Decision Making in Water Resources Policy and Management

Decision Making in Water Resources Policy and Management

An Australian Perspective

Barry T. Hart

Water Science Pty Ltd, Echuca and Monash University, Melbourne, VIC, Australia

Jane Doolan

University of Canberra, Canberra, ACT, Australia

ACADEMIC PRESS

An imprint of Elsevier

Academic Press is an imprint of Elsevier
125 London Wall, London EC2Y 5AS, United Kingdom
525 B Street, Suite 1800, San Diego, CA 92101-4495, United States
50 Hampshire Street, 5th Floor, Cambridge, MA 02139, United States
The Boulevard, Langford Lane, Kidlington, Oxford OX5 1GB, United Kingdom

Library of Congress Cataloging-in-Publication Data
A catalog record for this book is available from the Library of Congress

British Library Cataloguing-in-Publication Data
A catalogue record for this book is available from the British Library

ISBN: 978-0-12-810523-8

For information on all Academic Press publications visit
our website at https://www.elsevier.com/books-and-journals

www.elsevier.com • www.bookaid.org

Publisher: Candice Janco
Acquisition Editor: Louisa Hutchins
Editorial Project Manager: Tasha Frank
Production Project Manager: Maria Bernard
Cover Designer: Miles Hitchen

Typeset by SPi Global, India

Contents

PART 1   INTRODUCTION

PART 2   RECENT WATER RESOURCES POLICY AND MANAGEMENT CHANGES IN AUSTRALIA

PART 5   TRANSFER OF THE KNOWLEDGE—INTERNATIONAL CASE STUDIES

PART 6   CHALLENGES FOR THE FUTURE

Contributors

K. Auty
Commission for Sustainability and the Environment, Canberra, ACT, Australia; Honorary Professor with the University of Melbourne

C. Barlow
Australian Centre for International Agricultural Research, Canberra, ACT, Australia

S.A. Bekessy
RMIT University, Melbourne, VIC, Australia

N. Bond
Latrobe University, Wodonga, VIC, Australia

J. Brodie
James Cook University, Townsville, QLD, Australia

S.E. Bunn
Griffith University, Brisbane, QLD, Australia

I. Campbell
Rhithroecology Pty Ltd, South Balckburn, VIC, Australia

J. Catford
University of Southampton, Southampton, United Kingdom

J. Chong
University of Technology Sydney, Ultimo, NSW, Australia

A. Clarke
Department of Environment, Land, Water and Planning

D. Davidson
Water Science Pty Ltd, Echuca, VIC; Murray-Darling Basin Authority, Canberra, ACT, Australia

J. Doolan
University of Canberra, Canberra, ACT, Australia

R. Eberhard
Eberhard Consulting, Brisbane, QLD, Australia

C. Fitzpatrick
Independent Consultant

B. Fu
Australian National University, Canberra, ACT, Australia

C. Gippel
Griffith University, Nathan, QLD, Australia

C. Guest
Independent Consultant

S.H. Hamilton
Edith Cowan University, Joondalup, WA, Australia

B.T. Hart
Water Science Pty Ltd, Echuca, VIC; Monash University, Melbourne, VIC; Murray-Darling Basin Authority, Canberra, ACT, Australia

A. Horne
The University of Melbourne, Melbourne, VIC, Australia

A.J. Jakeman
Australian National University, Canberra, ACT, Australia

D. James
Consultant, Whale Beach, NSW

W.J. Kong
Chinese Research Academy of Environmental Sciences, Beijing, People's Republic of China

C. Leigh
Griffith University, Nathan, QLD, Australia

S.E. Loo
Department of Environment, Land, Water and Planning

R. McLoughlin
Commonwealth Department of Agriculture and Water Resources, Canberra, ACT, Australia

W. Meng
Chinese Research Academy of Environmental Sciences, Beijing, People's Republic of China

A.K. Parashar
CSIRO Land and Water, Canberra, ACT, Australia

G.M. Podger
CSIRO Land and Water, Canberra, ACT, Australia

C.A. Pollino
CSIRO Land and Water, Canberra, ACT, Australia

X.D. Qu
China Institute of Water Resources and Hydropower Research, Beijing, People's Republic of China

R. Rendell
RM Consulting Group, Bendigo, VIC, Australia

J. Schirmer
University of Canberra, Canberra, ACT, Australia

M.J. Selinske
RMIT University, Melbourne, VIC, Australia

R. Speed
Badu Advisory Pty Ltd, Brisbane, QLD, Australia

M.J. Stewardson
The University of Melbourne, Melbourne, VIC, Australia

P.-L. Tan
Griffith Law School, Brisbane, QLD, Australia

P.J. Wallbrink
CSIRO Land and Water, Canberra, ACT, Australia

J. Waterhouse
James Cook University, Townsville, QLD, Australia

A. Webb
University of Melbourne, Melbourne, VIC, Australia

J.A. Webb
The University of Melbourne, Melbourne, VIC, Australia

S. White
University of Technology Sydney, Ultimo, NSW, Australia

Y. Zhang
Chinese Research Academy of Environmental Sciences, Beijing, People's Republic of China

Preface

WHAT IS THIS BOOK ABOUT?

Contemporary water resources decision-making is an inherently complex undertaking. Every decision affects people in some way, whether as individuals, industries, communities, regional economies or the environments on which they depend. Water resource decisions can have profound impacts on freshwater, on marine and terrestrial ecosystems over large temporal and spatial scales, on agricultural industries reliant on irrigation, and also on the local communities dependent upon these industries. Given this, every decision involves trade-offs requiring careful balancing of environmental, social, cultural and economic impacts to develop a broadly accepted solution, which can then be adopted by governments at a point in time and practically implemented with communities. As a result, water reform is evolutionary in nature, with progress achieved through incremental decisions taken over time and at a pace at which the community can adapt.

Australia has been at the forefront of reforms in water resources policy, planning and management over the past three decades, particularly in the Murray-Darling Basin. Reform has occurred in all aspects of water management as Australian governments transitioned to a new environmentally and financially sustainable paradigm of water management. Over 30 plus years, this has involved the establishment of secure water entitlement systems, sustainable diversion limits, and a market-based approach to water allocation. It has required significant improvement in environmental protection and management, massive reforms in water pricing and a transformation in the governance of the water sector. Water reform has been hard, politically challenging and resource intensive, but the benefits have been great.

This book is unapologetically parochial in seeking to summarize the key aspects of these Australian reforms and to capture the learnings from them, particularly those elements that have contributed to improved decision-making. This collective review and analysis of the Australian experience should assist in advancing the decision-making process in water resources policy, planning and management, both in Australia and overseas.

As noted, the book is focused on the major water reforms in Australia over the past 30 years, both nationally and in the Murray-Darling Basin. Our purpose with the book is to present the key elements of these reforms and to analyze how they have led to improved water resources decision-making.

We believe these learnings will be of great interest to water managers in Australia and overseas. The situation that led to the recent water reforms in Australia is common to many other countries as they have to deal with financial, environmental and social issues associated with the development of their water resources. We recognize of course that water policy and management in each country will need to reflect the particular climatic, biophysical, social, economic and political aspects of that country. Nevertheless, there are many general principles and concepts that emerge from the Australian experience that could be useful to other countries as they grapple with their own water resource management problems.

There is increasing awareness that the management of water resources must be considered in a social, cultural, ecological, economic and political context. Gone are the days when it was only the hydrological aspects of water resources that constituted water resources management. Contemporary water resources decision-making now encompasses environmental, social, cultural and economic aspects of water resource planning and management, an extension of the so-called triple bottom line decision-making.

The book covers water resources decision-making at a number of levels: national planning and policy development; regional planning and policy development; regional and site-specific irrigation, environmental and urban water management; and individual farm business decision-making. Additionally, there are chapters covering the key components in the triple bottom line decision-making process, namely environment, economic and social. The increasingly important area of community and stakeholder engagement, including the involvement of Aboriginal and Torres Strait Islander peoples, is also covered.

This book complements other recent books covering aspects of decision-making in water resources planning and management (e.g., Garrick, 2015; Horne et al., 2017; Jakeman et al., 2016; Setegn & Donoso, 2015; Shrestha, 2015; Squires, 2016; Young & Esau, 2016).

BOOK STRUCTURE

The book is structured around five broad questions related to the water resources decision-making process:

- How has water resources policy and management changed in Australia over the past 30 years and what has driven these changes?
- What is the essential evidence base required in contemporary water resources decision-making and how has this been used to effect change in policy, planning and management?
- What are some of the key attributes of successful decision-making?
- What are the essential elements in exchanging/transferring this new knowledge to other countries?
- What are the future challenges in water resources decision-making?

THE DECISION-MAKING PROCESS

Doolan and Hart (Chapter 1) provide a brief introduction to water management in Australia, including an overview of the history of water management in Australia since European settlement. It focuses particularly on the last 30 years, during which Australian governments have been undertaking a program of major water reform, transitioning to a new paradigm of environmentally and financially sustainable water resource management providing high economic value to the nation.

RECENT WATER RESOURCES POLICY AND MANAGEMENT CHANGES IN AUSTRALIA

One overriding conclusion that comes out of the past 30 years of water resources reforms in Australia is that these changes generally occur over extended periods of time (evolution rather than revolution) and are assisted by crisis (e.g., floods and drought).

This section starts with a long-term perspective capturing the history of water resources planning, policy and management in Australia (focused on the Murray-Darling Basin) in the 100 plus years since

Federation in 1901. Guest (Chapter 2) provides a fascinating account of the political interactions between the basin states (Victoria, New South Wales, South Australia and more recently Queensland) and the Commonwealth government in water resources policy and management in the Murray-Darling Basin.

The next four chapters in this section of the book cover the changes in planning and management of water resources for irrigation (rural Australia), for the major urban centers and for the environment.

Fitzpatrick (Chapter 3) reviews the water reforms in Victoria over the past 30 years, starting with the conversion of poorly specified water rights into legally binding volumetric water entitlements, followed by the establishment of a water market.

Rendell (Chapter 4) provides a personal view of the changes that have occurred in the irrigation districts in the southern part of the Murray-Darling Basin over the past 30 years. He focuses particularly on how individual farmers and farm businesses have had to modify their decision-making practices to take into account factors such as climate variability, changes in agricultural practice, the water market and government policies.

Chong and White (Chapter 5) review the water reforms that have occurred in Australia's major cities. This commenced with the corporatization of the major urban water utilities in the 1990s, and since then has involved focus on supply-demand planning, the augmentation of supply options with desalinization plants, and increased emphasis on the integrated whole of water cycle management.

Bunn (Chapter 6) summarizes the significant water reforms that have been undertaken over the past three decades to address overallocation and improve the health of Australia's freshwater ecosystems. The environment is now a major player in water resources management in Australia, particularly in the Murray-Darling Basin.

EVIDENCE BASE REQUIRED FOR WATER RESOURCES DECISION-MAKING

Contemporary water resources decision-making now must go beyond so-called triple bottom line (TBL) decision-making—balancing environmental, social and economic aspects—to also include cultural or indigenous aspects. And as far as possible this decision-making should be evidence based. This does not mean that trade-offs and judgements will not have to be made, but simply that the better the evidence that underpins the decision-making process, the more likely it is that better decisions will be made.

This section of the book contains seven chapters that contribute to the essential evidence base required for contemporary water resources decision-making.

Stewardson et al. (Chapter 7) review the eco-hydrological methods used in Australia to assess the environmental flow regimes required firstly as environmental input to water allocation decisions and secondly as a guide in outline form to recent reforms in three key complex policy areas in Australia environmental water management.

James (Chapter 8) discusses the application of economic analysis (e.g., cost-benefit analysis, economic impact analysis, and optimization of economic, social and environmental outcomes) in the planning and management of water resources, with particular reference to the Murray-Darling Basin Plan.

Bekessy and Selinske (Chapter 9) review three frameworks for understanding social-ecological systems and seven decision tools for modeling and analyzing social-ecological systems, and then describe pathways to mainstream the use of these approaches.

Schirmer (Chapter 10) identifies key principles for assessing and managing the social effects of water reform. The development and implementation of the Murray-Darling Basin Plan are used to highlight three important principles: the need for good public participation; ensuring robust methods are used to assess the social outcomes of water reform throughout the development and implementation; and ensuring social effects are considered when selecting and designing the policy instruments used to enact water reform.

Tan and Auty (Chapter 11) discuss the importance of effective community engagement as an essential component of any water resources reform process. They identify five key elements that underpin an effective community engagement process: engaging in places where people are on familiar territory and care about them; taking time to listen and share information; accepting that compromises will need to be made; effectively planning and organizing the engagement process; and showing leadership in process.

Pollino et al. (Chapter 12) discuss the need for more integrative approaches in water resource decision-making, and in particular focus on multidisciplinary approaches that are central to tackling real-world, complex science and social-political problems, where acceptable solutions often range across disciplinary boundaries.

EXAMPLES OF POLICY CHANGE—AUSTRALIAN CASE STUDIES

This section of the book contains three case-study chapters that cover recent reforms in three complex water policy topics in three different areas in Australia.

Hart and Davidson (Chapter 13) provide an analysis of the key decision-making aspects in the development of the Murray-Darling Basin Plan, with particular focus on the decision-making steps employed in developing the sustainable diversion limits (SDL), the key policy initiative that underpins the Basin Plan.

Loo and Clarke (Chapter 14) discuss the process used to develop a new Victorian Waterway Management Strategy that replaced the 2002 Victorian River Health Strategy. They argue that clear management arrangements, a dedicated fund source, political authorization, and partnership with communities were key factors that underpinned the successful development and implementation of this new policy.

Eberhard et al. (Chapter 15) review the management of water quality in the catchments affecting the Great Barrier Reef (GBR), an internationally recognized World Heritage site. They note that efforts by the Commonwealth and Queensland governments to improve water quality over the last 15 years have not delivered measurable improvements to the health of inshore reef ecosystems. They conclude that greater effort is need to overcome constraints to current management approaches and to employ the additional policy measures required to help sustain the GBR into the future.

TRANSFER OF NEW KNOWLEDGE OVERSEAS

This section contains three case-study chapters covering experience in exchanging Australian water resources knowledge with Southeast Asia (Mekong, India and China). These show that, although every country has its own suite of problems and issues, elements of the Australian experience have been useful in assisting countries in developing water management tools, policies and institutional arrangements.

Campbell and Barlow (Chapter 16) identify six challenges in the international transfer of knowledge between developed and less-developed countries, gained largely from their experiences in the Mekong. These include: differences in the biophysical environment; lack of knowledge and local capacity; differences in socioeconomic conditions; and differences in culture (e.g., importance of age/seniority, agreement and social networks).

Pollino et al. (Chapter 17) report on a three-year project aimed at using Australian water resources expertise (particularly river system modeling) to assist India in building capacity and managing water at a Basin scale—the Brahmani-Baitarni River basin. A number of challenges were experienced during the project and these are discussed.

Gippel et al. (Chapter 18) discuss Australia's assistance in developing a framework for a national program to assess the health of China's rivers; there is currently no consistent nationwide assessment of river health in China. The framework was shaped by Australian experience, and tailored to suit the resources and expertise available in China, the biophysical characteristics of China's rivers, and the key management issues.

CHALLENGES FOR THE FUTURE

As the chapters in this book show, much has been achieved in Australia over the past three decades in improving the sustainable management of the nation's water resources. This has included: transforming the water allocation process; improving the protection and management of environmental water; reforming the pricing of water services; modernizing institutional arrangements; ensuring meaningful community and stakeholder engagement; and improving water information and knowledge.

This thrust towards more integrated water resources planning and management in Australia has also involved the development of improved adaptive decision-making processes that are focused on triple bottom line decision-making. However, there is still more to be done.

The decade-long Millennium drought (1997–2009) was a major wakeup call to water resource planners and managers in Australia. It highlighted our vulnerability to drought, a phenomenon that will become even more prevalent as climate change becomes more severe. There were a number of lessons learned that have already resulted in changes in water resource management to mitigate the impacts of water scarcity.

Chapter 19 considers the six biggest challenges to water management in Australia over the next three decades to 2050: *climate change, population growth, water-energy interactions, community expectations and demands, maintaining affordability, and maintaining the impetus for and commitment to future water reform.* These challenges are discussed from the perspective of how they will influence future decision-making in rural, urban and environmental water management. These will form the basis for the next wave of water reform in Australia.

B.T. Hart, J. Doolan

REFERENCES

Garrick, D. E. (2015). *Water allocation in rivers under pressure: Water trading, transaction costs and transboundary governance in Western US and Australia.* Cheltenham, UK: Edward Elgar Publishing.

Horne, A., Stewardson, M., Webb, J. A., Richter, B. D., & Acreman, M. (Eds.), (2017). *Water for the environment: From policy and science through to implementation.* New York: Elsevier Publishing.

Jakeman, A. J., Barreteau, O., Hunt, R. J., Rinaudo, J. -D., & Ross, A. (Eds.), (2016). *Integrated groundwater management: Concepts, approaches and challenges:* Springer. (open access), 762 pp., http://link.springer.com/book/10.1007%2F978-3-319-23576-9.

Setegn, S. G., & Donoso, M. C. (Eds.), (2015). *Sustainability of integrated water resources management: Water governance, climate and ecohydrology:* SpringerLink [Online].

Shrestha, S. (Ed.), (2015). *Managing water resources under climate uncertainty: Examples from Asia, Europe, Latin America and Australia:* SpringerLink [Online].

Squires, V. R. (Ed.), (2016). *Ecological restoration: Global changes, social aspects, and environmental benefits.* New York: Nova Publishing.

Young, M. D., & Esau, C. (Eds.), (2016). *Transformational change in environmental and natural resource management.* New York: Routledge, Taylor & Francis Group. 210 pp.

Author Biographies

Dr. Kate Auty is presently the ACT Commissioner for Sustainability and the Environment and she was previously appointed as the Victorian Commissioner for Environment and Sustainability. Kate continues as an Honorary Vice Chancellor's Fellow with the University of Melbourne and chairs the University's Melbourne Sustainable Society Institute Advisory Board. Other roles include membership on the Australian Urban Research Infrastructure Network advisory board and the Murray Darling Basin Authority's Advisory Council on Social, Economic and Environmental Sciences. Kate recently retired as the chair of the National Electronic Collaborative Tools and Research advisory board.

Dr. Chris Barlow spent 9 years as manager and technical advisor to the Fisheries Program of the Mekong River Commission in Phnom Penh and Vientiane. Since 2009 he has been with the Australian Centre for International Agricultural Research, managing the Fisheries Program, which administers research and development partnerships across Southeast Asian and Pacific countries.

Dr. Sarah Bekessy leads the Interdisciplinary Conservation Science research group at RMIT University. She is interested in the intersection between science and policy in environmental management and is currently involved in a range of interdisciplinary research projects. She is an ARC Future Fellow running a project titled "Socio-ecological models for environmental decision making." She also runs a project funded by the Myer Foundation titled "Reimagining the suburb: planning for biodiversity in the urban fringe." She is also a theme leader of the ARC Centre of Excellence for Environmental Decisions and leads projects in two National Environment Science Program Hubs (Threatened Species Hub and Clean Air and Urban Landscapes Hub).

Dr. Nick Bond is professor of freshwater ecology in the School of Life Sciences and Director of the Murray-Darling Freshwater Research Centre at La Trobe University. He has wide-ranging interests in the ecology and management of aquatic ecosystems, with a particular focus on the use of quantitative methods to understand the influence of landscape-scale drivers on ecosystem structure and function. He has worked extensively on approaches to river health assessment in Australia and in China as part of the AECDP Program.

Dr. Jon Brodie is an environmental scientist and manager with 35 years of experience working in these fields in the Pacific Islands, the Middle East, Southeast Asia and Australia. Jon's main focus for the last 30 years has been the Great Barrier Reef, while working in a management role as the Director of Water Quality at the Great Barrier Reef Marine Park Authority and as a research scientist and program leader at James Cook University. He led the writing teams preparing the Scientific Consensus Statements on water quality and the GBR in 2008, 2013, and 2016/17 for the Queensland and Australian governments. He is currently a professorial fellow at the ARC Centre of Excellence for Coral Reef Studies at James Cook University, Townsville.

Professor Stuart Bunn is the Director of the Australian Rivers Institute at Griffith University. His major research interests are in the ecology of river and wetland systems with particular focus on the science to underpin river management. Stuart has extensive experience working with international and Australian government agencies and with industry on water resource management issues. He has served as an Australian National Water Commissioner and as a Director of Land and Water Australia, and currently chairs the science advisory committees for the Murray-Darling Basin Authority and Healthy Waterways Limited.

Dr. Ian Campbell completed his MSc on the biological assessment of water pollution, and his PhD on the ecology of a group of aquatic insects. After 25 years teaching and conducting research on stream ecology at Monash University, Ian joined the Mekong River Commission as the Senior International Environment Specialist. He later worked as an environmental adviser at the Ok Tedi mine in PNG and as Principal River Scientist at consulting company GHD, before establishing his own environmental consulting company, Rhithroecology Pty Ltd.

Dr. Jane Catford is a lecturer in biological sciences at the University of Southampton, UK. She previously worked as a research fellow at the University of Melbourne, Australian National University and University of Minnesota. A plant ecologist, Jane focuses on community assembly and invasion ecology. She works in both terrestrial and freshwater ecosystems, and likes to link ecological theory and ecosystem management.

Joanne Chong is a research director at the Institute for Sustainable Futures (ISF) at the University of Technology Sydney. She is an environmental economist and policy analyst with extensive experience in the urban water sector, including designing, monitoring and evaluating sustainability and environmental management policies, programs and projects in Australia and internationally. Prior to joining ISF, Joanne worked for the Australian Government Department of Environment and Heritage and as a senior research economist at the Productivity Commission. Jo has also spent time as a volunteer and consultant for the Asia Regional Environmental Economics Programme of the World Conservation Union (IUCN).

Dr. Amber Clarke received her PhD from Monash University for research investigating the patterns and determinants of macroinvertebrate diversity in headwater streams. Amber also completed a master of environmental science and bachelor of environmental science (honors) at Monash University. The focus of her professional career to date has been leading the development of state policy to improve the environmental condition of waterways across Victoria. Amber is currently the Director, Catchments at the Department of Environment, Land, Water and Planning. She oversees programs for waterway health, environmental water, water quality, catchment planning, catchment governance, and integrated catchment reporting.

Dianne Davidson is an agricultural scientist who has worked primarily in the irrigated horticultural and wine industries throughout Australia and internationally for 40 years. She has considerable expertise in water and soil management as well as wide experience working and living in regional Australia. She has worked in natural resource management and served on many government advisory boards and committees and has a close involvement with the University of Adelaide. Di has strong business management skills and manages her own irrigated farms in South Australia. She well understands community concerns surrounding water availability and cost. Di has been a board member of the Murray-Darling Basin Authority since 2009 and was involved in the development and implementation of the Murray-Darling Basin Plan.

Professor Jane Doolan is a Professorial Fellow in Natural Resource Governance at the University of Canberra. She has extensive senior leadership experience working in sustainable water resource and environmental management, having provided policy advice to governments on issues such as urban and rural water supply and security, national water reform, and water sector governance. She has driven important initiatives in river health, environmental water allocation and catchment management, and the management of water during drought and climate change. Her career encompasses intergovernmental policy development and negotiations, particularly in relation to the management of the Murray-Darling Basin. Previously held positions include Commissioner with the National Water Commission and Deputy Secretary for Water in the Victorian Department of Environment and Primary Industries. She is currently the chair of the Murray-Darling Freshwater Research Centre and a director of Western Water Authority.

Rachel Eberhard is an environmental scientist with 25 years of experience in planning, brokering and evaluating natural resource management policy, programs and partnerships. She has worked as a sole operator consultant (12 years) and Queensland government scientist (7 years) and as a PhD scholar is researching collaborative water governance in the Great Barrier Reef and Murray-Darling Basin. Rachel has had a close involvement in Great Barrier Reef policy and planning since 2003. She led the establishing of the Reef Water Quality Partnership that facilitated Reef Rescue and laid the foundation for subsequent investments in agricultural practice change incentive in the GBR.

Campbell Fitzpatrick worked in the water industry for over 30 years and in senior executive positions for close to 20 years. He has been responsible for leading the development and implementation of water resource policy within Victoria. Campbell was actively involved in all aspects of the implementation and development of Victoria's water entitlement and water market arrangements, working closely with governments, water authorities and water users. He is currently providing strategic advice to the water sector.

Dr. Chris Gippel is adjunct senior research fellow, Australian Rivers Institute, Griffith University, Brisbane, and the Director of Fluvial Systems Pty Ltd, based in Newcastle, Australia. His PhD used theory, laboratory experiments and field observations to evaluate the use of turbidity as a surrogate for suspended sediment concentration. In 2013 he worked as a Visiting Fellow at Changjiang Water Resources Protection Institute to study the health of Poyang Lake. He has also been a Visiting Fellow at the College of Water Resources and Hydropower Engineering, Wuhan University on three occasions, in 2013, 2014, and 2015. Chris has worked in Australia, China, United States, United Kingdom, Europe, Japan, Laos, Pakistan, and Peru. His main current research interests are in assessment of river and lake health, assessment of environmental flow requirements, prediction of river geomorphology, water balance, hydrological prediction and hydraulic modeling for ecological and geomorphological objectives.

Dr. Chris Guest, as a career public servant and an economist, has had a long interest in water policy and politics, an interest that grew as governments sought to tackle the problems of overallocation of water. He represented New South Wales in the negotiations for the National Action Plan for Water Quality and Salinity, and later in the negotiation of the National Water Initiative. In 2007 he joined the Commonwealth Government team developing the new Water Act, and later assisted in the development and implementation of the first Murray-Darling Basin Plan.

Professor Barry Hart is director of an environmental consulting company, Water Science Pty Ltd. He is also emeritus professor at Monash University, where previously he was director of the Water Studies Centre. He has established an international reputation in the fields of ecological risk assessment, environmental flow decision making, water quality and catchment management and environmental chemistry. He is well known for his sustained efforts in developing knowledge-based decision-making processes in natural resource management in Australia and Southeast Asia. Prof. Hart is currently a board member of the Murray-Darling Basin Authority and a nonexecutive director of Alluvium Consulting Australia Pty Ltd. He is also deputy chair of the Scientific Inquiry into Hydraulic Fracturing of Onshore Unconventional Reservoirs in the Northern Territory, which commenced in Dec. 2016.

Dr. Serena Hamilton is a research fellow at the Centre for Ecosystem Management at Edith Cowan University, Perth. Her research interests include modeling socioecological systems, ecohydrology and decision support tools for natural resources management. Her recent research has focused on integrated modeling for improving understanding of system linkages and management of water resources and dependent ecosystems.

Dr. Avril Horne is a research fellow in the Department of Infrastructure Engineering at the University of Melbourne, Australia. She has worked on water resource policy and management projects in consulting, government and academia. She has experience across environmental flows, water trade, water allocation mechanisms and hydrology.

Professor Tony Jakeman is director of the Integrated Catchment Assessment and Management Centre at the Fenner School of Environment and Society at the Australian National University. He is leader of the National Centre for Groundwater Research and Training Program on Integrating Socio-economics, Policy and Decision Support (groundwater.com.au). In 2012 he was presented the Ray Page Lifetime Achievement Award by Simulation Australia. In 2011 he was awarded the Silver medal of Masaryk University. He is on the 2015 highly cited researcher list by Thomson Reuters (top 1%). In 2016 he was elected a Fellow of the American Geophysical Union.

Dr. David James has more than 40 years of academic, consulting and research experience in Australia and internationally, specializing in environmental and natural resource economics. He was a Special Commissioner with the Australian Resource Assessment Commission and has acted as adviser on numerous government technical committees. For the last 20 years he has served as a Resource Person with the Economy and Environment Program for Southeast Asia. In 1988 he received a UNEP Global 500 Award for his work in environmental impact assessment incorporating economic principles. Currently he is a member of the Murray-Darling Basin Authority's Advisory Committee on Social, Economic and Environmental Sciences.

Professor Kong Weijing is a professor in the Water Environment Institute of the Chinese Research Academy of Environmental Sciences. His key research interests involve watershed planning and management and wetland vegetation. His current work mainly focuses on freshwater ecoregion delineation and riparian vegetation landscape succession under natural restoration.

Professor Meng Wei is the academician of the Chinese Academy of Engineering and dean of the Chinese Research Academy of Environmental Sciences. His research mainly focuses on water pollution control theory and technology, and environmental science of estuaries and coastal ecosystems. He has made unique and pioneering academic contributions to understanding drainage basin water pollution capacity, watershed ecological quality assessment, management of the ecological health of estuarine and coastal zones, and pollution control measures, as well as assessment of ecological security at times of water pollution incidents.

Dr. Catherine Leigh is a research fellow and lecturer at the Australian Rivers Institute and Griffith School of Environment, Griffith University. She was previously a key participant in the international Intermittent River Biodiversity Synthesis and Analysis project, based at IRSTEA in Lyon, France, and in the Australian National Water Commission's Low Flows Project. Her expertise is widely sought by national and international committees and government, industry and research agencies, including the Australian Centre for International Agricultural Research, SeqWater, International Water Centre, Chinese Academy of Sciences and the European Committee for Standardization. Her work focuses on understanding the responses of biota to natural and human-induced changes in river flow, including river drying, by using ecosystem, community and food web science methods to investigate biodiversity patterns and ecosystem dynamics.

Dr. Sarina Loo is a senior public servant at the Department of Environment, Land, Water and Planning in Victoria. Over the past 10 years, her responsibilities have included policy, strategy, investment, monitoring, evaluation, reporting and governance in the areas of waterway health and catchment management. Sarina has a PhD in freshwater ecology with her research focusing on invasive species distribution modeling. She has been lecturer of sustainability at Swinburne University of Technology and an independent consultant to government and private business. Sarina is an inaugural fellow and past board member of the Peter Cullen Trust.

Baihua Fu is a research fellow at the Australian National University. She has a background in water quality and ecohydrology. Her recent research interests include integrated assessment for water resource management, exploratory modeling and scenario analysis to inform decision making, and uncertainty assessment of environmental models. She has also been involved in several integration projects in the Murray-Darling Basin regulated systems.

Richard McLoughlin is the Assistant Secretary for Water Resources in the Australian Government Department of Agriculture and Water Resources, where he is responsible for a wide range of national water policy matters, the National Water Infrastructure Development Program and Commonwealth involvement in management of the Lake Eyre Basin and Great Artesian Basin. He currently leads on national water policy reforms and international engagement in the Department, focused on strategic relationships with China, India, Indonesia and the United States. Richard also spent an extended period as chief executive officer of the National Water Commission in addition to his existing roles. Richard holds a BSc (Hons) and MSc from the University of NSW and is a graduate of the Australian Institute of Company Directors and the Executive Fellows program of the Australian and New Zealand School of Government.

Amit Parashar is a principal research scientist at CSIRO. He draws upon 18 years of professional experience in Australia and abroad. Starting from a background in computer science and information systems, he combined expertise in information systems with environmental management. Building on further education in business administration (MBA) he has strongly developed skills in organization change, Amit has been working in the nexus of information systems, organization and people, with a focus on delivering sustainable organizational change. He is currently based in New Delhi in India.

Geoff Podger is a senior river modeler and senior principal scientist at CSIRO. Geoff has significant experience in hydraulic and hydrologic modeling, with 22 years of experience in hydraulic and hydrologic modeling, river basin management and planning. Geoff has been involved in the implementation and development of numerous hydraulic and hydrologic computer models. He plays a significant part in developing Australian hydrological models and is the project director for the Sustainable Development Investment Portfolio, which has a focus on transboundary water modeling in South Asia.

Dr. Carmel Pollino works as a principal research scientist at CSIRO. She has 15 years of experience working in environmental flows, risk assessment, integrated river basin planning and ecological modeling. She has experience working on water quality and quantity issues, considering ecological outcomes within broader systems, including complex governance, stakeholder and cultural contexts. Carmel has worked with Australian agencies in developing methods for and in evaluating the outcomes of basin planning, and these methods have recently been applied in India.

Dr. Qu Xiaodong is associate professor of the Department of Water Environment, China Institute of Water Resources and Hydropower Research and the Director of Laboratory of Freshwater Ecosystem Assessment and Civilization. Xiaodong's PhD concerned spatial-temporal dynamic modeling and bioassessment of the macroinvertebrate communities in the Xiangxi River system. His main current research interests include quantitative and statistical analysis on community response and species abundance patterns, river health assessment methods, and ecological modeling implemented to provide sustainable ecosystem management.

Rob Rendell is an agricultural engineer with the RM Consulting Group (RMCG). He has worked in northern Victoria for over 40 years covering all aspects of irrigated agriculture. Rob has a practical irrigation farm background and has worked at both the farm business and technical irrigation level across many industries. He was involved in the introduction of laser technology in the 1980s, salinity management and groundwater management planning in the 1990s, and was the architect of the plan to modernize the Goulburn Murray Irrigation District in recent times. Rob is heavily involved in the implementation of the Murray-Darling Basin Plan and its socioeconomic impacts.

Dr. Jacki Schirmer is an associate professor at the University of Canberra, working jointly between the Health Research Institute and the Institute for Applied Ecology. She manages the Regional Wellbeing Survey, the largest survey to focus on examining how changes in rural and regional areas of Australia affect the wellbeing of people and communities. Much of her work examines the interaction between natural resources and human health and wellbeing. In particular, her research focuses on understanding how changes in the management of resources such as water, agricultural land, forests and fisheries affect the wellbeing of people and communities who depend on these resources, and how we can understand the social impacts of natural resource management change. She collaborates with a wide range of government, nongovernment and industry organizations to work on developing policies and programs that support the health and wellbeing of people during periods of natural resource policy change.

Matthew Selinske is a PhD candidate at RMIT University investigating the social dimensions of conservation programs. He is specifically interested in the behavioral implications of incentivized conservation initiatives and social-ecological modeling. Matthew's previous research focuses on the effectiveness and social outcomes of voluntary stewardship programs on private lands in South Africa.

His professional experience includes landscape restoration in the US Midwest, urban habitat restoration and community engagement in New York City, and 6 years of protected area management and primate conservation in West Africa. Matthew's research is supported by RMIT University and the Australian Research Council's Centre of Excellence for Environmental Decisions.

Robert Speed is a principal of Badu Advisory, a strategic management company that focuses on water resources management, and is a PhD candidate at the Fenner School of Environment and Society at the Australian National University. Robert was previously the project director of the River Health and Environmental Flows in China Project. Robert has qualifications in environmental science and law.

Professor Michael Stewardson has focused his research over the past 24 years on interactions between hydrology, geomorphology and ecology in rivers (http://www.findanexpert.unimelb.edu.au/display/person14829). This has included physical habitat modeling, flow-ecology science, and innovation in environmental water practice. Michael has participated in Australia's water reforms through advisory roles at all levels of government. More recently, his research has focused on the physical, chemical and biological processes in streambed sediments and their close interactions in regulating stream ecosystem services. He leads the Environmental Hydrology and Water Resources Group in Infrastructure Engineering at The University of Melbourne.

Professor Poh-Ling Tan is the International Water Centre's Professor for Water Law and Governance at Griffith University. Her research and teaching focuses on water reform and governance particularly in the intersections of law, social and biophysical sciences. She collaborates with multidisciplinary teams of researchers and has significant experience working with Traditional Owners, water agencies, stakeholders and general communities in different Australian contexts.

Dr. Peter Wallbrink is the research director at CSIRO. Peter has long experience in developing tools and technologies to support basin scale planning, management and use of water, as well as in developing information platforms, standards and architectures to underpin planning and management processes. He is currently leading a significant portfolio of projects in South Asia with the aim of linking water management decisions to livelihood outcomes. Peter has published over 110 peer-reviewed articles in a diverse range of discipline areas, such as soil erosion, sediment generation and redistribution dynamics, land use impacts on water quality, river hydrology, and best practice in applying water management modeling.

Jane Waterhouse is an environmental scientist with 20 years of experience working in the Great Barrier Reef in water quality science, catchment management, monitoring and evaluation, science synthesis and regional planning. Jane is an independent consultant and also maintains a research capability as a research fellow in the Catchment to Reef Group in TropWATER at James Cook University. She is currently involved in several projects related to water quality science and management in the Great Barrier Reef and the Pacific Islands.

Dr. Angus Webb is a senior lecturer in Environmental Hydrology and Water Resources at the Melbourne School of Engineering, University of Melbourne. Much of his research centers on the monitoring and evaluation of major environmental flows programs in Australia, including leading the Lower Goulburn River project under the Commonwealth government's Long-Term Intervention Monitoring Project. He has a particular interest in developing models of ecological responses to flow regimes that can inform planning and adaptive management of environmental flow programs. This includes research into methods that can make better use of existing knowledge in the literature, in the minds of experts, and in major (and often unanalyzed) sets of environmental monitoring data.

Professor Stuart White is the director of the Institute for Sustainable Futures at the University of Technology Sydney and leads a team of researchers across a range of aspects of "creating change towards sustainable futures." He is a member of the International Water Association Strategic Council and deputy chair of the Efficient Urban Water Management Specialist Group. He has led research and has written and spoken on integrated water supply demand planning widely in Australia and internationally over the last 25 years. In 2012 he was awarded the Australian Museum Eureka Prize for Environmental Research.

Professor Zhang Yuan is the deputy director at the Water Environment Research Center at the Chinese Research Academy of Environmental Sciences, and an adjunct professor and PhD supervisor at Beijing Normal University. His key research involves ecosystem conservation and the restoration of rivers and lakes, and the influence of land use and water resource utilization on river health at a basin level.

INTRODUCTION

WATER RESOURCE POLICY, PLANNING AND MANAGEMENT IN AUSTRALIA—AN OVERVIEW

J. Doolan*, B.T. Hart[†,‡]

University of Canberra, Canberra, ACT, Australia* Water Science Pty Ltd, Echuca, VIC, Australia[†] *Monash University, Melbourne, VIC, Australia*[‡]

CHAPTER OUTLINE

INTRODUCTION

Australia is the driest inhabited continent in the world. It is renowned for its highly variable climate, with a history of recurrent droughts often punctuated by large floods. The long, dry periods have always been a major challenge for communities and governments as they sought to develop and use the nation's water resources to provide secure water supplies for cities and towns and promote economic growth.

Australia is a federation of six states and two major territories. Under the Australian Constitution, state and territory governments have responsibility for land and water management. Water in Australia is owned by the Crown and its management is vested in state governments. The Commonwealth government's role is one of oversight, facilitation, and investment, ensuring that the national interest is met, particularly in transboundary river and groundwater basins, like the Murray-Darling Basin (MDB), the Great Artesian Basin, and the Lake Eyre Basin.

In these major transboundary systems, the relevant state and territory governments work with the Commonwealth government to develop ongoing, formal management arrangements covering water sharing and use and other related matters. The most famous of these is the MDB involving the states of Victoria, New South Wales, South Australia, and Queensland, the Australian Capital Territory (ACT), and the Commonwealth government. The MDB is often seen as the showcase for Australian water reform. It has been the area of greatest water use, biggest environmental challenges, and highest economic value, and its management has required the states and the Commonwealth government to work together for over 100 years on issues of water sharing between states, river management and key aspects of catchment and land management (see Chapter 2).

Once the higher order issues of water sharing between states have been agreed upon, states then manage land and water within their own individual legislative frameworks. Each state has its own system of water entitlements and water planning, frameworks for environmental protection and water industry regulation. In multijurisdictional basins, where required, these systems have been made compatible. When national approaches to water management have been desired in the national interest, they have been achieved through the agreement of all governments through the Council of Australian Governments (COAG).

Each state has established institutional arrangements to deliver water supply, sewerage, drainage, and irrigation services to local communities. In a number of states, these services are delivered through local authorities—either dedicated urban and rural water authorities or in some cases through local government.

Australia is a highly urbanized society, with approximately 90% of its 24.2 million people[1] living in cities and towns, and around 60% living in cities of greater than 1 million people. In 2014–15, a year in which rainfall was 10% lower than the national average, bulk water extraction across the country was 16,700 GL. Of this, 75% was used for irrigation, 19% for urban use, and 6% for mining and power generation (Bureau of Meteorology, 2016a).

This chapter provides a brief introduction to water management in Australia, including an overview of the history of water management in Australia since European settlement. It focuses particularly on the last 30 years during which Australian governments have been undertaking a program of water reform, transitioning to a new paradigm of environmentally and financially sustainable water resource management providing high economic value to the nation.

This chapter also sets the scene for the rest of the book. Individual chapters go into more detail about specific elements that have contributed to the implementation of the reform agenda.

PHASES IN AUSTRALIAN WATER MANAGEMENT
THE EARLY YEARS—THE "BUILD AND SUPPLY" PHASE

Since 1901, when Australia became a federation, up until the late 1970s, the focus of governments was on the construction of storage and delivery systems to provide water security for growing cities and towns, and to develop irrigation areas and mining industries. At the start of the 20th century, the combined storage capacity of all large dams across Australia was 240 GL. This grew to 7200 GL by 1950 and to

---

[1]Estimated as at December 1, 2016 (http://www.abs.gov.au/ausstats/abs@.nsf/0/1647509ef7e25faaca2568a900154b63? opendocument).

84,800 GL by 2005 (Australian Bureau of Statistics, 2010). This dam building was often undertaken in the aftermath of drought and funded, predominantly, by governments. As a consequence, Australia now has the highest per capita surface water storage capacity of any country in the world.

This period between federation and the late 1970s was, effectively, a major "build and supply" phase (Doolan, 2016a). However, even in this early phase where the focus was on infrastructure construction, serious attention was also paid to developing legislative frameworks and management systems to properly allocate and administer the resources that were then under control. There were early laws governing water allocation and management, which retained ownership of water with the Crown. Water systems were constructed using the contemporary understanding of the hydrology of the water systems involved, and water allocation systems were put in place that were meant to reflect the actual water availability at that time.

While this early phase provided a firm foundation for water resource management at the time, by the 1980s, at the end of this development phase, a number of new issues were emerging. These included large government debt, poor pricing policies, service delivery challenges and widespread environmental damage.

The high level of investment in water infrastructure had left governments with a significant legacy of debt and an ongoing future requirement for maintenance and refurbishment. While the gains in water storage had provided a level of water security, there was still increasing competition for these water supplies as towns and the irrigation industries grew. Every drought raised issues within communities and calls for more storage (Keating, 1992). However, by the 1980s, there was limited potential to increase supply in regions of high water demand, due to a shortage of cost-effective, large-scale dam sites.

With increasing competition for water, problems arose with the existing water rights systems. In many areas, existing water rights were not based on a good understanding of resource availability and the full suite of demands, leading to a number of systems being overallocated. This resulted in detrimental impacts to both the environment and downstream users. Additionally, water rights were tied to land, so the only way to effectively transfer water between users was through land purchase (National Water Commission, 2011a).

In urban areas, there were problems with service delivery. Many towns across the country were provided with water that did not meet drinking water quality standards, often by small local authorities that were not financially sustainable (Industry Commission, 1992). Water pricing was generally based on property value and did not reflect the costs of supply or water use. As a result, water use in cities was highly inefficient. In the 1980s, water use in Melbourne was greater than 180 kL per capita per year and greater than 160 kL per capita per year in Sydney (Water Services Association of Australia, 2003). In other areas, it was difficult to tell what the per capita use was because metering was not widespread. Urban wastewater discharges were also a major pollution source into rivers and marine environments.

In irrigation areas, water was being used inefficiently to produce low-value crops. For example, in 1991, over half of Australia's irrigated land was used for fodder production. The profitability of irrigated agriculture was low, with many farms considered to be marginal businesses, and irrigators still highly susceptible to drought. Moreover, irrigation authorities were highly dependent on government subsidies, with only part of their revenue being funded by irrigators (Industry Commission, 1992).

In addition to these issues, by the 1980s, it was clear that water management was also causing widespread environmental damage. Obvious problems included high salinity in the River Murray and extensive land salinization, which had major socioeconomic costs. In 1981, the mouth of the River Murray closed for the first time since European settlement as a result of upstream water extraction. In 1991, a toxic algal bloom extending over 1000 km occurred in the Darling River. Algal blooms were also occurring in Perth in Western Australia, and in the Gippsland Lakes in Victoria. There was extensive erosion and sedimentation in rivers and wetlands.

Many of these environmental issues had flow-on economic and social impacts on agricultural production, regional tourism, and community wellbeing.

THE REFORM ERA

The issues of environmental degradation, government debt, poor pricing for water services, and service delivery challenges caused governments to closely examine the way water resources were being managed in their jurisdictions. In the mid-1980s, in Victoria, a bipartisan review of water management across the state was undertaken by the Parliamentary Public Bodies Review Committee, which developed a long-term reform agenda covering almost every aspect of water resource management (see Chapter 3).

The MDB governments worked together with the Commonwealth government to first develop a strategy for salinity management in 1988. They then undertook a water audit in 1995, which showed that water extraction had reduced the median annual flow to the sea to 27% of that occurring naturally (Murray-Darling Basin Ministerial Council, 1995). This clearly showed that water use in the Basin was unsustainable and prompted a landmark decision by all governments in 1995 to "cap" water extraction across the MDB, generally limiting water diversions to 1993 levels of development (Murray Darling Basin Commission, 1996). This decision created the drivers for the evolution of more accurate water accounting, widespread water trading, and an interstate water market in the MDB. At the national level, governments worked together on strategies for managing water quality (ARMCANZ/ANZECC, 1994) and providing water for the environment (ARMCANZ/ANZECC, 1996).

Despite these efforts, by the early 1990s, governments had collectively recognized that overall water management across Australia was providing a poor return to the national economy and was causing significant environmental damage along the way. They agreed that the entire system of managing, using and funding water had to transition from the environmentally damaging, government-funded, inefficient, "build and supply" philosophy to a new system that was both economically and environmentally sustainable, providing a secure basis for future investment and capable of yielding high returns to the community (Doolan, 2016a).

In 1994, the COAG signed off on a national water reform framework (COAG, 1994)—a highly challenging, comprehensive policy agenda covering all elements of water management, which was intended to move Australia into a new phase of sustainable water management. It was reviewed in 2004 and extended as the National Water Initiative (NWI) (COAG, 2004). This was reendorsed by all governments again in 2014.

KEY ELEMENTS OF AUSTRALIAN WATER REFORM

The underlying objectives of the national water reform agenda were to transform the management of water in Australia to increase the productivity and efficiency of Australia's water use and ensure the health of river and groundwater systems while servicing the needs of rural and urban communities (COAG, 1994, 2004).

Doolan (2016a) reviewed the initial reform efforts of state governments, the collaborative efforts of the MDB governments and the national COAG and NWI reform frameworks, and identified four key overarching areas of water reform that have been pursued collectively over the past 30+ years, together with advances in two enabling areas. These are illustrated in Fig. 1 and include the following.

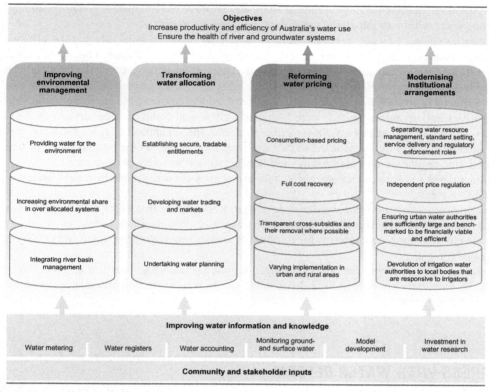

FIG. 1

Objectives and key elements of water reform in Australia over the past 30 years.

From Doolan, J. (2016a). The Australian water reform journey—An overview of three decades of policy, management and institutional transformation. *Australian Water Partnership. Canberra.*

Transforming water allocation: This involved moving from the old, administrative method of water allocation that assumed no environmental limits to the resource to a new system that works within sustainable resource limits, is market-based and provides economic value to individual water entitlement holders and the nation overall.

Critical steps in this process included, first, the conversion of existing ill-defined water rights into secure, legally defined, tradeable entitlements, which relate to the volume of water available, and, second, the establishment of diversion limits for surface and groundwater systems to protect the environment and the rights of existing users. These key steps were undertaken in a range of water-planning processes. Setting diversion limits effectively capped highly developed and overcommitted systems, which then created the driver for widespread water trading. The development of water trading rules and water markets was undertaken gradually in stages to minimize any adverse social and environmental consequences.

Improving environmental management: A market-based system of water allocation needs to be underpinned by a sustainable water resource base. This was achieved through providing a legally recognized share of water for the environment in all water systems, addressing overallocation in some water systems to increase the environmental share and improve environmental condition and, finally, managing water within an integrated catchment management context.

Reforming pricing of water services: Reforming pricing involved implementing the principles of consumption-based pricing, full cost recovery, and removal of cross-subsidies. This was aimed at promoting efficient and sustainable use of water resources and assets and improving the financial viability of water businesses by providing adequate revenue streams to fund service delivery.

Modernizing institutional arrangements: This involved transforming old institutions and local water authorities into organizations that were financially viable and could provide services to their communities efficiently and effectively. Key actions included the separation of roles for water resource management, service delivery standard setting and regulation as well as the establishment of independent economic regulation of water pricing and benchmarking authority performance.

Throughout the process of reform, these four key areas were underpinned by work undertaken in two enabling areas.

Ensuring community and stakeholder engagement: Recognizing that every aspect of water reform requires trade-off decisions that affect individuals, water users, communities, industries, regional economies and environments, mechanisms for community and stakeholder engagement were built into every facet of reform.

Improving water information and knowledge: Improving metering, monitoring, modeling, water accounting, and water knowledge to underpin advances in water planning and management.

Each of these areas is described in more detail in Doolan (2016a).

PROGRESS WITH WATER REFORM

Over the 30+ years since the 1980s, work has been undertaken in the areas described above in every state and territory, facilitated by the Commonwealth government. Enormous progress has been made in shifting toward a model of sustainable water management. This has been documented in regular reports by the National Water Commission (e.g., National Water Commission, 2011b, 2014) and summarized in Doolan (2016a).

Legally defined, secure, tradeable entitlement regimes are in place in all Australian states but one. Water markets have been established in a number of systems across Australia and the value of water to the economy is being revealed through these markets. The most mature of Australian water markets is in the MDB, where in 2016 the estimated total value of all water entitlements in the southern part of the basin was greater than $13 billion (Aither, 2016). In this system, the total value of commercial trade in 2015–16 was $660 million (Aither, 2016) and at least 30% of water available in any one year has been sold since 2007 (Grafton & Horne, 2014).

Water-planning processes have been undertaken in all states. Through these processes, undertaken with community consultation, caps for consumptive use have been set and rules established for environmental and consumptive use. Most states now have more than 80% of water used managed under some form of plan or equivalent (National Water Commission, 2014).

There are clear, statutory-based provisions of water for the environment made through all water plans or planning processes. In most systems, these are "rules-based," placing constraints on the take of consumptive users. However, in a number of regulated systems, this is supplemented with environmental water entitlements with the same characteristics as other entitlement holders.

Where the latter is the case, there is considerable discretion in how, where and when this water can be used (Doolan et al., 2017). In Victoria and at the national level, formal statutory institutions have been established with the responsibility to make these choices and to use environmental entitlements to achieve the best environmental outcomes possible. Currently, the Commonwealth Environmental Water Holder holds a range of water entitlements with varying reliabilities, equivalent to 1703 GL (long-term average annual yield) as at Sep. 30, 2016 (Department of Agriculture and Water Resources, 2016).

Progress on restoring overcommitted systems to a more sustainable level of extraction has been slow because of the significant economic and social implications of taking water from the consumptive pool to provide to the environment. Most success has occurred when it has been possible for governments to purchase entitlements or to invest in more efficient water supply and irrigation systems and practices, saving water that can be converted to environmental entitlements. In the MDB, the Commonwealth government has a major long-term initiative to restore the Basin to a more sustainable balance. This involved setting a new sustainable diversion limit (SDL) for the Basin and the investment of around $13 billion in irrigation modernization, on-farm efficiencies and buying entitlements on the market to implement the new balance (see Chapter 13). As a result of this, all catchments in the MDB should be returned to environmentally sustainable levels of diversion by 2024.

In terms of water pricing, significant progress has been made in metropolitan systems where there is full cost recovery (although there is still some debate about what this actually means) and generally a return on assets. In regional urban systems, there is generally full cost recovery except in some remote areas. In irrigation systems, reform has been slower, but for the most part prices cover the provision of services and asset maintenance and refurbishment. In some states, water prices include components that cover planning and management costs and/or some environmental externalities. Some form of independent economic regulation of pricing is in place in every state.

Water services are now generally of a high standard, and the performance of water utilities is nationally benchmarked annually on a range of service parameters (Bureau of Meteorology, 2016b). Drinking water provided by urban water utilities is monitored against national guidelines and standards and is consistently safe and of a high quality.

This progress has been accompanied by considerable advances in metering, monitoring, measurement, reporting, and forecasting and significant institutional reform. Governance has been significantly improved with the separation of policy, service delivery, and regulatory functions, and the consolidation of water utilities into authorities of a size that is financially sustainable and able to deliver quality services efficiently to their local communities.

THE MILLENNIUM DROUGHT (1997–2009)

The progress outlined here has been achieved despite the fact that this period included the Millennium drought. This drought commenced in late 1996 and lasted through to 2009, affecting much of southern Australia (van Dijk et al., 2013). It was the longest, most severe drought on record, producing harsh conditions affecting the more densely populated southeast and southwest of the continent, including the MDB and Australia's largest cities: Perth, Adelaide, Melbourne, Hobart, Canberra, Sydney, and Brisbane (South Eastern Australian Climate Initiative, 2011). The 10th year of the drought (2006) was an extraordinary year, with the lowest inflows on record for most of the MDB and southern Australia. This produced conditions that were well outside the boundary settings that, based on historical records, had been used in the design of water planning and entitlement systems, river operations, and water supply management (DELWP, 2016).

The impacts on communities, regional economies, and the environment were extremely severe. Towns were on the highest level of water restrictions. In the MDB, production of annual crops like rice and cotton was reduced by 99% and 84%, respectively, between 2002 and 2009, while production of perennial crops fell by 32% between 2003 and 2007 (van Dijk et al., 2013). There were consequential losses in employment, reductions in household income, downturns in local business, reduced recreational and tourism opportunities as local lakes dried up, and a general break in social cohesion in rural areas, with resulting increases in mental health problems and suicides.

From an environmental perspective, the combined effect of river regulation and drought meant that streamflows in some systems were around 80% less than would have occurred under natural conditions (SEACI, 2011). This resulted in widespread deaths of river red gums, reductions in waterbird numbers, threatened species extinction, and the large Ramsar-listed lakes at the downstream end of the MDB at imminent risk of widespread irreversible acidification (Doolan, 2016b). Moreover, climate change predictions suggested that similar conditions could occur more frequently in the future (CSIRO, 2008).

While the impact of the drought was undoubtedly severe, in general, the economic and social consequences of the drought were considered to be less harsh as a result of the reforms that had been put in place (Frontier Economics, 2007; Productivity Commission, 2010).

During the drought, the water market became a critical tool, enabling the scarce water that was available to move to where it had highest value. Because of the market, irrigators survived consecutive years of drought, though with varying levels of impact. They could decide whether to plant a crop with the possibility that they would have to buy more water, or to sell their seasonal water allocation to realize cash. If the farm was financially unviable, they could permanently sell their water entitlements.

As a consequence, use of the water market increased significantly during the drought. Since 2007–08 about 30% of water allocated in any water year has been sold on the temporary allocation market (Grafton & Horne, 2014). For the most part, water moved to support high-value perennial plantings and horticulture. It has been reported that, without trading, the dairy industry would have fared much worse than it did during the years of the drought, and many horticultural farms in Victoria and South Australia would not have survived the extremely dry conditions (Productivity Commission, 2010). The market also revealed the economic value of water, which then created a drive for greater water efficiency. In the critical period between 2005–06 and 2008–09, although water availability for irrigation dropped by 53%, the gross value of irrigated agricultural production only fell by 29% (National Water Commission, 2011a).

In urban systems, demand was managed through a combination of restrictions, significant water conservation and behavioral change programs. In southeast Queensland, this resulted in reducing water consumption by 60% from predrought volumes to an average of 125 L per person per day (Turner et al., 2016). However, as the drought continued, most major cities and many towns had to augment their water supplies (see Chapter 5). As they did this, they looked to increase the diversity of their supplies and make more efficient use of existing regulated systems to be more effective in a drier future. Water utilities increased their use of recycled water and stormwater, using these as a substitute for potable water. A number built system interconnections to make more efficient use of existing supply systems. Some augmented with groundwater or used managed aquifer recharge. Several in the MDB bought permanent water entitlements on the water market from irrigators. Finally, most of the capital cities built desalination plants as a climate-independent water source. For the most part, as a result of the reform process, most of these strategies and augmentations were funded through water pricing with the result that water prices increased rapidly (Doolan, 2016b).

While the environment suffered significant degradation during the drought, once again it was not quite as severe as it would have been without the reforms. The provision of water for the environment provided some means of ameliorating the dry conditions. Environmental entitlements were used to protect drought refuges, prevent catastrophic events, and save species from extinction. As a result of environmental watering, a number of serious biological losses were avoided (Mount et al., 2016). In a number of rivers, environmental passing flows had to be reduced because of the extremely dry conditions. However, when this did occur, it was to provide water only for critical human needs and was undertaken in ways that minimized as far as possible the impact on the environment. In Victoria, in a number of cases, there was some form of compensation either through water payback at a future time or the provision of funds from urban water authorities for complementary works (DELWP, 2016).

One of the major initiatives taken in response to the drought was the Commonwealth government's long-term initiative to reset the balance in the MDB. The drought showed that, despite previous efforts of governments to improve environmental flows in the River Murray, the Basin was still overallocated. The Commonwealth government provided around $13 billion to assist in establishing a new sustainable balance for the Basin. This involved establishing a new authority to develop a Basin Plan, which set a new SDL for the catchments within the Basin and the Basin overall (see Chapter 13).

Funding was then provided to acquire water for the environment to meet the SDL through buying water entitlements from willing irrigators on the water market and through investment in irrigation modernization and on-farm efficiencies. The latter aimed to assist irrigation industries and communities to adjust to the new lower diversion limit and maintain production with less water, increasing their drought resilience at the same time. This whole initiative has been highly controversial within the Basin communities, with considerable community outrage occurring over the process, the actual final SDL, the science and the socioeconomic analyses behind the establishment of the SDL, the final volume of water to be recovered from irrigators, and the means by which it is being recovered. The Basin Plan is a major initiative, which illustrates both the difficulty of water reform and what can happen when the process goes awry. However, it also demonstrates some of the advances in analyses, information, and community engagement that this has necessitated to get it back on track.

The Millennium drought severely tested the water management framework built through the Australian water reform agenda and showed it to be robust. In general, the economic and social consequences of the drought were less severe as a result of water reforms that had been put in place. While governments shifted their priorities during the drought to deal with water scarcity, for the most part the actions they took either accelerated much of the water reform agenda or were consistent with it.

THE PROCESS OF WATER REFORM

While there has been significant progress over 30 years in Australian water reform, none of it has been simple or easy—nor has it led to perfect outcomes. Every advance under each of the key elements of reform had impacts on individuals, industries, communities, regional economies, and environments. Often, these advances are taken during periods of water scarcity when communities are already experiencing hardship. Because of this, community and stakeholder engagement has been absolutely critical to the reform process. Water reform affects their futures and this needs to be understood and respected.

Each step forward in the reform process has required trade-offs to be made and settlements to be negotiated to enable a politically feasible way forward at that point in time. Where community and stakeholder engagement has not been undertaken well, proposed reforms have frequently foundered. The reality is that each step takes time to develop resources, and community acceptance and political capital take time to set up and time to consolidate. Each step then throws up new issues, which drive the next steps in the reform process. The pace of reform can only move as fast as community acceptance allows. The result is that water reform is an evolution of continuous improvement, occurring in a series of steps over decades.

Recognizing this, water reform in Australia over the past 30+ years has not just been the work of governments. Reform has involved communities, industries, local authorities, NGOs, and academics, with leaders emerging in all these areas over the decades. Often, the drive comes from governments, but just as frequently the impetus can come from stakeholder groups or communities concerned about issues or advocating for different outcomes.

The previous discussion has provided a high-level summary of the progress that has been made in Australian water management over the past 30+ years. Part 2 of the book provides more detail on this evolution from four key perspectives.

Part 2 starts with a long-term perspective capturing the history of water resource planning, policy and management in the MDB in the 100+ years since Federation in 1901 (Chapter 2, Guest). It outlines the key steps in reform and how they were achieved in a multigovernment, transboundary political setting, culminating in the recent Commonwealth government initiative to reset the MDB. This chapter provides a fascinating story of the political interactions between the Basin states (Victoria, New South Wales, South Australia, and more recently Queensland) and the federal government in water resources policy and management in the MDB.

The next four chapters in this section of the book cover the changes in water allocation and planning and the impacts of water reform for irrigation, major urban centers, and for the environment.

Fitzpatrick (Chapter 3) reviews the evolution of water entitlement frameworks and water markets in Victoria. The establishment of robust, secure statutory entitlement and water-planning processes is one of the critical steps in water reform. Secure water entitlement systems provide the foundation stone for the provision of water services, the protection of water environments, the operation of water markets, and future investment in water-related industries on which regional development is based. Ongoing community trust and support was highlighted as vital in underpinning these long-term reform processes.

Rendell (Chapter 4) provides a personal view of the changes that have occurred in the irrigation districts in the southern part of the MDB over the past 30 years. The focus is particularly on how individual farmers and farm businesses have modified their decision-making practices to take into account factors such as climate variability, changes in agricultural practice, the water market, and government policies.

Chong and White (Chapter 5) review the water reforms that have occurred in Australia's major cities. This commenced with the corporatization of the major urban water utilities in the 1990s, and since then has involved focus on supply-demand planning, the augmentation of supply options with the construction of desalinization plants, and increased emphasis on the integrated whole of water cycle management. They also speculate on some of the additional changes likely to occur in the future.

Bunn (Chapter 6) summarizes the significant water reforms that have been undertaken over the past three decades to improve the health of Australia's freshwater ecosystems and address overallocation. This started with the formal recognition of the environment as a user of water in the mid-1990s, following which governments introduced major changes to water legislation and planning to improve environmental outcomes.

DECISION MAKING IN WATER REFORM

Every step in the evolution of water reform requires decisions to be made that could potentially impact individuals, industries, communities, regional economies, and the environment in different ways. Because of this, decision making in water management in Australia generally involves a detailed process that includes community and stakeholder input to consider the issues, examine options, and understand trade-offs, to allow a broadly accepted position to be reached that can then be feasibly adopted by governments and practically implemented with communities. These decisions represent difficult trade-offs that balance social, economic, and environmental issues and sectoral impacts. Decision-making processes are run across all aspects of water resource management, and at local, state, MDB, and national scales, depending on the issue.

However, the most notable of these processes, where there are frequently competing views and hard trade-offs to be made, are those involving the following areas.

Water planning and allocation: where the aim is to provide a clear basis for the allocation of water entitlements for consumptive use, provide water for the environment, and develop implementation and operational arrangements. These processes are critical because they define the characteristics of entitlements held by consumptive users and the environment, establish whether and by what means water could be returned to the environment, and set up the conditions for water markets. Within a local water system, catchment or aquifer, they:

- Identify the nature and behavior of surface water and groundwater resources, resource availability under different climate scenarios and the environmental values and their water requirements,
- Involve key stakeholders and communities,
- Clarify and define the rights of existing users, including their relationship to water availability,
- Identify the agreed share of water for the environment,
- Cap water available for consumption,
- Develop implementation arrangements for the seasonal allocation of water for consumptive and environmental use, system operation, environmental obligations, and if relevant, market operation and investment programs.

The planning and allocation process in undeveloped water systems, while conceptually similar, is different from the processes used for highly developed or overcommitted systems, where there is real competition for water resources and real wealth and investment at stake. In undeveloped water systems, an environmentally sustainable diversion limit is set, the rights of any existing users are recognized, and a process by which additional consumptive water can be allocated is established.

In highly developed systems, the first stage of water planning generally establishes and defines the rights of existing consumptive users and caps the system so that no more water can be taken out, thereby allocating the residual water to the environment. Capping these systems means that no additional new water entitlements can be issued. This then creates the preconditions for a water market where water in the consumptive pool is able to be traded among new and existing users and enable water to move to higher-value uses. However, capping does not improve environmental sustainability—it simply stops the system getting worse. The second stage of water-planning processes in highly developed or over-allocated systems is to decide on whether to recover water from the consumptive pool to enhance the environment. These decisions can involve very difficult trade-offs for communities and can be highly controversial: the Murray-Darling Basin Plan process of the MDB is the most notable instance of this (see Chapter 13). Basins where this has occurred have set out pathways to achieve a more sustainable balance over time. The pathways can involve: (a) investment in efficient water and/or irrigation systems and on-farm efficiency with the water saved going to the environment, (b) purchase of water entitlements for the environment on the market, or (c) agreement to reduce entitlements at a rate of change to which the regional communities and economies can adapt over time.

Water-planning processes are undertaken at regular intervals for a water system as information improves, communities and industries become more sophisticated in their understanding of water management, the patterns of water use change and/or new risks to water resources become apparent. Each new iteration of a water plan builds on the knowledge of the previous plan and the experience of communities and industries in implementing it. With new knowledge and deeper experience, there are significant opportunities to improve system management rules, understand the surface water and groundwater connectivity, review the balance between environment and consumptive use and identify ways of moving closer to optimizing between environmental, social, and economic outcomes. Australian governments have developed a set of national water-planning guidelines which are used to guide state-based water allocation and planning processes (COAG, 2010).

Water service provision and pricing: the aim here is to work through types of water services to be provided to communities, the levels of service to be provided and the price to be paid for these services. These will include both urban and irrigation services and communities. In urban areas, they generally include discussions on the way to supply water in the future and will identify options for the next augmentation of water supply. These processes are of great interest to the communities involved and are critical to water authorities providing services as they establish the prices they are authorized to charge for the next pricing period, which is generally 5 years. These processes are undertaken regularly, are highly consultative and overseen by independent price regulation authorities who set guidelines for how they should be undertaken (e.g., the Victorian water pricing framework and approach; Essential Services Commission, 2016).

Environmental objective setting and catchment management: the aim here is to understand the environmental, social, and economic values provided by water environments and their catchments; understand trade-offs between environmental condition and various human uses of both catchment lands and water; and set objectives for environmental condition to provide the values that the community wants to see in the future. These processes often provide flow-related information to feed into water-planning processes. They also cover other critical elements of environmental condition, for example, river and aquifer water quality, habitat condition and catchment land use and management. These types of planning processes include: catchment management planning, waterway management planning, setting environmental protection standards, and integrated urban water management. This is where water management is set within the broader context of integrated catchment or natural resource management and integrates with water quality regulation undertaken by environmental protection agencies.

Regardless of the purpose, contemporary decision-making processes in Australian water management have a number of important and consistent attributes. These include:

- Comprehensive processes for community and stakeholder engagement to ensure that differing views are heard and taken into account, to provide sound information to all, and to ensure transparency in decision making.
- A triple bottom line (TBL) approach where the suite of economic, social, and environmental implications is recognized and understood as trade-off decisions are made.

- An evidence-based approach, using the best available science and socioeconomic analysis to underpin TBL decision making.
- An adaptive process, including regular review cycles to incorporate new knowledge, experience and emerging issues.

These four attributes are critical to successful decision making. And all four are referred to and required in some way by both the COAG water reform framework (1994) and the National Water Initiative (2004). Jurisdictions have had to report on their decision-making processes in their regular reports on implementation of the national water reform program. Undertaking decision making in this way is resource intensive and can take significant time (e.g., in Victoria, the water entitlement conversion processes took up to 3 years to complete in some systems; in New South Wales, many water plans took at least 2 years to prepare). However, these features build confidence in the planning and water management process, and the end result is one that is generally robust, understood by community and stakeholders, and able to be endorsed by governments and implemented.

These key features of Australian water planning and management have had significant implications for the state of water information and knowledge in Australia. Each step of the water reform process has been accompanied by investment in improved information, knowledge, evaluation methodologies, data management, and decision support systems. Advances have been made in water measurement, water monitoring, understanding of groundwater systems, and environmental flows, in socioeconomic analyses and in the development of water registers and water accounting—all to provide accurate, reliable, and relevant information to support decisions on water management not only by governments, but also by regional water authorities in their delivery of water services and by individual irrigators in the farm enterprises. These information bases are publicly available and underpin the operation of water management systems and the water market.

The preceding discussion provides a brief overview of how decision making has occurred during the 30+ years of Australian water reform. We have noted that contemporary water resources decision making should encompass TBL decision making, which, as far as possible, should be evidence based to enable informed trade-offs and judgments to be made. There is generally an assumption that the better the evidence that underpins the decision-making process, the more likely it is that better decisions will be made.

Part 3 of the book contains seven chapters that outline key elements of the essential TBL evidence base and the decision-making process in contemporary water resources decision making, particularly in relation to water allocation and planning.

Stewardson et al. (Chapter 7) review the largely eco-hydrological methods used in Australia to assess the environmental flow regimes required as a key input to water planning and allocation processes. These assessment methods are used to establish: the level of human impact on the river, estuary, groundwater-dependent, and wetland ecosystems; the environmental water requirements of the system; and potential environmental risks if the environmental water allocation scenarios fall short of the required environmental water.

James (Chapter 8) discusses the application of economic analysis in water planning and allocation, with particular reference to the Murray-Darling Basin Plan. The chapter covers the application of tools such as: cost-benefit analysis; valuing the environment; cost-effectiveness and cost trade-off analysis; economic impact analysis (input-output models and computable general equilibrium models); impact analysis at local community scale; and optimization of economic, social and environmental outcomes.

James notes the importance of ensuring that the economic analysis is focused at the appropriate level, for example, national, regional, and local.

Chapters 9 and 10 then cover the more social aspects of water resources decision making.

Bekessy and Selinske (Chapter 9) first review three frameworks for understanding social-ecological systems (Ostrom's social-ecological framework; the driver-pressure-state-impact-response framework; and the structured decision-making framework). They then review seven decision tools for modeling and analyzing social-ecological systems (agent based modeling, backcasting and forecasting, Bayesian belief networks, bioeconomic modeling, game theory modeling, social network analysis, and scenario planning and analysis). Finally, they describe pathways to mainstream the use of these approaches, and argue that more effort needs to go into the use of social-ecological modeling in water resources decision making.

Schirmer (Chapter 10) identifies key principles for assessing and managing the social effects of water reform, an often socially contentious exercise and one that is often a secondary consideration in the design and implementation of water reform. The development and implementation of the Murray-Darling Basin Plan is used to highlight three important principles: the need for good public participation; ensuring robust methods are used to assess the social outcomes of water reform throughout the development and implementation; and ensuring social effects are considered when selecting and designing the policy instruments used to enact water reform.

Tan and Auty (Chapter 11) discuss the importance of effective community engagement as an essential component of any water resources reform process. They analyze the community engagement processes adopted by the Murray-Darling Basin Authority (MDBA) during the 4-year period leading up to the successful adoption of the Basin Plan in 2012. They identify five key elements that underpin an effective community engagement process: engaging in places where people are on familiar territory and in places that people care about; taking the time to listen and share information; accepting that compromises will need to be made; seriously organizing and planning the process; and showing leadership in settings.

Pollino et al. (Chapter 12) discusses the need for more integrative approaches in water resource decision making. Integrative research applies to the multidisciplinary approaches that are central to tackling real-world, complex science, and social-political problems, where acceptable solutions often range across disciplinary boundaries. Recent literature on integration theory is reviewed, and three case studies from the MDB are used to demonstrate different approaches to integration, where each example has a different purpose and scale of application.

Part 4 of the book then provides three detailed case studies that outline recent reforms in three specific complex policy areas in Australia.

Hart and Davidson (Chapter 13) provide an analysis of the key decision-making aspects in the development of the Murray-Darling Basin Plan, with particular focus on the decision-making steps employed in developing the SDL, the key policy initiative that underpins the Basin Plan. This required a TBL approach to rebalance the overallocated Murray-Darling river system, specifically to provide more water for the environment with minimal impact on communities. They also provide an assessment of the updated decision-making process used to review and change the SDL set for the northern MDB.

Loo and Clarke (Chapter 14) discuss the process used to develop a new Victorian Waterways Management Strategy (VWMS), which provides the policy framework in which rivers, wetlands, and estuaries are managed in Victoria, including setting environmental condition objectives to support community values and integrating environmental flow management with other aspects of waterway management. It outlines the process of reviewing and refreshing the 2002 Victorian River Health Strategy. New management issues addressed in the VWMS included: wetland, estuary, and urban waterway management, Traditional Owner involvement, management of invasive species, and recreational use of waterways. They considered that clear management arrangements, a dedicated fund source, political authorization and partnership with communities were key factors that underpinned the successful development and implementation of this new policy.

Eberhard et al. (Chapter 15) review the management of water quality in the catchments affecting the Great Barrier Reef, an internationally recognized World Heritage site. In overviewing the efforts of the Commonwealth and Queensland governments to improve water quality over the last 15 years, they note that these efforts have failed to deliver measurable improvements to the health of inshore reef ecosystems. In this period, four phases of bilateral water quality planning and programs have developed scientifically robust targets and reporting systems. They conclude that, despite the strong science and partnerships that support reef policy and programs, greater effort is needed to overcome constraints to current management approaches and to employ the additional policy measures required to help sustain the GBR into the future.

These last two chapters move the focus away from water allocation and planning to environmental objectives and catchment planning and are examples of how water reform has integrated with the broader processes of integrated catchment and natural resource management.

THE AUSTRALIAN EXPERIENCE—INTERNATIONAL APPLICABILITY

The situations that led to the water reforms in Australia are common to many other countries. While the Australian water reform journey is not finished, elements of this experience to date have proved useful to other countries as they tackle their own water management challenges. Part 5 describes three instances where this has been the case.

Campbell and Barlow (Chapter 16) describe the involvement of working in the transboundary Mekong basin and identify six challenges in the international transfer of knowledge between developed and less-developed countries. These include: differences in the biophysical environment; lack of knowledge and local capacity; differences in socioeconomic conditions; and differences in culture (e.g., importance of age/seniority, agreement, and social networks).

Pollino et al. (Chapter 17) report on a 3-year project aimed at using Australian water resources expertise to assist India in building capacity and managing water at a Basin scale. The project was largely technical, focusing on the Brahmani-Baitarni River Basin located in the eastern part of India, and used Australian expertise in river system modeling to build capacity in central and state government engineers in India. A number of challenges were experienced during the project and these are discussed.

Gippel et al. (Chapter 18) discuss Australia's assistance in developing a framework for a national program to assess the health of China's rivers. There is currently no consistent nationwide assessment of river health in China. The framework was shaped by Australian experience, and tailored to suit the resources and expertise available in China, the biophysical characteristics of China's rivers, and the key management issues.

A key message from all of these is that while the Australian experience can be helpful in providing guidance and tools to other countries, nothing is ever directly transferable and will always require adaptation to suit the physical systems, capability, and problem requirements of the home country.

WATER REFORM IN AUSTRALIA—THE FUTURE

In Australia, substantial progress has been made over the past 30+ years in shifting toward a model of sustainable water resource management. However, there is still much to do in the next phase of Australian water reform. In each of the key reform areas outlined here, there is still work to be done. The challenge of climate change means that the water entitlements and markets regime and the new environmental water management arrangements will need to continue to be effective in a future predicted to be much drier in parts of the country already dealing with water scarcity. Water use efficiency will need to continue to be improved in all sectors: urban, irrigation, and environment. Population growth coupled with climate change will create serious challenges for our cities and towns. In these urban environments, the role of water services in contributing to livability will be a major focus in the future, as will making the best use of all water sources, including recycled and stormwater. In irrigation, the drive will continue for increased efficiency in water use and distribution, higher productivity from water use and achieving financial and environmental sustainability.

Across the water sector, the issues of energy use and carbon emissions are a significant problem currently under consideration. All these issues will raise questions about water services standards and maintaining affordability. Finally, the water sector will need to improve their connection with communities, including indigenous communities, to ensure they meet their needs in the future in a rapidly changing environment.

Chapter 19 of the book discusses these challenges in more detail and outlines some of the ways in which water resource managers will have to respond to address them in the future. These will form the basis of the next wave of water reforms in Australia.

REFERENCES

Aither (2016). *Water markets report*. 2015-16 review and 2016-17 outlook. Melbourne.

ARMCANZ/ANZECC. (1994). *National water quality management strategy*. Commonwealth of Australia Canberra.

ARMCANZ/ANZECC. (1996). *National principles for the provision of water for ecosystems*. Sydney: Sustainable Land and Water Resources Management Committee, Subcommittee on Water Resources.

Australian Bureau of Statistics. (2010). Available at: http://www.abs.gov.au/ausstats/abs@.nsf/Lookup/by Subject/1370.0~2010~Chapter~Water storage %286.3.6.2%292016.

Bureau of Meteorology. (2016a). *Water in Australia 2014–15*. Melbourne: Bureau of Meteorology.

Bureau of Meteorology (2016b). *National performance report 2014–15: Urban water utilities, part A*. Melbourne.

Council of Australian Governments. (1994). The Council of Australian Governments' Water Reform Framework. In *Extract from Council of Australian Governments: Hobart 25 February 1994 Communique*.

Council of Australian Governments (COAG). (2004). *Intergovernmental Agreement on a National Water Initiative between the Commonwealth of Australia and the Governments of New South Wales, Victoria, Queensland, South Australia, the Australian Capital Territory and the Northern Territory*. Canberra.

Council of Australian Governments. (2010). *National water initiative policy guidelines for water planning and management*. Canberra.

CSIRO. (2008). *Water availability in the Murray-Darling Basin: Summary of a report to the Australian Government from the CSIRO Murray-Darling Basin Sustainable Yields Project*. Canberra.

DELWP. (2016). *Managing extreme water shortage in Victoria—Lessons from the Millennium Drought.* Melbourne: Department of Environment, Land, Water and Planning.

Department of Agriculture and Water Resources. (2016). *Progress of registered water recovery towards sustainable diversion limits at 30 September 2016.* Available at: http://www.agriculture.gov.au/SiteCollectionDocuments/water/register-water-recoveries.pdf. [Accessed 29 November 2016].

Doolan, J. M. (2016a). *The Australian water reform journey—An overview of three decades of policy, management and institutional transformation.* Canberra: Australian Water Partnership.

Doolan, J. M. (2016b). *Building resilience to drought—The Millennium Drought and water reform in Australia.* Canberra: Australian Water Partnership.

Doolan, J. M., Ashworth, B., & Swirepik, J. (2017). Planning for the active management of environmental water. In A. Horne, A. Webb, M. Stewardson, M. Acreman, & B. Richter (Eds.), *Water for the environment*: Elsevier.

Essential Services Commission. (2016). *Water pricing framework and approach: Implementing PREMO from 2018.* Melbourne: Essential Services Commission.

Frontier Economics. (2007). *The economic and social impacts of water trading: Case studies in the Victorian Murray Valley: Report for the Rural Industries Research and Development Corporation.* Canberra: National Water Commission and Murray–Darling Basin Commission.

Grafton, R. Q., & Horne, J. (2014). Water markets in the Murray-Darling Basin. *Agricultural Water Management, 145,* 61–71.

Industry Commission. (1992). *Water resources and waste water disposal report no. 26.* Canberra: Australian Government Publishing Service.

Keating, J. (1992). *The drought walked through—A history of water shortage in Victoria.* Melbourne: Department of Water Resources Victoria.

Mount, J., Gray, B., Chappelle, C., Doolan, J., Grantham, T., & Seavy, N. (2016). *Managing water for the environment during drought—Lessons from Victoria, Australia.* San Francisco: Public Policy Institute of California.

Murray Darling Basin Commission. (1996). *The cap.* Canberra: Murray Darling Basin Commission.

Murray-Darling Basin Ministerial Council. (1995). *An audit of water use in the Murray-Darling Basin.* Canberra: Murray-Darling Ministerial Council.

National Water Commission. (2011a). *Water markets in Australia: A short history.* Canberra.

National Water Commission. (2011b). *The National Water Initiative – Securing Australia's water future: 2011 assessment.* Canberra: National Water Commission.

National Water Commission. (2014). *Australia's Water Blueprint: National reform assessment 2014.* Canberra.

Productivity Commission. (2010). *Market mechanisms for recovering water in the Murray-Darling Basin: Final report.* Melbourne.

South Eastern Australian Climate Initiative. (2011). *The millennium drought and the 2010/11 floods.*

Turner, A., White, S., Chong, J., Dickinson, M. A., Cooley, H., & Donnelly, K. T. (2016). *Managing drought: Learning from Australia.* Sydney.

van Dijk, A. I. J. M., Beck, H. E., Crosbie, R. S., de Jeu, R. A. M., Liu, Y. Y., Podger, G. M., et al. (2013). The Millennium Drought in southeast Australia (2001–2009): Natural and human causes and implications for water resources, ecosystems, economy, and society. *Water Resources Research, 49,* 1040–1057.

Water Services Association of Australia. (2003). *Submission to the House Environment Committee to examine the future sustainability of Australian cities.*

RECENT WATER RESOURCES POLICY AND MANAGEMENT CHANGES IN AUSTRALIA

MANAGING THE RIVER MURRAY: ONE HUNDRED YEARS OF POLITICS

C. Guest
Independent Consultant

CHAPTER OUTLINE

INTRODUCTION

The Murray-Darling Basin has always had a central place in Australian politics (see Chapter 13 for a map of the Murray-Darling Basin). When he was Prime Minister, Paul Keating described the Murray-Darling as[1]:

> Australia's greatest river system, a basic source of our wealth, a real and symbolic artery of the nation's economic health, and a place where Australian legends were born. Nowhere is the link between the Australian environment, the Australian economy, and Australian culture better described.

Conflict over the control and sharing of the waters of the River Murray was one of the most contentious issues at Federation in 1901. The River Murray Waters Agreement, signed by New South Wales (NSW),

---

[1]Keating, P. Environment statement launch speech, 21 December 1992, Adelaide.

Decision Making in Water Resources Policy and Management. http://dx.doi.org/10.1016/B978-0-12-810523-8.00003-3

Victoria, South Australia and the Commonwealth governments in 1914, was the political resolution to the conflict.

The Agreement has evolved through 15 iterations to the current Murray-Darling Basin Agreement, and is possibly Australia's longest standing intergovernmental compact. Its history can be seen as an allegory of the way in which the roles of Commonwealth and state governments, public policy instruments, social values, and the health of the river system have changed over the past 100 years.

THE ROAD TO THE FIRST RIVER MURRAY WATERS AGREEMENT

Water politics at the end of the 19th century was a troubled cocktail of the vast differences in the endowments of nature between the three states, their competing economic interests in the use of the water, and uncertainty about their legal entitlements to water.

Always sitting behind South Australia's position was the chronic anxiety that it was the "weaker state and the one lowest down on this great river system" (SA Parliament, 1904). Being at the end of the River Murray, it was vulnerable to whatever the upstream states did, along with being burdened by the driest climate in Australia and few economic opportunities. Not only that, but the state's interest in the river was for the transport of merchandise upstream and of commodities downstream ("navigation"). The considerable variability of the volume of water in the river meant navigation was a risky enterprise. Steamers were often stranded for months during droughts.

South Australia argued it had riparian rights to the water that would be naturally available from the river. The state's political case was that it was prepared to compromise the riparian rights, if NSW and Victoria constructed locks on the Murray and Darling rivers that would hold water back and assist the passage of boats.

South Australia's claim to riparian rights was tenuous, as it would have been difficult to enforce them between the separate self-governing states. In practice, South Australia's claim would have denied the other states' use of the water. New South Wales and Victoria were never going to agree to that and, as the upstream states, held the superior hand in any political conflict.

In the early 20th century, Victoria's irrigation development was extensive, with more in prospect. The area under irrigation was 125,000 ha, vastly greater than the 700 ha in each of South Australia and NSW. Victoria's annual water use was estimated to be 515 GL, while the other two states diverted 30 GL in total (IRC, 1902). So far as irrigators were concerned, water not used was "wasted" and that was a "flagrant injustice." It followed the objective should be to "irrigate the Murray dry" (MRMCL, 1902).

Victoria, however, was in the weakest legal position. As well as South Australia's claimed riparian rights, New South Wales asserted ownership of the water in the Murray under the *Imperial Act 1855*, which placed the border between the two states on the southern bank of the river, giving NSW ownership of the riverbed.

To prosecute its case, Victoria appealed to the "rights" of vested interest. As the first to take water for irrigation, a far more profitable use than navigation, its position was that the state's irrigation developments warranted secure access to the waters of the River Murray. This was a political argument, which pointed to Victoria's need for a political settlement.

For its part, NSW devoted attention to River Murray matters only as and when needed. With limited irrigation, no river trade, and the Murray being a long way from the capital, Sydney, the river was not prominent in the state's political life. This meant NSW did not bring the same fervor to the issue as the other two states. However, when NSW did become involved, it did so with the sense of being the largest, and on that basis, the most politically powerful government.

The impasse was a central issue that Federation was designed to overcome. Arguments about rights to water "delayed Federation and occupied more debating time in the federal conventions than any other issue" (Clark, 2003). The outcome in the Constitution was a compromise. Section 98 gave the Commonwealth power to regulate the use of the River Murray for interstate navigation. However, to meet NSW and Victoria's interests in ensuring their autonomy over water use, Section 100 prohibited Commonwealth interference with the rights of states and their residents to the "reasonable use of water for conservation or irrigation" (conservation at that time meant the storage of water). So the resolution of the River Murray question was held over for another day.

In 1902, in the depths of the Federation drought, irrigators convened the Corowa Water Conference (MRMCL, 1902). They wanted governments to build storages and distribution channels to underpin irrigation, and called on the Commonwealth Government to "take over" responsibility for the River Murray. When Prime Minister Edmund Barton rejected the resolution, the obligation to act turned to the state ministers at the Conference, who announced they would establish an Interstate Royal Commission on the "conservation and distribution of the waters of the Murray River Basin."

While it would be another 12 years before a River Murray Waters Agreement was ratified, its key elements can be traced to the report of the Interstate Royal Commission. They were:

- A settlement that recognized the interests of all three states without reference to their claimed legal rights.
- A volumetric entitlement flow to South Australia determined by the estimated needs of navigation.
- Water sharing between New South Wales and Victoria determined by their relative contributions to the river.
- Autonomy for the upper states in the use of tributaries.
- During droughts each state shared the reduced water proportionately.
- A program of building locks and dams for regulating and storing water.
- A "River Murray Commission" comprising representatives of the three states being established to administer the arrangements.
- A principle for cost sharing.

Following the Royal Commission, Premiers negotiated and signed four River Murray Waters Agreements, in 1903, 1906, 1907, and 1908. Equal funding shares and the need for each parliament to ratify the Agreement were both adopted in 1903. The requirement for decisions to be unanimous, which has exercised a fundamental influence on the way the Agreement has worked, was settled in 1907. One of the principal differences between the Agreements was the amount of water allocated to South Australia. This swung between the 1700 GL recommended by the Royal Commission and adopted in the 1903 and 1908 Agreements and 2120 GL in the 1906 and 1907 Agreements.

None of the four Agreements was put to a vote in the parliamentary debates on ratification. In Victoria, there was a sharp division between two opposed perspectives. For the minority "business view," it was *"worthwhile for [the] state to concede a great deal to bring about a settlement"* (Murray, 1903). The majority view was that the state should not enter into any Agreement that had the effect of *"compelling the people of Victoria to part with their birth right in the waters flowing upon Victorian soil."* [2]

---

[2]The Argus newspaper, 4 November 1903, Melbourne.

The Agreement was a low priority for NSW, which meant it was open to finding reasons not to submit a bill to parliament. South Australia chose to wait until the other states ratified.

Despite what must have been immense frustration, the commitment to reaching a settlement through a political process, rather than a legal resolution, was never compromised. Even South Australia, which repeatedly referred to the legal recourse available to it in the face of the resistance by the upper states to its demands, did not pursue a legal remedy. The commitment to a political process was due not only to uncertainty about the outcomes of legal action, but also to the belief that, with a political process, managing the conflict remained in the bailiwick of governments.

The failure of the states to implement a settlement showed that the experience of recurring serious droughts, on top of the need for certainty to underpin Victorian irrigation development, South Australia's navigation interest, the need for storage capacity to sustain irrigation and the clear economic benefit of this being a joint enterprise between NSW and Victoria, were not enough. More was needed. And that came in the form of Commonwealth money.

The 1914 Agreement was reached principally as a result of the Commonwealth contributing £1 million (only slightly less than the states themselves were each providing), and to a lesser extent, on it mediating the negotiations. In these ways, the Commonwealth saw itself as carrying out its responsibility for nation building, which was to strengthen the Federation by resolving the conflict over water between the states, and allowing them to increase their populations and generate prosperity.

In every way the 1914 Agreement delivered the framework recommended by the 1902 Royal Commission, and refined through the following four Agreements (RMWA, 1914). The new element was the Commonwealth money. On the political arithmetic, the Commonwealth funding bridged the gap between the cost of the extensive system of locks South Australia wanted, which exceeded its funding contribution, and what NSW and Victoria were prepared to pay to secure a resolution to their River Murray problem. For their part, Victoria and NSW were also funding the project from which they would benefit, a major storage on the upper River Murray. As South Australia was getting the locks it wanted without having to pay their full cost, it accepted an annual entitlement of 1547 GL.

The 1914 Agreement established the River Murray Commission, which the Commonwealth would chair, with the expectation that it would be a mediator among the states, as the party with no direct economic interest in the river. Decisions required unanimity. As a political compact among equals, a consensual process for trying to reconcile the competing interests in the River Murray water was the only arrangement that could have been agreed.

THE BUILDING YEARS (1917–73)

In the context of the desire for reconstruction and recovery after the First World War, Prime Minister Billy Hughes saw the River Murray Scheme as "a great national work of the utmost importance to land settlement and the development of Australia generally" (Premiers' Conference, 1920). The largest project in the scheme was the Upper Murray Storage, at the junction of the Murray and Mitta Mitta rivers. By early 1920, the storage, which became known as the Hume Dam, was only at an "elementary stage." Hughes wanted faster progress, and to this end tabled some significant reform proposals at the Premiers' Conference of May 1920.

By then, the cost of the works in the Agreement had increased from the initial budget of £4.7 million to an estimated £8–10 million. Hughes offered to assist states to meet the additional costs by removing the Commonwealth's £1 million funding cap and becoming an equal partner with the states for capital expenditure.

There were two conditions to the offer. One was that the River Murray Commission be responsible for originating, designing, and managing projects, instead of the individual states. The other was to replace the unanimity requirement for decisions. According to Littleton Groom, the Commonwealth President of the Commission at that time, the need for unanimity meant the "Commission cannot go any faster than any state will allow it." Hughes proposed that a three-quarters majority be required for all decisions.

Persuaded by the funding offer, the Premiers accepted the reforms. Ratification by the Commonwealth, Victorian and South Australian Parliaments was prompt and untroubled. However, harmony ended in NSW. After its passage through the Legislative Assembly, the bill went to the Legislative Council on 23 Dec., the last day of the session. It appears the debate finished shortly before 4 a.m. on Christmas Eve, not usually a time for "sober thought."

Joseph Carruthers led the Opposition response. His concern was NSW being outvoted under the three-fourths rule. He asked the Council, "What mercy are we likely to get in a period of extreme difficulty?" (NSW Parliament, 1920). Arguing that above all else the state's interests needed protection, Carruthers moved that the Agreement be amended to require unanimity for decisions about water shares in times of drought. The motion became a rallying call for expressions of the importance of protecting NSW interests and was carried.

Without any debate, the Council also agreed that state governments remain the construction authorities, and so rejected another key element of the 1920 Agreement. The Legislative Assembly supported the Council amendments, again without debate, and with this brought the 1920 Agreement to an end.

State Ministers met with the Commonwealth in May 1923 to renegotiate the Agreement (Ministers'Conference, 1923). Once again, negotiations were taking place under the pressure of a drought; in Mar. 1923 the River Murray had stopped flowing at Swan Hill. In every other way, much had changed in the 3 years. Hughes was no longer Prime Minister, having been replaced by the conservative Nationalist, Stanley Bruce. What was more, the need to expedite construction work had dissipated—"great progress" was being made on the Hume Dam.

With one key exception, the 1923 Agreement reversed the previous Agreement. States would continue to be design and construction authorities and the requirement for unanimity was restored. On the other hand, the Commonwealth agreed to remain an equal funding partner. This was becoming more and more significant, as the estimated cost of the scheme had increased to £12 million.

Having restored the governance status quo, now underpinned by Commonwealth funds, the business of the Agreement returned to the capital works program, with extended attention being paid to the capacity of the Hume Dam. The initial design capacity of 1233 GL was reviewed at a conference of ministers in Aug. 1924. The Commonwealth, NSW and Victorian governments wanted the dam to be 1850 GL. South Australia believed construction should proceed to 2500 GL. South Australia's fear was that the lower capacity would be absorbed by upstream developments, and believed its allocation would only gain additional protection with a larger dam.

Under the yoke of unanimity, ministers could only agree that construction should proceed for a capacity of 2500 GL, with the "ultimate capacity" to be determined by a conference of ministers in 3 years' time (Conference of Ministers, 1924). In practice, the River Murray Commission found that the decision left too much uncertainty about what was to be constructed, and in 1926 requested a final decision. Ministers endorsed a capacity of 2500 GL.

In early 1930, the darkest days of the Great Depression and government fiscal stringency, the capacity was reduced back to 1500 GL (the size of the dam when completed in 1936). At the same conference, Ministers also asked for a report from the River Murray Commission on the advisability of constructing works or allocating water to protect the lower Murray against high levels of salt.

In response, the Commission reported: *"There appears to be no question but that, in the future, when the [flow freshes] in the Murray River are stored in the Hume Reservoir, and the upper riparian states are making fuller use of their tributaries, conditions on the lower Murray will be much worse than they have been in the past"* (RMC, 1931).

To tackle salinity, the options were constructing barrages at the Murray Mouth or granting additional water to South Australia. Neither NSW nor Victoria was prepared to consider more water for South Australia, so the Commission recommended five barrages, costing £549,000. It also proposed that nine navigation locks not be constructed. Only South Australia dissented, while suggesting that it might agree if the barrages were accepted as part of the River Murray Scheme.

Ministers considered the Commission's advice in Jul. 1934, and decided to proceed with the five barrages and that the remaining navigation locks not be built. They also agreed the Hume Dam could be increased to 2500 GL, if supported and funded by the states.

The 1934 Agreement was a turning point for South Australia's navigation interest. In speaking to the bill, the South Australian Minister of Agriculture said that with a total net cost of the new works of £1.2 million, South Australia's share would be £300,000. This, he said, was very favorable for his state, given the £549,000 cost of the barrages. He did not provide any explanation for the remaining locks not being built, and no one took up the issue (SA Parliament, 1934).

The era of navigation was over. With the effluxion of time and change in circumstances, something that had once been of great moment "stepped gently to the rear" (Murray, 2006).

Another severe drought, which left the Hume Dam nearly dry, struck in the middle of the Second World War. The consequences were more serious than ever, because the dam had stimulated extensive new irrigation developments, particularly in Victoria. In Oct. 1943, Victoria and NSW wrote to Prime Minister John Curtin, asking that a conference of ministers be convened to consider increasing the capacity of Hume to 2500 GL, even though under the 1934 Agreement the matter was one for states to agree and fund.

At the meeting of ministers in Jan. 1944, Victoria was the most ardent supporter of the idea. They argued that increasing the Hume volume was one of the "surest and safest ways of adding to the population" and an increased population was required "to defend the Commonwealth." New South Wales, on the other hand, claimed not to be getting any benefit from the existing Hume Dam, but then called on the Commonwealth to fund a portion of the required expenditure. Consistent with its position from the beginning, South Australia's support was on the condition that the additional water was used for meeting the state's entitlements under the Agreement, and as a reserve in dry years (Conference of Ministers, 1944).

When Ministers met again in May 1945, the Commonwealth advised that it was prepared to share the cost of the project, subject to the introduction of controls over the purposes for which irrigators could use water and over catchment management. The Commonwealth argued that, while "Murray Valley works" were of a "national character," development work by the states was "largely independent and uncoordinated" and needed to be "comprehensively controlled" by an "enlarged" River Murray Commission or a new authority (Conference of Ministers, 1945).

The Commonwealth was also concerned about the build-up of silt in the Hume catchment as a result of deafforestation and erosion. The catchment was the responsibility of NSW and Victoria, who acted independently of each other. To address the problem, the Commonwealth wanted the Commission to exercise control over the catchment.

The states resented the conditionality of the funding offer, which they said was not in the spirit of the Agreement. New South Wales objected to every part of the Commonwealth position, arguing that each of the proposed controls was already being implemented by the states and the problems were being overstated. Victoria was "opposed to giving any outside authority power to say how Victoria shall use any waters allotted to her as a result of the River Murray Agreement…it is our job and we are going to maintain it." The Victorian Minister declared that the Commonwealth should simply provide the money and leave everything else to the states.

Matters remained dormant until a Premiers' Conference in Aug. 1946 when Prime Minister Ben Chifley presented a considerably diluted position. The Commonwealth would contribute one quarter of the cost of the Hume expansion, subject to only one condition, that the River Murray Commission be given the power to take legal action where breaches occurred in relation to the protection of the Hume catchment.

New South Wales resisted the proposal, and the debate then played out through correspondence over the following 2 years. The resolution was that NSW and Victoria report to the Commission on the condition of the catchment. The Commission would have the power to inspect the catchment and advise the relevant state if actions were required. The state would then be obliged to remedy the position "in so far as may be practicable" (RMWA, 1949). And so, for a modest change, the states secured Commonwealth funding for the Hume Dam expansion.

The final decision on the Hume Dam, taken in 1954, was to expand further its capacity to 3000 GL, the present-day size. The increase was a consequence of the Snowy Mountains Hydro-Electric Scheme, which involved channeling water from the Tooma and Snowy Rivers to the Murrumbidgee and Murray Rivers. While the decision to expand to 3000 GL was straightforward, the Snowy Scheme triggered the first serious test of relations between governments under the Agreement.

In early 1956, South Australia learned that NSW and Victoria intended to share the additional water that would be diverted into the Murray system, and were negotiating with the Commonwealth an agreement to this effect. South Australia was excluded from the negotiations. So, for the first and only time, it pursued a legal remedy, taking out a High Court writ for a declaration that the *Snowy Mountains Hydro-Electric Power Act 1949–56* was beyond the power of the Commonwealth and therefore invalid.

The writ served its purpose. Prime Minister Bob Menzies promptly convened a series of meetings, under the aegis of the Agreement, to negotiate a resolution. The outcome recognized the additional water as being subject to the Agreement, and gave South Australia its share in times of restrictions.

This, however, was not the end of the affair for South Australia and its Premier, Tom Playford. For Playford, the Snowy episode demonstrated that South Australia needed to look after itself, which meant having its own water storage. Without that, he repeatedly told South Australians, the increasing diversions by the upstream states meant water to his state would be "only a trickle" by 1970.[3]

In Mar. 1960, Playford announced that South Australia would build a dam at Chowilla. The site for the proposed dam wall was 10 km downstream of the border with Victoria. The dam would be huge and shallow, with a capacity of 5900 GL and cover 520 km^2.

---

[3]The Advertiser newspaper, 4 March 1960, Adelaide.

At Menzies' insistence, the proposal was referred to the River Murray Commission for advice. The Commission concluded that the Murray needed additional regulation to give greater security during droughts, and that Chowilla was the most effective and economical means of doing this (RMC, 1961). And so in 1963 the governments signed an Agreement to proceed with Chowilla.

The Chowilla project started to unravel very early in its life. While the forecast cost was $28 million, the lowest tender bid in Apr. 1967 was $68 million. In addition, there were high levels of salinity at the dam site. Chowilla was intended to hold South Australia's water entitlements, leaving the water in the Hume Dam for Victoria and NSW. Having to supply a flow from Hume to manage salinity at Chowilla would considerably reduce the benefit of the new dam to Victoria and NSW.

In the light of these problems, the River Murray Commission investigated alternative sites and recommended Dartmouth on the Mitta Mitta River. Dartmouth would store water from the Victorian high country snowfields, the largest single catchment for the Murray. A new Agreement endorsing Dartmouth and deferring Chowilla "until the contracting governments agree that work shall proceed" was signed in early 1970. The Commonwealth Minister, David Fairbairn, made clear the meaning of the deferral: it was Dartmouth or "nothing" (Aust Parliament, 1970).

Playford's successor as Premier, Steele Hall, signed the Dartmouth Agreement, despite the continued strong desire for Chowilla in South Australia. To garner support, Hall secured an increase in the state's yearly entitlement under the Agreement, from 1547 GL to 1850 GL. However, this was not enough. The bill to ratify the Agreement was defeated in the South Australian Parliament and the government lost the subsequent election.

At first the new Premier, Don Dunstan, tried to secure support from other governments for a commitment to both Dartmouth and Chowilla. However, in Aug. 1971 he announced that these efforts had "failed." Having been rebuffed, he told the Parliament he was forced to seek ratification of the Agreement endorsing Dartmouth. Although the debate on the bill was long and bitter, it passed comfortably on the government's numbers.

Following some delays to deal with an increase in the estimated cost of Dartmouth, work commenced in Feb. 1973, and the dam was completed in 1979 with a capacity of 3850 GL.

With Dartmouth Dam, the Agreement's building years ended. It was a period of great achievements. Over 60 years, an extensive network of river infrastructure had been built and a body of practice had evolved for managing the assets and sharing the water. Both were the product of the Agreement and its processes. In these ways, the Agreement fostered and underpinned the industries, employment, communities, and settled ways of life of an Australian food bowl.

TACKLING WATER QUALITY AND ENVIRONMENTAL DEGRADATION (1973–2006)

SALINITY

During a severe drought in the mid-1960s, evidence started to emerge that the River Murray was facing a water-quality problem. The issue was salinity, which impacted adversely on agricultural productivity, the river's ecological health and urban consumers. A milestone study, initiated by the River Murray Commission, found half of the salinity load in the Murray was from "man's activities," and the "greatest source" was the "irrigation schemes" (RMWP, 1973).

In early 1973, Prime Minister Gough Whitlam met with the NSW, Victorian, and South Australian Premiers to consider what to do about salinity (Whitlam, 1973). They were able to agree that the answer lay in giving the Commission a new "supervisory" role in relation to water quality. While South Australia wanted the Commission to have greater "control," NSW and Victoria believed a new role should be confined to "coordination" and "advice." "Supervision" was the compromise on the day and with that the work of defining the new role was handed to officials.

It took 9 years before an Agreement was reached in 1982 to give effect to the outcome of the 1973 meeting. A significant reason for the delay was the resistance by NSW officials to any new water qual-ity role for the Commission on the basis that the responsibility should stay with states. A serious drought in 1980–81 led to high salinity levels and the closure of the Murray Mouth for the first time in European history, and showed something had to be done. However, the Commission's new role was modest. It was to monitor water quality, and investigate and recommend corrective measures. No new funding was provided and the Commission's staff establishment remained at 11.

Recognizing more needed to be done, in Nov. 1985 Ministers met in Adelaide to set a new course for the governance and scope of the Agreement. The headline decision was to establish a Ministerial Council, to "wrest control from the engineers." To address the range of the causes of water-quality problems, the scope of the Agreement was broadened from the river to the Basin landscape. Ministers from the environment and agriculture portfolios would join water ministers on the new Council. Sub-sequently, the Ministerial Council established the Murray-Darling Basin Commission to replace the River Murray Commission, with a membership that mirrored the Council. Provision was also made for Queensland to become a party to the Agreement.

Water-quality issues challenged the Agreement. Making decisions about building structures that served economic development was far easier than agreeing on ways to tackle water quality. Addressing the prob-lem would entail costs, by way of constraints on states' autonomy, through restrictions on land uses, irrigation practices and vegetation clearing, the economic costs of limiting irrigation, and the financial costs incurred on salt-interception schemes. No state wanted to give up its autonomy, impose restrictions on key stakeholders or allocate their budgets to programs with benefits outside the jurisdiction.

For these reasons, the 1982 and 1985 reforms were light handed. An advisory water-quality role, and a Ministerial Council and Commission with wider memberships, were all to the good. But the requirement for unanimity and the absence of additional powers and funding meant that, while the scope of conversations in the governance bodies was greater, nothing in the reforms provided the capacity actually to do things. Their effectiveness required "goodwill" on the part of the states.

INTRODUCTION OF THE "CAP"

Over the summer of 1991–92, an event occurred that crystallized the Basin's water-quality problems. The Darling and Barwon Rivers in NSW experienced the largest ever-recorded toxic algal bloom. By generating a single, strong image of what was going wrong with the rivers, the algal bloom had a sub-stantial political impact. Research by a NSW agency found that the principal cause of the bloom was a lack of flow caused by upstream extractions for irrigation.

Water consumption in the Basin had reached 11,000 GL per year. Median annual flows from the Basin to the sea were only 21% of what would have occurred under natural conditions. The lower reaches of the Murray were experiencing severe drought-like flows in over 60% of years, compared with 5% under natural conditions (Close & McLeod, 2000).

Research found that a healthy river system depends on a minimum sustaining flow and is likely to exhibit significant stress if the flow regime is below around two-thirds of the natural level (Dole, 2002). By the early 1990s, the River Murray flows were considerably less than this. As a result, native fish were in severe decline, and wetlands and red gum forests were suffering from a reduction in the frequency of flooding.

The evidence of environmental stress, and the finding this was linked to low river flows resulting from irrigation diversions, caused a shift in the question being asked. The issue moved from addressing salinity through mostly landscape measures, to finding the right balance between the extraction of water for irrigation and the need for water to be left in the river to support a healthy riverine environment.

The articulation of the problem in this way pushed the evolution of thinking to perhaps the single most important decision made by the Ministerial Council. The decision was to limit or "Cap" diversions of surface water for irrigation. Within the Ministerial Council process, the origin of the decision lay in a paper presented by South Australian Minister John Klunder in Jun. 1993. Based on work by Commission staff, the paper argued that the Basin's surface waters were overcommitted and environmental health was declining. What was required, the paper concluded, was *"consistent and comprehensive political direction."* In response, the Ministerial Council asked for a report from the Commission *"on the issue of no further regulation and diversion arrangements which would exacerbate the deteriorating flow regimes"* (MDBMC, 1993).

The South Australian paper had been delivered at the same time as the Council of Australian Governments (COAG) commissioned a report to recommend a strategic framework for water reform, addressing among other things "unsustainable natural resource use." The report found extensive evidence that rivers were overallocated and stressed and existing policies were not going to *"halt the process of natural resource degradation, particularly in the Murray-Darling Basin"* (Neal Report, 1994).

In response, COAG announced that *"arrangements would be instituted and substantial progress made by 1998 to provide a better balance in water resource use including appropriate allocations to the environment in order to enhance/restore the health of river systems"* (COAG, 1994).

Despite the clear signal from COAG, preparing the report for the Ministerial Council was fraught. A year after the Council's request, its premise was still being debated. The minutes of Commission meetings record Commissioners arguing that "it was wrong to say now that diversions should be limited," and expressing concern about the impact on "current water users and their capital investments" (MDBC, 1994). However, after 2 years of debate, they agreed to recommend to Ministers that surface water diversions be limited to 1993–94 levels. New irrigation development could still occur, but would require that water be obtained by improving water-use efficiency or purchasing water.

When the issue came to the Ministers in Jun. 1995, there was a new Minister from NSW at the table, following the defeat of the Coalition Government by Bob Carr's Labor Party. The change was critical to the outcome. The Minister discarded his predecessor's resistance, ignored his officials' advice not to support a Cap "at this time," and with other Ministers agreed that "diversions must be capped" (MDBMC, 1995).

Not only was the Cap important of itself, a consequence was that it decisively brought the Agreement back to focusing on water, compared with the integrated catchment management paradigm that had until then underpinned the thinking of the Ministerial Council and the Commission.

As is often the case, the implementation of the Cap was even more difficult than the decision. A 5-year review conducted in 2000 found that while Victoria and South Australia were complying with the Cap, Queensland and the Australian Capital Territory were yet to establish Caps. New South Wales was exceeding the Cap in three major river valleys. Looking to the future, the review described the Cap as an "essential first step in providing for the environmental sustainability of the river system of the Basin." Clearly, it had concluded more was needed. For one thing, the review recommended that groundwater be managed "within the spirit of the Cap" (MDBC, 2000).

In being set at 1993–94 levels of development, the Cap was as politically palatable as it could be, while still establishing a benchmark for water extractions. Burden sharing between the states was determined by drawing a line in the sand at existing development. However, the corollary of setting the Cap at a historic level of use was that it bore no relation to achieving any particular environmental objectives. In the absence of other measures, achieving stronger environmental outcomes would require the Cap being set at a lower level, and there was no policy framework for doing that.

ENVIRONMENTAL FLOWS AND THE LIVING MURRAY INITIATIVE

On the night before a Murray-Darling Basin Commission meeting in Mar. 2001, Senator Robert Hill, the Commonwealth Environment Minister, wrote to the President of the Commission seeking an increase in environmental flows to keep the Murray mouth open and deliver environmental benefits to the Coorong. Senator Hill also made his case at the Ministerial Council. The Council directed the Commission to prepare an options paper on increases in environmental flows, alternatives for providing the water, their costs and benefits, any institutional issues, and consultation arrangements.

Environmental flows posed difficult scientific, policy, and political questions. By considering increased environmental flows, the Ministerial Council was acknowledging that the Cap was not enough. If this was the case, then the questions were the levels of river health to aim for and the amount of additional water required, taking account of any adverse social and economic consequences, and finding ways of securing the additional water.

The Commission was asked to examine three options for annual environmental flows: 350, 750, and 1500 GL. However, states could only identify projects yielding 335 GL, at an unfunded cost of $445 million. In the face of this, work then focused on 350 and 500 GL scenarios, but proceeded slowly.

Unexpectedly, Prime Minister John Howard announced in Aug. 2003 that he would propose to COAG a "new $500 million fund to fight the River Murray's decline," to which the Commonwealth would contribute $125 million. Howard wanted governments "to recognize the need for a historic 'coming together' to help the ailing Murray."[4] In the face of state hostility, the Commonwealth agreed to increase its contribution to $200 million—a significant departure from the enshrined principle of equal funding—and with that the new fund was approved.

Three months later, the Ministerial Council made the "First Step" decision of The Living Murray (TLM) project. An additional 500 GL would be recovered and managed to improve environmental outcomes at six "icon" sites on the river. The water would be sourced from engineering works, water savings and the purchase of water licenses. South Australia urged that still more needed to be done, and looked to a longer-term target of 1500 GL. Outside government, the Wentworth Group of Concerned Scientists was demanding 4000 GL.

One state official from the time believes that without COAG's decision and the Commonwealth funding, "The Living Murray would never have happened." By taking the issue to First Ministers

---

[4]Australian Associated Press, 23 August 2003.

at COAG, Howard gave the issue the highest level of leadership and channeled substantial new funds into the mix.

By 2005, TLM was underway and the Cap was close to being fully implemented. And yet all was not well. Drought, a recurring trigger for change, was setting in. As also in the past, South Australia was restless. At the Sep. 2005 meeting of the Ministerial Council, the South Australian Minister for the River Murray presented a paper on "Revitalizing the Murray-Darling Basin Initiative."

The paper argued there were uncertainties about the volume of water TLM would find. At the time, projects expected to yield 240 GL had been identified, only half of the 500 GL target, and 169 GL of the 240 GL was from Victoria. Apart from asking Council to acknowledge publicly the difficulties being encountered in implementing TLM, the Minister's central proposition was that water-savings projects would not yield the volume of water required for the environment. Other ministers were not prepared to embrace these ideas, and simply acknowledged the "challenges" of delivering the First Step decision (MDBMC, 2005).

At the same time as the Agreement was wrestling with the Cap and TLM decisions, the Commonwealth started to take a more prominent role in water reform, through the *National Action Plan for Water Quality and Salinity* in 2000 (NAP, 2000) and the *National Water Initiative* in 2004 (NWI, 2004). Both overlapped with the Agreement. The former delivered onground, funded, catchment-based programs for water quality. The latter presented a detailed blueprint for the goals and processes of water planning. Yet the Ministerial Council and Commission did not have a significant role in either of them.

A NEW ERA

On Australia Day 2007, the pressure of the worst-ever recorded drought, the politics of the time, and a conception of national leadership by the Commonwealth, buttressed by its fiscal strength and a commitment to water reform by John Howard, triggered a proposal for the Commonwealth to take over water management in the Murray-Darling Basin and end the Murray-Darling Basin Agreement.

In a speech to the National Press Club, John Howard presented "*a $10 billion 10-point plan to improve water efficiency and to address the overallocation of water in rural Australia, particularly in the Murray-Darling Basin.*"[5] The proposed expenditures were conditional on governance being put on what he described as a "proper national footing." This required "an end to the parochial pursuits of state interests," which would always be present, he said, as the states had "competing interests." "We could," Howard continued, "muddle through as the states have been doing, but, frankly, that gets us nowhere." Instead, "decisive action" was needed.

The National Plan for Water Security would rein in overallocation in the Basin through a "sustainable Cap" on surface and groundwater extractions. The new sustainable Cap would be central to a "strategic plan for the Basin." The new cap was to be complemented by investments to recover water and deliver the reductions in water use required to achieve it.

A major part of the investment program was to line and pipe delivery channels and increase on-farm irrigation efficiency. Almost $6 billion was budgeted for these investments, which were forecast to save 3000 GL of water. Up to $3 billion would be allocated to "adjust water entitlements" in the Basin.

---

[5]Address to the National Press Club, 25 January 2007, Canberra.

This, Howard said, was "the Commonwealth assuming responsibility for a problem created by the states." His Government, he promised, "stands ready to provide structural adjustment and, if necessary, to purchase water allocations in the market."

And then came the crunch for the states, and the Agreement, the point to which all the criticism and spending proposals were leading. Howard announced he would *be writing to all relevant State and Territory leaders requesting that they refer to the Commonwealth their powers of water management over the Murray-Darling Basin.*

New South Wales was the first state to declare a position on Howard's proposal. The day after the speech, the Premier, Morris Iemma, announced he was "100% in favor of a national approach to water."[6] There was an expectation Labor would face a difficult time in the upcoming Mar. state election, and support was intended to "take the politics out of the issue." In addition, NSW was regarded as having the most serious overallocation of water and so stood to receive more of the $10 billion than any other state.

The response by Victoria's Premier, Steve Bracks, was to say he wanted "to protect Victoria's opportunities and rights," and would only support the "takeover as long as Queensland and NSW irrigators were forced to sacrifice water instead of Victorian ones."[7]

A "disappointed and surprised" South Australian Premier, Mike Rann, welcomed the additional funding, but was concerned that his state would be surrendering control of the river to the eastern states, because their politicians dominated the Commonwealth Parliament. For this reason, he urged an independent commission, rather than a Commonwealth Government entity, be established to manage the Basin.[8]

Like Bracks, the Queensland Premier, Peter Beattie, emphasized he would "protect the rights of our farmers." However, unlike his Victorian counterpart, he kept the way open to joining with Howard. As he put it, while Queensland had "real and specific problems with the Prime Minister's plan," he was "optimistic" he could "work these issues through with the Prime Minister" (Qld Parliament, 2007).

To find a way forward, Beattie and Rann presented a proposal for the new body responsible for the "strategic plan for the Basin." The Commonwealth would appoint the chair and two members and the states would appoint the other two. Members would be experts and not jurisdictional representatives. The new authority would advise a Commonwealth Minister. If the minister overruled the advice, the Minister's reasons for doing so had to be tabled in the Commonwealth Parliament. When the Commonwealth accepted the Premiers' proposal, the two states moved behind the plan. Only Victoria held out.

Toward the end of Jul. 2007, just days before Bracks announced his resignation from Parliament, he wrote to Howard to tell him that Victoria would not refer the powers the Commonwealth was seeking. The Commonwealth then resolved to proceed to the extent possible under its own Constitutional powers. The core elements of the resulting Water Bill were (Aust Govt, 2014):

- The establishment of a Murray-Darling Basin Authority—this was to be an independent, expert entity, under the governance of a five-person authority.
- The Murray-Darling Basin Authority would produce a 10-year Basin Plan.

---

[6]Daily Telegraph newspaper, 26 January 2007, Sydney.
[7]The Australian newspaper, 26 January 2007, Sydney.
[8]The Age newspaper, 26 January 2007, Melbourne.

- The Basin Plan would specify *sustainable diversion limits* (SDLs) for all of the surface water and groundwater catchments in the Basin.
- The SDLs were to be given effect through the states' water resource plans—this was a critical difference with Howard's Australia Day proposal, and reflected that the states retained their water management powers.
- The Basin Plan would include an environmental watering plan.
- A Commonwealth Environmental Water Holder would be established as a statutory office, with responsibility for managing the water recovered through the efficiency and buyback programs, to achieve improved environmental outcomes in the basin.
- The Commonwealth Minister could make water charge and water market rules, on the advice of the Australian Competition and Consumer Commission.
- The Bureau of Meteorology was tasked with collecting, holding, and reporting on information about water resources.

When the bill came to the Commonwealth Parliament, it enjoyed bipartisan support. The resulting Water Act 2007 was the first-ever Commonwealth water legislation (Aust Govt, 2014).

The Murray-Darling Basin Agreement, the Ministerial Council and the Commission all remained in place, unchanged alongside the Water Act and the new Murray-Darling Basin Authority, in parallel worlds as it were. While the Commission and its staff continued to serve the Ministerial Council, the Authority had no resources supporting its functions.

Not only were the two sets of arrangements addressing the same issues of overallocation and environmental health, there was no connection between them. Although the implementation of the Basin Plan was to be through the water resource plans of the states, there was no formal means of engaging with the states on the Basin Plan. Without that, securing compliance by states with the Plan was going to be problematic.

After 11 years in government, the Howard government was defeated at the Nov. 2007 election. An early priority of the new Labor Government was to address the unfinished business of the Water Act reforms.

The first step was a Memorandum of Understanding with the states. Under the Memorandum, the Murray-Darling Basin Authority and the Commission office would be "brought together as a single institution, to be known as the Murray-Darling Basin Authority" (MDBA), which would be a Commonwealth entity. This required the states to transfer the functions of the Commission office to the new body. The latter would then have two functions: developing, implementing, and monitoring the Basin Plan, and the functions set out in the Agreement.

A new Ministerial Council, comprising only water ministers, was to be established as the governance body for the Agreement and to provide advice to the Commonwealth Minister on the Basin Plan. The Murray-Darling Basin Commission would be abolished and replaced by a Basin Officials Committee (BOC). The BOC would have the same role as the Commission in relation to the Agreement and would also provide advice on the Plan. The roles of the MDBA regarding the Agreement would continue to be funded by the states and the Commonwealth. The Commonwealth would fund the Basin Plan functions.

The Commonwealth agreed to fund "priority projects" in each state. Victoria secured up to $1 billion for its Food Bowl Modernisation Project, reportedly settled at breakfast between the new Victorian Premier John Brumby and Prime Minister Kevin Rudd on the morning of the COAG meeting.

Basin governments, including Victoria, signed the Memorandum in Mar. 2008 (MOU, 2008). The substance of the Memorandum was then developed into the text for referral legislation (IGA, 2008), which all Parliaments passed at the end of 2008.

Work on the Basin Plan took 4 years, double the time contemplated at the time of the Water Act. The delay was partially due to the very hostile reaction to the first draft of the Plan (the "Guide") (MDBA, 2010; Chapter 14). Subsequently, a new Chair, Craig Knowles (a former NSW natural resources minister), one new Authority member and a new Chief Executive Officer were appointed to the Murray-Darling Basin Authority.

A revised draft Basin Plan was provided to the Ministerial Council at the end of May 2012 for their comment. Of the matters raised collectively by Ministers, perhaps the most important was a proposal for a mechanism to give some flexibility to the size of the reduction of the SDL. The draft Plan was designed to achieve an annual reduction of 2750 GL in surface-water diversions. South Australia was looking for a reduction of 3200 GL, while Victoria and NSW wanted 2100 GL. The latter's concern was that a reduction of 3200 GL would impose excessive socioeconomic costs, while South Australia believed a 2100 GL decrease would have insufficient environmental benefits.

The solution was a clever but complicated mechanism. To meet the Victorian and NSW concern, the volume of water to be recovered for environmental purposes could be reduced by implementing works and measures that decreased the amount of water required to yield equivalent environmental outcomes. To address South Australia's concern, the recovery of water for the environment could be increased, without worsening socioeconomic outcomes, by improving the efficiency of irrigation water use.

The MDBA revised the draft Plan to include the adjustment mechanism. Following two further rounds of comments and responses, the Ministerial Council endorsed the Plan, and the Commonwealth Minister for the Environment, Tony Burke, then tabled it in the Commonwealth Parliament. On 29 Nov. 2012, the House of Representatives voted 95 to 5 in favor. And so the first Basin Plan became law (Aust Govt, 2012).

In the midst of the negotiation of the Basin Plan, financial tensions emerged in relation to the Agreement. First NSW and then South Australia reduced their funding contributions. The result was a reduction in the budget, from $107 million in 2011–12 to $90 million in 2013–14. At the time of the NSW decision to cut funding, the NSW Minister for Primary Industries was critical of the amount of water the Basin Plan proposed withdrawing from use, which she claimed would "devastate irrigation-reliant towns."[9] She also pointed to the Murray-Darling Basin Authority's "bloated bureaucracy."[10]

Six months later, the South Australian Treasurer announced that the state would be reducing its contribution. He explained the decision by saying, "[w]e don't see why South Australia should share more than its burden for the administrative expenses of the Murray-Darling Basin Authority when New South Wales have made a similar reduction."[11]

The controversy about joint program funding was probably inevitable. With the Commonwealth taking responsibility for and funding basin-wide water planning, it was only a matter of time before states began to think about reducing their contributions to the Agreement. Whatever the reasons for the two states' funding decisions, the consequence has been to reduce the scope of what the Agreement can deliver.

---

[9]Sun Herald newspaper, 17 June 2012, Sydney.
[10]The Land newspaper, 21 June 2012, Sydney.
[11]The Australian newspaper, 21 December 2012, Sydney.

CONCLUSION

The Murray-Darling Basin Agreement has been a remarkably enduring framework for managing states' competing interests. One reason for this is the Agreement has provided a shared governance structure to which the parties can turn in the face of pressures and tensions. The importance of having a process to go to, one that governments themselves had set up and having the credibility of experience and achievement, should not be underestimated. However, good processes are not sufficient for securing outcomes in the face of competing interests.

The record of the Agreement has been that, as autonomous entities accountable to their electorates, states joined in cooperative action when there was a benefit from doing so. The requirement for unanimity has meant there needs to be a clear line of sight between participation and benefits for all states. Where this has not been the case, help from the Commonwealth, the pressure of circumstance or influential individuals have been needed to secure action. Of these, Commonwealth financial assistance has been far and away the most decisive.

Without Commonwealth money, the Agreement has struggled to meet its challenges. This was the case with the problems of salinity and environmental deterioration. Given states' reluctance to compromise their autonomy or bear costs, progress was slow. The most significant decision was adopting the Cap on consumptive water use in 1995, by which burden sharing between the states and for industry was determined by drawing a line in the sand at then existing levels of diversions.

Triggered by the crisis conditions of the Millennium drought, the recent reforms have occurred as water use reached the limits of availability, the environment became increasingly stressed and of concern to the community, and the Commonwealth had the political confidence and fiscal strength to assume leadership in the Basin.

The challenge now is for the Commonwealth and states to develop complementary roles that will deliver a healthy working river system within the framework of the Basin Plan (Chapter 13).

REFERENCES

Aust Govt. (2012). *Water Act 2007—Basin plan 2012, extract for the federal register of legislative instruments* [p. 245]. Canberra: Australian Government, Office of Parliamentary Counsel

Aust Govt. (2014). *Water Act 2007 (with amendments)* [p. 571]. Canberra: Australian Government, Office of Parliamentary Counsel

Aust Parliament. (1970). *House of representatives hansard* (p. 84). Canberra.

Clark, S. (2003). The Murray-Darling Basin: Divided power, co-operative solutions? *Australian Resources and Energy Law Journal, 22*, 67.

Close, A. F., & McLeod, A. J. (2000). The Cap as public policy in natural resource sharing. In: *Xth world water congress, Melbourne, 12–17 March.*

COAG, (1994). *Council of Australian Governments, 'Communiqué'.* Hobart.

Conference of Ministers. (1924). *Resolutions adopted at conference of ministers representing the governments of the commonwealth and the states of New South Wales Victoria and South Australia, Hume Reservoir, 9 August.*

Conference of Ministers. (1944). Upper Murray storage—Increase in capacity. In: *Report of conference between ministers representing the commonwealth of Australia and the states of New South Wales, Victoria and South Australia, Melbourne, 19 January.*

Conference of Ministers. (1945). River Murray waters. In: *Report of proceedings of conference of ministers representing the commonwealth and the states of New South Wales, Victoria and South Australia on proposals to increase the storage capacity of the Hume Reservoir and to widen the inlet channels to Lake Victoria, Melbourne, 21–22 May.*

Dole, D. (2002). Managers for all seasons. In D. Connell (Ed.), *Unchartered waters.* Canberra: Murray-Darling Basin Commission.

Guest, C. (2017). *Sharing the water: One hundred years of River Murray politics.* Canberra: Murray-Darling Basin Authority.

IGA. (2008). *Council of Australian governments, agreement on Murray-Darling Basin reform.* Canberra.

IRC. (1902). *Interstate royal commission on the River Murray. Report of the commissioners, Sands and McDougall.* Melbourne.

MDBA. (2010). *Guide to the proposed basin plan. Vol. 1 Overview.* Canberra: Murray-Darling Basin Authority.

MDBC. (1994). *Minutes of meeting 29.* Canberra: Murray-Darling Basin Commission.

MDBC. (2000). *Overview report—Review of the operation of the cap.* Canberra: Murray-Darling Basin Commission.

MDBMC. (1993). *Minutes of meeting 12.* Canberra: Murray-Darling Basin Ministerial Council.

MDBMC. (1995). *Minutes of meeting 16.* Canberra: Murray-Darling Basin Ministerial Council.

MDBMC. (2005). *Minutes of meeting 38.* Canberra: Murray-Darling Basin Ministerial Council.

Ministers' Conference. (1923). *Conference of Commonwealth and State Ministers, Melbourne, May–June, Albert J Mullett, Melbourne.*

MOU. (2008). *Memorandum of understanding for Murray-Darling Basin reform.* Canberra: Council of Australian Governments.

MRMCL. (1902). Murray river main canal league. In: *Official report of the Corowa water conference, George Hamilton, Berrigan* [pp. 51, 59–60].

Murray, S. (1903). *Press clipping—note to the minister, 4 November.* Accessed at the Victorian Archives.

Murray, L. (2006). *The Wedding at Berrico, collected poems* [p. 381]. Melbourne: Black Inc.

NAP. (2000). *National action plan for salinity and water quality.* Canberra: Council of Australian Governments.

Neal Report. (1994). *Report of the working group on water resources policy to the Council of Australian Governments.* Canberra.

NSW Parliament. (1920). *New South Wales Legislative Council Hansard* [p. 4138]. Sydney.

NWI. (2004). *Intergovernmental agreement on a National Water Initiative.* Canberra: Council of Australian Governments.

Premiers' Conference. (1920). Report of the resolutions. In: *Proceedings, and debates, May, Albert J Mullett, Melbourne.*

Qld Parliament. (2007). *Queensland legislative assembly Hansard* (p. 428). Brisbane.

RMC. (1931). *River Murray commission, 'Report of special investigation'.* Melbourne.

RMC. (1961). *River Murray commission, 'Report by and resolutions of the river Murray commission on Chowilla Dam'.* Melbourne.

RMWA. (1914). *River Murray waters agreement 9 September, Schedule, River Murray Waters Act 1915 (Commonwealth).*

RMWA. (1949). *River Murray waters (Amendments) Act 1949, (Commonwealth) second further amending agreement, Clause 8.*

RMWP. (1973). *River Murray working party 1973, Interim report* [p. 5]. Canberra: Australian Government Printing Service.

SA Parliament. (1904). *South Australian House of Assembly Hansard* (p. 387). Adelaide.

SA Parliament. (1934). *South Australian Legislative Council Hansard* (p. 1640). Adelaide.

Whitlam. (1973). *Meeting on problems associated with the use of river Murray Waters between the premiers of New South Wales, Victoria and South Australia and the prime minister, 'Summary record of discussion', 2 March, Canberra.*

EVOLUTION OF WATER ENTITLEMENTS AND MARKETS IN VICTORIA, AUSTRALIA

3

C. Fitzpatrick
Independent Consultant

CHAPTER OUTLINE

INTRODUCTION

Formal, legal, managed systems of water entitlements and markets have been successfully implemented in southeast Australia over the past 30 years. This chapter focuses on how a system of water entitlements and markets was progressively implemented and subsequently evolved in the state of Victoria. Some insights into solving the political problems of transition are also provided.

The evolution of markets in Victoria is an exemplar "of principled pragmatism—in which the principles of sound economic management are well defined and respected, but in which they are applied in a pragmatic and sequenced way which takes account of local circumstances and political economy, and in which the focus is on moving in the right direction, and on the art of the possible" (Briscoe, 2011).

Under the Australian constitution water management is the responsibility of the states, with each state having its own water law. Alfred Deakin's Irrigation Act of 1886 laid the foundations of Victoria's water law (DCE, 1991). Deakin's law overrode existing common law rights and vested water resources in the Crown. Both New South Wales and South Australia also legislated to vest water resources in the Crown. Victoria's and Australia's water entitlement and water trading are therefore based on statutory law rather than riparian rights and common law such as occurred in the United States and England.

In the time before the major reforms, management of Victoria's irrigation areas was undertaken by the Victorian State Rivers and Water Supply Commission, which was created in 1905. This agency was also responsible for providing oversight of some 375 nonmetropolitan local water authorities responsible for town water supply and sewerage services. In the Melbourne metropolitan area, the Melbourne and Metropolitan Board of Works was responsible for providing water supply, sewerage, and some drainage services (DWR, 1986a).

Victoria shares the waters of the River Murray with New South Wales and South Australia. A water-sharing agreement was reached in 1914, followed by legislation in 1915, and the River Murray Commission was established to construct and operate works on the River Murray and to assist in sharing water between the three states (Paterson, 1987a; Chapter 2).

CHANGING THE OBJECTIVES OF WATER RESOURCE MANAGEMENT

Substantive water reform relies on long-term leadership and political commitment and this cannot be left to chance.

Water management became a matter of political interest in Victoria in the early 1980s at the same time that Australia was pursuing a major microeconomic reform agenda (Productivity Commission, 1996). The government was providing ongoing subsidies for the operation of the irrigation and non-metropolitan sector, environmental opposition was emerging to proposals to build increasingly costly new dams to meet the growing demands for irrigation water, and it was becoming clear that the institutional structure of the sector needed to be modernized (DWR, 1992).

As set out in the second reading speech, the *Parliamentary Committees* (*Public Bodies Review Committee*) *Act 1980* provided the mechanisms to establish the Parliamentary Public Bodies Review Committee (PPBRC) with powers to review the efficiency, effectiveness, structure, and role of Victoria's public bodies (Vic Parliament, 1979). The committee consisted of eight parliamentarians drawn from both the government and the opposition.

One of the first actions of the PPBRC was to review the 375 legally separate water sector public bodies that included the State Rivers and Water Supply Commission (the state's dominant irrigation authority) and each constituted water, sewerage, drainage, and river improvement trust or authority in Victoria, but excluded the Melbourne and Metropolitan Board of Works (Vic Parliament, 1980). In conducting the review, the PPBRC made site visits, commissioned technical reports, called for public submissions and held public hearings (Vic Parliament, 1984a, 1984b).

An important focus of the Committee was to consider what should be the contemporary objectives and strategies for water-resource management. At the time objectives expressed in legislation were to develop the land and water resources of the State, to obtain the best beneficial use of them in production, and to achieve this in an efficient manner, resulting in a direct financial return to the State. The PPBRC found a clear pattern of government decisions over almost 70 years that had consistently favored the objective of closer settlement and increased production ahead of the other objective of direct financial return and economic efficiency. This had consistently resulted in the relaxation of economic efficiency measures and the introduction of substantial government subsidies.

The PPBRC considered that modern management objectives were required to enable the water sector to effectively respond to emerging challenges. In coming to this conclusion, the Committee was aware that Victoria was approaching the technically practicable limit to the diversions from rivers that could be obtained.

At the irrigator level the Committee was aware that:

- Water is one of the most important inputs into the activities of irrigation farmers, but farmers had little control on allocation, which was largely controlled administratively rather than by the needs of the irrigator;
- Different types of irrigated production require different quantities of water for optimal production, but the essentially arbitrary nature of water rights meant that the opportunities for an irrigator to grow or experiment with different products was arbitrarily limited;
- Incentives were needed to reward farmers who make efficient use of water;
- The system of water rights prevented the reallocation of water from areas of lower productivity to those of higher productivity.

The opportunities for reform at both the state level and the farm level helped generate the necessary community and therefore political support to reform the sector, despite many vested interests.

The PPBRC recommended five organizational strategies—flexibility, commercially based, customer oriented, future oriented and accountability— to reform the management of the water sector in the state. These strategies unambiguously shifted the focus of management from the objective of developing the state's water resources for productive use, which had resulted in ongoing government subsidies of urban and irrigation water services, to a strong focus on the commercial delivery of these services based on user-pay principles. They underpinned all water reform that followed and, although rarely referenced, continue to underpin the management of the Victorian water sector today. They also provided the basis for devolving water and wastewater service delivery responsibilities to statutory water authorities.

In subsequent years, care has been taken to mitigate the effects of implementing a commercial approach to irrigation reforms. However, the well-being of local communities' dependent on public irrigation systems has been an ongoing concern and is still a major political consideration in managing ongoing transitional pressures in the rural water sector.

Interestingly, a consequence of establishing the PPBRC was that a significant number of parliamentarians from both the government and opposition developed a deeper understanding of the state's water resource management issues and shared views about how the arrangements should be improved. As a result, despite inevitable short-term political disagreements, long-term water reform in Victoria survived changes in government.

A major outcome of the PPBRC review was the establishment of the Department of Water Resources on 1 Jul. 1984 (*Water (Central Management Restructuring) Act (1984)*). It was established to provide independent policy advice to the Minister and be responsible for coordinating the policies and programs of all agencies in the water sector that is to lead the reform of the water sector. Dr. John Paterson, the first Director General of the Department of Water Resources, was responsible for leading the government's water reform agenda.

The formation of the Department of Water Resources was the starting point in Victoria for separating institutional responsibilities for policy and service delivery and for establishing independent regulators. This process was completed with the establishment of an independent economic regulator, the Essential Services Commission at the start of 2002, and the passage of the *Safe Drinking Water Act (2003)*.[1]

The reform of Victoria's water entitlement regime and the development of water markets were seen as important parts of the broader microeconomic reform of the water sector.

ESTABLISHING WATER ENTITLEMENTS BASED ON THE CONCEPTS OF PROPERTY RIGHTS

In the early 1980s a significant literature was emerging about the importance of introducing stronger economic incentives into the water sector, including the introduction of water markets. Dr. Paterson recognized that the starting point for reforming water-resource allocation processes was to properly define rights to water using a property rights framework. He argued that any system of economic incentives that rests on a disjointed foundation of basic law is inevitably second best and is highly productive of unforeseen and unwanted consequences and that mere improvements in the transactional framework will generally be insufficient to rectify them (Paterson, 1987b).

At the time a multitude of arrangements existed that specified rights to water (DWR, 1987). Catchments that supplied water to Melbourne were vested in the water authority, but the vesting provisions did not specify the volume of water that could be taken. Nonmetropolitan water authorities generally obtained water through works approval Orders in Council. These Orders permitted works to harvest water but did not specify how much and under what circumstances water could be taken. The control of irrigation storages was vested in irrigation authorities. Again, the rights to water were implicit rather than explicit.

In each case, the poorly specified rights to water failed during times of scarcity because they did not define shares of flow, were not tradeable and had scant regard for environmental needs. The failures led to political interventions, which were difficult for governments to resolve and had unpredictable results.

Paterson proposed that the legal foundation of water-resource management must capture the subject that is the terrestrial path of the hydrological cycle, in a single hierarchical vesting framework. He also strongly advocated that the approach of defining rights by partitioning runoff, storage capacity, and losses simultaneously satisfies the requirements for the exhaustive and unique partitioning of the resource, and a legal basis for a consistent set of nonattenuated rights.

Paterson wrote (Paterson, 1987b):

> Attention to the definition of rights offers enormous opportunities for enhancement of the technical efficiency of water allocation decisions. Better definition of rights is a pre-condition to creation of more efficient allocative frameworks, but property rights and efficient markets are mere technical servants of broader social purposes.

Therefore, the initial focus of the reform of Victoria's water allocation arrangements was on the definition of the underpinning rights to water.

---

[1]http://www.parliament.vic.gov.au/static/www.legislation.vic.gov.au-statbook.html.

The task of clarifying rights to water was generally supported by the entities that held poorly specified rights. Some feared that their vaguely specified rights could be eroded by a range of factors, including by water abstractions upstream of their offtakes and by centralized administrative decisions. Devolving responsibility and accountability for managing water entitlements to entitlement holders was supported by the entitlement holders, but required the center to give up administrative and operational influence. Strong leadership was required from the Minister and Department to manage the transition.

The then Department of Water Resources subsequently commenced a comprehensive review of Victoria water legislation with a view to introducing a simplified, briefer and more systematic legislative framework. The review process involved significant public and stakeholder consultation consistent with the earlier approach taken by the PPBRC. A series of discussion papers formed the basis for initial consultation. They highlighted difficulties with the existing legislation and proposed a possible new management framework.

Water-resource allocation was just one of the many issues considered by the review of the legislation. The first Water Law Review Discussion Paper (DWR, 1986c) found that, overall, the existing framework for allocating water resources and planning for future allocations lacked coherence and that an overhaul was long overdue. The paper proposed a new framework to systematically deal with bulk water allocations, environmental water allocations and the introduction of transferable water entitlements (TWEs) for irrigation water. Further technical detail about the specification of bulk water entitlements was provided in a subsequent issue paper (DWR, 1987).

A hierarchical structure of rights was proposed, with supply authorities to hold enduring unattenuated bulk water entitlements (or wholesale rights), which were to be defined in terms of shares of catchment run off, storage capacity, release capacity, and evaporative and seepage losses. They were to be designed to be explicit, exclusive, enforceable and transferable, and when added up they were not to exceed the divertible physical resources in a catchment.

The *Water Act* (*1989*) (Vic Parliament, 1989), prior to subsequent amendments, provided significant safeguards to ensure the bulk entitlements had similar attributes as property rights and to protect their integrity. The safeguards include: the unauthorized use of water was made an offense; a formal statutory process needed to be followed to amend bulk entitlements; limitations were placed on the discretion of the Minister when considering applications to amend bulk entitlements; and finally, either house of parliament could disallow a proposed amendment of a bulk entitlement.

A key safeguard was that amendments to bulk entitlements require the bulk entitlement owner to apply to the Minister for their entitlement to be amended. This meant that their agreement to the proposed amendment was required. A further safeguard was that the Minister could not approve an application to amend a bulk entitlement if this would cause significant adverse effects on other authorized uses of water or the environment.

As well as these safeguards, the *Water Act* (*1989*) allowed the Minister to step in under extreme circumstances to "qualify rights" on a temporary basis if they declared that a water shortage exists and they are satisfied that there is or will shortly be insufficient water available to satisfy any rights to water. This power has been used during severe droughts to reduce passing flow obligations to ensure town water demands under severe water restrictions could be supplied.

An amendment of the *Water Act* (*1989*) in 2005 enabled the Minister to permanently qualify rights if a long-term water assessment revealed that changes in hydrology caused a disproportionate impact on some entitlements. The long-term water resource assessments are to be conducted every 15 years.

This amendment eroded the conceptual basis of the property-right framework by providing a mechanism amending rights every 15 years through a political process and in effect gave government the right to confiscate water from rights holders without compensation.

The establishment of bulk entitlements was fundamentally important in implementing Victoria's new devolved water service delivery arrangements, because they transferred the responsibility for managing the supply risks caused by climate variability and droughts from a centralized allocation process administered by the government to the water authorities. This was an essential prerequisite for loosening the water bodies from previous bureaucratic command-and-control arrangements and placing them on a more commercial footing. It also greatly increased the accountability of the water authorities to their customers and reduced their ability to transfer water resource management costs to government.

Private end users such as irrigators were to hold retail rights that were effectively a share of the bulk water entitlements. The retail rights were defined indirectly as a share of the total amount of water available in the superior bulk entitlement each year, thus also transferring risks caused by climate variability to the end user.

MANAGING THE TRANSITION TO NEW ENTITLEMENTS

Paterson recognized that, while it is not too difficult to design appropriate management arrangements from first principles, the design problem became distinctly nontrivial when implementation of the solution must also satisfy a complex and internally contradictory set of entitlements, many of which are poorly defined (Paterson, 1987b).

The approach taken to this dilemma was to provide a choice. Water authorities had the option of retaining their poorly specified rights if they wished or apply to convert these to the new form of entitlements. Enabling voluntary conversion was critical for ensuring sufficient stakeholder support for the proposed new legislation. The Act was subsequently amended in 1995 to enable the Minister to request that an Authority apply to convert their existing rights to a bulk water entitlement, but this power proved to be unnecessary.

Providing legislative provisions to enable temporary and permanent trading of both retail irrigation rights and bulk water entitlements proved more controversial, with many irrigators fearing that water trading would result in water moving out of some areas, causing social hardships, while others saw that water trading would provide opportunities for them to grow their enterprises.

The new water bill was passed in 1989, but not without vigorous debate in Parliament and many amendments. Despite the amendments, the *Water Act (1989)* provided the legal basis for reforming the Victorian water sector, including its water entitlement and market arrangements. A key principle adopted when drafting the Act was that it was to provide an enabling framework, rather than prescriptive administrative detail. This proved to be an important design feature as the process of converting existing rights into the new form of explicit, exclusive, and enforceable property rights revealed the need for considerable flexibility to partition the resource.

Water rights to Victoria's share of Murray River water resources were the next to be converted. The process of converting the poorly specified water rights to legally binding water entitlements was undertaken in two stages—first a pilot in the Goulburn-Broken river systems and then in the River Murray.

GOULBURN-BROKEN PILOT

The Goulburn-Broken Valley in northern Victoria was chosen in 1991 to pilot the bulk entitlement conversion process because it included a wide range of different water supply types ranging from small communities that pumped directly from rivers, towns that operated their own reservoirs, towns that were supplied from supply systems operated by other authorities, a large irrigation system and hydro-electricity schemes. Also, water trading between irrigators in this area offered substantial benefits.

An open and participatory process was used with the conversion process being guided by the Goulburn-Broken Forum, made up of representatives from the water authorities applying to convert their entitlements, irrigator interests, environmental interests, and the Water Bureau, the successor body to the Department of Water Resources. The Forum adopted the principle that existing, poorly defined rights would be converted to explicit rights and the process of conversion would not result in new commitments (Water Bureau, 1995).

These principles were critical because they enabled the entitlement holders to participate without fear of having their rights reduced. It was possible to adopt this approach because water authorities recognized that, once the entitlements were properly defined, water could be reallocated through the market. The water authorities and irrigators saw that the market was a far superior mechanism to reallocate water than reallocation by administrative fiat, which they reasonably saw as the confiscation of a property right.

The principle that the conversion would not create new entitlements was equally important because it minimized ambit claims and protected existing environmental values. The open and participatory process was important because it enabled the Minister to direct people back into the process should they seek special favors. Arguing their case to their peers was a much tougher task than making ambit claims to the Minister.

A further important feature of the processes was that the water authorities were not expected to formally apply to convert their existing rights to bulk entitlements until the end of the processes after the detailed analyses had been completed and the Forum had agreed on how the available water was to be shared and the form of the entitlements to be issued.

A computer modeling package (called REALM[2]) was used to model the water-supply systems in the Goulburn-Broken catchment. It was used to simulate the current operation of both the urban and rural water-supply systems to define the security of supply and to evaluate proposed rules to be included in the bulk water entitlements. REALM is public-domain software that uses an advanced optimization routine and continues to be supported.

The Goulburn-Broken pilot program took 3 years to complete and resulted in 36 bulk entitlements being formally issued by the Minister. Copies of the bulk entitlements can be accessed from the Victorian Water Register (2016).

Two types of bulk water entitlements were defined. The first were called *source entitlements*, which specified rights to all or some of the following: the amount and rate that water that can be abstracted from a waterway, passing flow obligations, the share of the capacity of the storages that can supply the entitlement, the share of the capacity of the natural and constructed carriers that can be used to supply the entitlement, and a share of the evaporation and operational losses in the system used to supply the entitlement.

---

[2]http://www.depi.vic.gov.au/water/water-resource-reporting/surface-water-modelling/resource-allocation-model-realm.

The second type—*delivery entitlement*—were issued to urban water authorities supplied from large irrigation systems. These specified: the maximum daily and annual volume that can be taken at a specified point, the reliability of the entitlement and restriction rules, head works and delivery financial obligations, and water accounting and reporting obligations.

The bulk entitlements of the rural water authorities included water allocations to operate the irrigation systems and obligations to supply the retail entitlements tied to the bulk entitlements. An early decision was to specify irrigator rights in terms of a share of the whole regulated system rather than source entitlements to the individual harvesting sites and storages. This latter approach would have resulted in each irrigator holding many source entitlements to different sources and needing to place orders for deliveries of water from particular sources throughout the year. They would have faced a very complex task to optimize their entitlements, and water trading and water accounting would have been excessively complex.

This decision was made to simplify water trading. Only two tradeable rights were created, a *high reliability* right and a *low reliability* right. Both rights effectively provided a share of the total amount of unregulated water available in the system. This approach greatly reduced the number of types of tradeable entitlements and simplified trading transactions for both water users and administrators. For pragmatic reasons the relatively small water entitlements for urban water supplies in the regulated system were converted at a higher level of reliability than other high-reliability entitlements.

The customers of urban water authority customers do not hold water entitlements; instead, the water authorities have an obligation to supply their customers on a fee-for-service basis. During periods of water shortages, these authorities may impose water restrictions to manage demand.

The legislative framework also enabled the Minister to establish an operating framework for the bulk supply systems based on: bulk and environmental entitlements, the formal appointment of a storage or bulk supply system operator to operate the supply system and be accountable to the entitlement holders, storage/system management objectives determined by the Minister, and detailed water-supply system operating rules as these may affect the reliability of water supplied to the entitlements.

This framework includes formal processes for varying the legal instruments as well as consultative and reporting requirements. Changes can only be made with the agreement of all entitlement holders.

RIVER MURRAY SYSTEM

Water rights to Victoria's share of the River Murray water resources were the next to be converted. The water resources of the River Murray are shared between the states of Victoria, New South Wales, and South Australia in accordance with statutory rules in the Murray-Darling Basin Agreement (MDBA), which is now a Schedule of the Commonwealth *Water Act, 2007* (Aust Govt, 2013). The Agreement set aside water to operate the river, a reserve for the following year, and then shares the remaining water available to the states in accordance with prescribed rules. Each state is then free to allocate its share of water according to its own policies. Rights to water derived from Victoria's share of River Murray resources specified in the Agreement were converted to bulk water entitlements. No attempt was made to renegotiate the Agreement. Subsequently, the Murray-Darling Basin Plan (see Chapter 13) has defined the quantity of water available to be taken by NSW, Victoria, and South Australia from their state water shares, but not alter the state water shares themselves.

A similar approach to the Goulburn-Broken was again taken. The Murray Water Entitlement Committee was formed in Feb. 1996, chaired by a trusted local member of parliament and with representatives from water user and irrigation industry groups, water authorities, environment, and catchment management groups (Murray Water Entitlement Committee, 1997). They decided to "Cap" the volume of water that could be diverted at the levels that reflected 1993–94 levels of development and reliability of supplies. This meant that the conversion to bulk entitlements was based on average usage based on 1993–94 levels of development, rather than the capacity of their approved works. This meant that water authorities experiencing population growth quickly used the water market to supply growing demand.

The Committee achieved agreement to the water-sharing arrangements and the form of the bulk entitlements in mid-1999 (NRE, 1999) and the Minister issued bulk water entitlements soon after. Copies of these bulk entitlements can also be accessed from the Victorian Water Register (2016).

A combination of high- and low-reliability water shares were created on the Murray regulated supply system. Urban water authorities received rights that were effectively the same as the high reliability rights held by irrigators. Irrigators received the same type of high reliability entitlement irrespective of the type of crops they irrigated, with water users required to use the water market to adjust the amount and mix of entitlements they held to suit their needs.

All poorly defined bulk rights have now been converted to bulk water entitlements. In total, 186 bulk and environmental water entitlements have been issued to water corporations, thermal power companies, hydroelectric power companies, and the Victorian Environmental Water Holder. Environmental entitlements are effectively bulk entitlements established specifically to supply water to the environment.

The bulk entitlement framework is widely supported by water authorities and is now accepted as the foundation stone of property rights to water in Victoria. Entitlement holders are fiercely protective of their entitlements because they are valuable assets that are accepted as credit-worthy assets that can be mortgaged. The economic value of entitlements is revealed on a daily basis through the operation of the water market.

MANAGING THE EVOLUTION OF THE WATER MARKET

Interest in water markets, particularly in irrigation areas, was emerging prior to the reform of Victoria's property rights–based bulk entitlement regime. Water markets in Victoria developed in parallel to the conversion of the specified rights to bulk water entitlements.

The *Water Act* (*1905*) introduced the concept that each irrigation district should have a water right for all land commanded by works and suitable for irrigation. The administrative formula that determined the amount of water right provided was settled in 1963 (DWR, 1986b). The formula tied water allocations to the land and favored small areas of land over large areas. It effectively prevented intensive irrigation development on larger farms. It also committed volumes of water that were not available from existing storages, but were expected to become available when additional storages identified in a 10-year plan were completed.

This constraint was somewhat loosened by allowing the water rights on contiguous holdings in the same ownership to be amalgamated so all of the water could be used on one part of the property, on the condition that the water could be supplied from existing supply points. This enabled farms to increase their water holdings by buying their neighbors' land.

Temporary amalgamations and transfer of water were also permitted during the severe droughts of 1967–68 and 1982–83 between noncontiguous properties in the same ownership. This rule was continued after the 1982–83 drought in response to the needs of irrigators to source additional water to develop their properties. However, the process was obviously cumbersome and economists and some irrigators were interested in establishing a system of TWEs.

A review of the benefits and costs of enabling irrigators to transfer their water entitlements concluded that implementation of a TWE policy was justified on economic grounds since it would lead to improvements in the efficiency of irrigation water use (ACIL, 1984). The review also found that irrigators would benefit from the increased flexibility with which they could obtain, dispose of and allocate water, and that significant volumes of water would trade away from some local areas. This latter finding reinforced significant misgivings about TWE in some local irrigation communities.

Temporary transfer of water rights commenced in mid-1987 following an amendment to the *Water Act* (*1958*). This change in the legislation facilitated modest temporary trades of around 25,000 ML/year. Subsequently, the *Water Act* (*1989*) enabled both temporary and permanent trade of water rights and bulk water entitlements. Trade commenced in late 1991 and rapidly developed in the large interconnected irrigation systems in northern Victoria. Despite some initial resistance to the trading provisions of the *Water Act* (*1989*), almost all "commercial" farmers became active participants in the market, and generally supported the market.

Early trading rules were rudimentary but trading rules and administrative arrangements rapidly evolved as the potential of the market was tested. The early years of water trading have been described in detail (NRE, 2001).

Water-trading rules initially limited trading within gravity irrigation districts of the Goulburn-Murray Irrigation District supplied from the same water-supply system. The *Water Act* (*1989*) was amended in 1994, enabling trading rules to be relaxed to allow trading from an irrigation district to land outside the districts, provided no more than 2% of entitlements was traded out of a district in any year. This was introduced to assuage local concerns about the effects of water leaving their areas and was necessary to gain sufficient community support for the liberalization of the rules.

In 1995, a Cap was placed on the volume of water that could be diverted in the Murray-Darling Basin (MDBC, 1998). This provided a significant impetus to market activity. Victoria implemented the cap by tightening water allocation rules and water-trading rules.

Trading rules were also amended to allow urban water authorities to participate in irrigation water markets and to enable the temporary interstate trade of water. For the first time, environmental water allocations were traded to irrigators, further expanding the depth of the market (NRE, 2001).

The next step in the evolution of the market was a pilot project on interstate water trading that commenced in 1998 after amendments to the *Water Act* (*1989*). A new schedule to the MDBA was also needed to put in place the water-accounting arrangements needed to properly account for the transfer of water between the states. The creation of this scheduled required the agreement of the Commonwealth and the NSW, Victorian, and South Australian governments.

The liberalization of the trading rules enabled the volume of water traded on the temporary market to rapidly increase from around 25,000 ML/year to 200,000 ML/year and permanent trades to increase to around 25,000 ML/year. The level of both temporary and permanent trade was a surprise and tested the administrative, record keeping, and water accounting processes used at the time.

Despite the obvious benefits to both the buyers and sellers of water, the cumulative effects of water trading out of some areas was causing community concern and creating some local political pressure points. Unhelpfully, those areas where water was trading in were reluctant to talk about the economic benefits and development in their area enabled by the trading.

RESETTING REFORM DIRECTIONS

The Victorian Government commenced a process to reexamine the state's water management arrangements in Aug. 2003, some 20 years after the work of the PPBRC. A Green Paper identified that a key challenge for Victoria was to provide secure, reliable water supplies for our homes, our farms and industry while meeting the needs of the environment (DSE, 2003).

All aspects of Victoria's water management arrangements were considered, including Victoria's water entitlement, market, and planning frameworks. Issues and proposals to respond to the issues were documented in the Green Paper that formed the basis of a public consultation process that involved public submissions, meetings with stakeholders, and public hearings.

Opportunities and issues revealed by the operation of the water market since 1991 were identified and discussed. The Green Paper included proposals to improve the specification of irrigators rights to water to improve market operations and address concerns about the effects of water trading out of some parts of irrigation districts, stranding some irrigation delivery infrastructure.

The unbundling of the water entitlements held by irrigators into a water share not attached to land, a share of delivery capacity, and a license to irrigate land was proposed. The water share was to be expressed as a share to a water-supply system and lenders would be able to directly mortgage the water share. Water shares would be able to be traded or leased. The delivery share would be tied to a supply point and based on the infrastructure used to deliver water. Separate cost reflective tariffs were to be applied to the water share and the delivery share.

A water-use license would be required to irrigate land. License conditions would be limited to controlling the off-site impacts of irrigation. Both delivery shares and a water-use license would be required to irrigate land in the public irrigation districts.

The decision to create water shares that were not attached to land had the strong support of economists and policy makers, but was a concern to many in the irrigation community who feared that speculators would corner the market in water shares, forcing up prices. Many irrigators, although owning water shares, feared that increases in the value of water shares would ultimately make some irrigation farms unviable leading to painful local adjustment. This was seen to be a particular threat to irrigation districts, where the unit costs of service charges to recover the largely fixed operating and maintenance costs of shared delivery infrastructure would increase with reduced deliveries.

The Victorian Government released a White Paper in Jun. 2004 (DSE, 2004) supporting relatively open-water trading arrangements, and committing to actions to unbundle irrigators water entitlements and establishing a modern central register of all water entitlements. The *Water Act* (*1989*) was subsequently amended in 2005 to implement the actions in the White Paper with irrigators' water rights unbundled in northern Victoria on 1 Jul. 2007 and in southern Victoria on 1 Jul. 2008.

The government also put in place two short-term measures to respond to irrigator concerns. The first was to limit the amount of water entitlements that could be traded out of irrigation districts each year to 4% of the total entitlements in the district. The second was a rule that no more than 10% of the water entitlements could be owned by people who did not own land. This latter rule was designed to control speculators. These transitional measures were removed in 2014 and 2009, respectively.

The Victorian government also agreed that water markets did pose challenges to the management of long-lived assets, and responded by providing powers to the water authorities to reconfigure their delivery infrastructure, including the power to close channels where a majority of the customers on the channel had sold their water entitlements and no longer irrigated.

THE NEED FOR A CENTRAL REGISTER OF ENTITLEMENTS

The creation of the Victorian Water Register was a key reform initiative coming out of the White Paper and fundamental to protecting the integrity of Victoria's entitlements to water and expanding the water market. It is a public register of all water-related entitlements in Victoria (Victorian Water Register, 2016), and is the single point of truth for all entitlements, recording details about ownership, attributes of the titles, and encumbrances. It is the point of control for all changes to entitlements.

Unlike land registers, the water register also records water allocations to each entitlement as well as water use information over the year. This information is necessary to support trading of entitlements and temporary trading of water provided by the entitlements.

The register applies commercially accepted accounting standards to the processing of trades and rigorous audit trails for all transactions. It is customer focused with online capabilities. It facilitates the market, reduces transaction costs and times, and improves accuracy, transparency, and accountability.

The Register was progressively populated as retail water shares were unbundled. Major data cleansing was required, highlighting the lack of control provided by the previous systems. All retail and bulk water entitlements had been added to the water register by Jul. 2011.

STRESS TESTING THE NEW FRAMEWORK

Victoria experienced severe drought conditions between 1996 and 2010. This drought, called the "Millennium drought," was more severe than any previously recorded since European settlement of Victoria in 1835 (CSIRO, 2012). It was more severe than the droughts used for planning purposes to determine the yield and reliability of water-supply systems and for converting poorly defined entitlements to bulk water entitlements.

In effect the Millennium drought "stress-tested" Victoria's approach to water, entitlements, markets, and planning. In general, these arrangements proved very effective, but important improvements were also made to clarify rights in extreme circumstances (i.e., rights to pump dead storage of reservoirs), enhance retail entitlements to enable water holders to manage their own risks by "carrying over" their water allocations from the current year to the next year, and to amend water reserve policy of large water-supply systems to reduce the likelihood of zero allocations (DELWP, 2016).

The drought also allowed the water-sharing arrangements set out in the MDBA to be stress-tested. In the 2006–07 water year, inflows to the River Murray storages were the lowest on record (Kirby, 2012), and there were doubts that the water-sharing arrangements in the Agreement would be capable of supplying critical urban and stock and domestic water needs. The Prime Minister of the day responded by convening a water summit with the three southeastern state premiers on 7 Nov. 2006 to develop emergency water-sharing arrangements (see Chapter 2).

Drought conditions persisted in 2007–08 and 2008–09. Storage reserves were severely depleted and allocations to high-reliability irrigation water entitlements were just 43% in 2007–08 and 35% in 2008–09 in Victoria and 32% and 18% in South Australia, insufficient to meet the needs of horticultural crops that required reliable water supplies.

Despite this serious water shortage, disaster was averted on many fronts. Farmers with perennial crops, such as orchards, were able to minimize long-term damage by purchasing water on the market from, for example, farmers with pasture-based dairy farms. Dairy farmers were able to use the proceeds from selling water to purchase fodder. The market empowered farmers to make their own business decisions, avoiding the need for governments or their bureaucrats to use unpopular and inefficient administrative powers to reallocate water.

In 2008–09, of the 924 GL of water used by the farmers in northern Victoria, 44% was purchased on the water market, 27% was purchased from other sources, with over 200 GL traded in from other states. Market prices for water allocations reached $400 per ML. The interstate market developed strongly in this period and enabled crops requiring uninterruptable water supplies to be watered. The National Water Commission (NWC) found that most of the benefits of trading accrue in dry years, when the need to reallocate water is greatest. In the exceptionally dry years of 2007–08 and 2008–09, benefits in the southern Murray-Darling Basin were estimated at $1.05 billion and $1.2 billion, respectively (NWC, 2012).

The northern Victorian water market continued to evolve in the large, interconnected water supply systems where there were many potential buyers and sellers. Over 700 GL of high-reliability entitlements (or 30% of the 2338 GL of high reliability of entitlement available in northern Victoria) have been traded, with most of the entitlements transferred from public irrigation districts for use outside the districts (Cummins, 2016). Between 1991 and 2015, irrigators in the Mallee purchased 152 GL of high reliability entitlements mostly for new horticultural developments, and are continuing to purchase water. Urban water authorities in Victoria connected to these systems had purchased 62 GL (<3%) of high reliability water shares.

However, the Commonwealth government has been the largest participant in northern Victoria's interconnected water-supply systems in recent years. By mid-2015 it had purchased 488 GL of entitlements to provide environmental flows in the southern Murray-Darling Basin (Cummins, 2016). This is more than double the volumes purchased for commercial purposes.

MANAGING ADJUSTMENT PRESSURES

The cumulative effects of water trading have resulted in deliveries in the Goulburn-Murray irrigation district falling from around 2200 GL in 2004–05 to about 1170 GL in 2015–16. Irrigation deliveries have also fallen significantly in irrigation districts in New South Wales. The long-term reduction in deliveries is presenting serious asset management and financial challenges to the operators of these irrigation districts. The powers to compulsorily reconfigure irrigation delivery systems provided in 2005 had not been used by mid-2016 despite the growing "stranded asset" problem.

Despite the obvious benefits markets provide to both buyers and sellers of water, areas where there is net selling are undergoing significant structural adjustment. Irrigation-dependent communities in these areas have responded by challenging the benefits and operations of the water market. They have made strong political representations to introduce trading rules to restrict trade out of irrigation districts to both the Commonwealth and state governments. Governments continue to support the market and are considering how they can ease these adjustment pressures.

The Murray Darling Basin Authority (MDBA) now sets the water trading rules in the Murray-Darling Basin, and the Authority must obtain and have regard to advice provided by the Australian Competition and Consumer Commission (ACCC) (Aust Govt, 2016). The ACCC is an independent Commonwealth statutory authority responsible for promoting competition and fair-trading. The involvement of the ACCC limits the ability of stakeholders to assert political pressure to introduce barriers to water trading. The Commission provided their advice to the Murray-Darling Basin Authority in 2010 (ACCC, 2010) and trading rules have been incorporated into the Murray-Darling Basin Plan and commenced on 1 Jul. 2014. Despite the advice of the Commission that it was inappropriate to restrict trades associated with the water's past, present, or future intended use, the rules in the Basin Plan to guard against restrictive trading rules do not apply to trade by urban water authorities.

Water trading is available in small surface and groundwater supply systems across Victoria; however, much lower levels of activity occur because of the small number of potential buyers and sellers. Trading of bulk water entitlements in these systems held by urban water authorities has also been muted.

Victoria's water entitlement framework has had the effect of devolving much of the responsibility for planning from government to the entitlement holders. Individual entitlement holders are now responsible for managing the risks of water scarcity within their own enterprises and systems.

The entitlement framework supports the entitlement holder's risk-management strategy by enabling entitlement holders to "carry over" water from year to year, to access the water market and most importantly to enforce and protect the integrity of the entitlements so that the entitlement holders are confident that they will receive the benefit of their risk-management activities.

The Victorian government does carry political risks should entitlement holders run out of water and the water market fails. However, government is very constrained on what it might do. Other water users have very little water to reallocate at the point of market failure, and government intervention prior to market failure is not tenable. Often the only option available to government is to provide access to water allocated to the environment, if available, and to provide welfare support.

Victoria's water-planning framework now comprises planning undertaken by:

- Retail entitlement holders, which is a completely private affair but may be informed by information provided in sustainable water strategies;
- Environmental water holders and catchment management authorities who are the "caretakers of river health" (DEPI, 2013);
- Urban water authorities who prepare long-term water-supply demand strategies on a 50-year horizon, taking into account climate-change scenarios and population-growth forecasts consistent with guidelines issued by the government (DSE, 2011);
- Governments who prepare regional sustainable water strategies every 7–10 years that provide long-term water availability forecasts, build on local planning efforts, and consider water issues that cannot be solved by a water authority or catchment-management authority acting alone;
- Long-term water resource assessments of the resource base and river health every 15 years.

In northern Victoria, the Victorian planning approach is overlain by the Murray-Darling Basin Plan (Fisher, 2011). The Basin Plan has set a long-term environmentally sustainable level of take of water from its rivers of 10,873 GL, a reduction of 2750 GL from the consumptive water take using 2009 as the baseline year. The Basin Plan requires the states to prepare regional water-resource plans that implement the sustainable level of take. The Commonwealth Minister is to approve the plans and the MDBA is to ensure compliance to the plans. In this way, the Basin Plan may override state water law and the states' water entitlements.

The Commonwealth is responsible for reducing the level of take in the states. They have decided not to appropriate water entitlements and instead have committed to obtain the required amount of water by two mechanisms: purchasing water on the water market and investing in water infrastructure to reduce distribution losses. In doing so, they have respected Victoria's system of property rights and established a very strong precedent that will make it exceedingly difficult for Victoria to use the *Water Act (1989)* to permanently qualify entitlements. In deciding to purchase water entitlements that exist under state law, the Commonwealth has used a market approach rather than an administrative approach to achieve a sustainable level of take. In retrospect, the objectives of the Basin Plan could have been achieved with much less administrative and legal complexity if a greater reliance was placed on the markets at the outset and more attention was paid to assisting local irrigation communities to adjust.

KEY LEARNINGS AND REFLECTIONS

The development of Victoria's water entitlement, water market, and planning arrangements has been nearly 30 years in the making. The key objective has been to introduce economic incentives into managing the competition for the state's increasingly scarce water resources. A principled but pragmatic approach has been used to manage the politics of transition. Importantly, political leadership spanning many election cycles and many Ministers, a leadership process that depended on ongoing community trust and support, sustained the reform process.

The establishment of a hierarchy of unattenuated water entitlements modeled on the concept of property rights provided the basis for the introduction and evolution of water markets and a revolution in the approach to planning. Converting existing poorly specified and inferior rights into the new volumetric entitlements through a voluntary process created the initial set of entitlements. Open participatory processes were used to judge whether the proposed new entitlements were fair and reasonable when considered together. This encouraged entitlement holders to participate and support the process, because they saw the benefits of holding secure enduring entitlements over the old inferior rights. Their support and trust was critical in sustaining the community and therefore political support for water reform.

The new entitlement regime provided the foundation stone for major institutional reform of the Victorian water sector based on devolving water and waste service delivery responsibilities to statutory water authorities.

Tightly restricted water trading was initially introduced, but restrictions were progressively relaxed as confidence in the market increased. The expanding and evolving water market quickly revealed the real dollar value of water entitlements, further enhancing the support for the new entitlements arrangements. It also revealed that the existing entitlement recordkeeping and trading processes were inadequate.

A modern register of water entitlements was created with a customer focus and with online capabilities. It applies commercially accepted accounting standards and controls rigorous internal audit trails for all transactions, to protect the integrity of the entitlements. It has sped up transaction times and enabled transaction fees to be reduced. It has helped to maintain entitlement owners' and therefore the community's confidence in the entitlement system.

Irrigation communities recognized that the introduction of water markets in the irrigation districts would move water from low-value uses to high-value uses, often in different locations. Furthermore, they recognized that the value of their entitlements would increase, placing further pressure on low-value uses. Trading rules were introduced in response to these concerns, to slow the rate that water could be traded out of districts to manage the adjustment process in local communities. These rules were sufficient to sustain local support for the market. The market proved particularly effective in managing the effects of drought.

The development of the entitlement regime and water markets has changed the roles and responsibilities for water-resource planning. The primary responsibility for planning now rests with entitlement holders—irrigators, urban water corporations and environmental water holders. The focus of governments' planning role is now on issues that cannot be dealt with by an entitlement holder alone, and with the provision of public goods and services associated with the state's water resources.

Water deliveries in some irrigation districts have fallen by more than 30% due to the progressive relaxation of trading rules, and the Commonwealth entering the water market to purchase water for environmental flows. This has creating significant concern in many irrigation-dependent local communities, resulting in demands that more restrictive trading rules be reintroduced. To date the Victorian government has resisted these calls, but has so far failed to come up with effective programs to assist local irrigation communities undergoing rapid adjustment.

The establishment and ongoing effectiveness of water markets depends on sharing the benefits with water entitlement owners and on the ability of local communities to respond to the unavoidable adjustment pressures that result. Using consultative processes, transitional measures and carefully timed reform steps is critical for maintaining sufficient community and political support. Ultimately, the resilience of Victoria's water entitlement and water markets is measured by the robustness of the community's faith and trust in the arrangements. That is the only way the necessary political support can be maintained.

REFERENCES

ACCC. (2010). *Water trading rules: Advice development.* Canberra: Australian Competition and Consumer Commission.

ACIL Australia. (1984). *Summary report transferability of water entitlements: Report to Department of Water Resources, Melbourne.*

Aust Govt. (2013). *Murray-Darling Basin Agreement.* Canberra: Federal Register of Legislation, Australian Government.

Aust Govt. (2016). *Water Act 2007 (with amendments).* Canberra: Office of Parliamentary Council, Australian Government.

Briscoe, J. (2011). *Water as an economic good: Old and new concepts and implications for analysis and implementation. In* P. Wilderer (Ed.), *Treatise on water science.* Amsterdam: Elsevier Science. http://www.sciencedirect.com/science/referenceworks/9780444531995#ancv0010.

CSIRO. (2012). *Climate and water availability in south eastern Australia*. Canberra: CSIRO. http://www.seaci. org/publications/documents/SEACI-2Reports/SEACI_Phase2_SynthesisReport.pdf.

Cummins. (2016). *Water market trends—Trends in northern Victorian water trade 2001–2015: Report by Tim Cummins and Associates to Melbourne: Department of Environment, Land, Water and Planning*. http:// waterregister.vic.gov.au/images/documents/Water%20Market%20Trends%20Report.pdf.

DCE. (1991). *Water Victoria: The next 100 years*. Melbourne: Department of Conservation Forests and Land.

DELWP. (2016). *Millennium drought report*. Melbourne: Department of Environment, Land, Water and Planning. http://delwp.vic.gov.au/water/millennium-drought-report.

DEPI. (2013). *Victorian waterway management strategy*. Melbourne: Department of Environment and Primary Industries. http://www.depi.vic.gov.au/water/rivers-estuaries-and-wetlands/strategy-and-planning.

DSE. (2003). *Securing our water future—green paper for discussion*. Melbourne: Department of Sustainability and Environment.

DSE. (2004). *Securing our water future together*. Melbourne: Department of Sustainability and Environment.

DSE. (2011). *Water supply demand strategies*. Melbourne: Department of Sustainability and Environment. http:// www.depi.vic.gov.au/water/governing-water-resources/water-corporations/water-supply-demand-strategy.

DWR. (1986a). *Second annual report 1985-86*. Melbourne: Department of Water Resource Victoria.

DWR. (1986b). *Irrigation management study*. Melbourne: Department of Water Resources Victoria.

DWR. (1986c). *Report no. 1 water law review discussion paper*. Melbourne: Water Resource Management Report Series, Department of Water Resource Victoria.

DWR. (1987). *Security for major water allocations*. Melbourne: Department of Water Resources Victoria.

DWR. (1992). *Water Victoria: A scarce resource*. Melbourne: Department of Water Resources Victoria.

Fisher, D. (2011). *A sustainable Murray-Darling Basin: The legal challenges*. Canberra: Australian National University. http://press-files.anu.edu.au/downloads/press/p115431/html/ch24.xhtml?referer=109&page=33#toc-anchor.

Kirby, M. C. (2012). *The economic impact of water reductions in the Millennium Drought in the Murray-Darling Basin*. Canberra: CSIRO. https://publications.csiro.au/rpr/download?pid=csiro:EP12033&dsid=DS1.

MDBC. (1998). *Water audit monitoring report 1996/97*. Canberra: Murray Darling Basin Commission.

Murray Water Entitlement Committee. (1997). *Sharing the Murray*. Melbourne: Department of Natural Resources and Environment.

NRE. (1999). *Entitlements to the Murray—outcome of work to define how Victoria's River Murray water is to be shared*. Melbourne: Department of Natural Resources and Environment.

NRE. (2001). *The value of water a guide to water trading in Victoria*. Melbourne: Department of Natural Resources and Environment.

NWC. (2012). *Impacts of water trading in the southern Murray-Darling Basin between 2006-07 and 2010-11*. Canberra: National Water Commission. http://archive.nwc.gov.au/library/topic/rural/impacts-of-trade-2012.

Paterson, J. P. (1987a). *The river Murray and Murray Darling Basin agreements: Political economic and social foundations*. Melbourne: Department of Water Resources Victoria.

Paterson, J. P. (1987b). *Law and water rights for improved water resource management: DWR staff paper 01/87*. Melbourne: Department of Water Resources.

Productivity Commission. (1996). *Stocktake of progress in microeconomic reform*. Canberra: Productivity Commission. http://www.pc.gov.au/research/supporting/microeconomic-stocktake.

Vic Parliament. (1979). *Hansard*. Melbourne: Victorian Parliament. http://www.parliament.vic.gov.au/images/ stories/historical_hansard/VicHansard_19791121_19791127.pdf.

Vic Parliament. (1980). *Public bodies review committee first report to parliament*. Melbourne: Victorian Parliament. http://www.parliament.vic.gov.au/papers/govpub/VPARL1980-81NoD9.pdf.

Vic Parliament. (1984a). *Twelfth report to parliament future structures for water management volume 4 final report, irrigation and water resource management*. Melbourne: Victorian Parliament.

Vic Parliament. (1984b). *Salinity committee final report on water allocations in northern Victoria*. Melbourne: Victorian Parliament. http://www.parliament.vic.gov.au/papers/govpub/VPARL1982-85NoD48.pdf.

Vic Parliament. (1989). *Water Act 1989*. Melbourne: Victorian Parliament. http://www.parliament.vic.gov.au/static/www.legislation.vic.gov.au-statbook.html.

Victorian Water Register. (2016). *Bulk entitlements*. Melbourne: Department of Environment, Land, Water and Planning. http://waterregister.vic.gov.au/water-entitlements/bulk-entitlements.

Water Bureau. (1995). *The bulk entitlement conversion process*. Melbourne: Water Bureau, Department of Conservation and Natural Resources.

AGRICULTURE IN NORTHERN VICTORIA (AUSTRALIA) OVER THE PAST 20–30 YEARS: FACTORS INFLUENCING DECISION MAKING BY INDIVIDUAL FARMERS

4

R. Rendell

RM Consulting Group, Bendigo, VIC, Australia

CHAPTER OUTLINE

Decision Making in Water Resources Policy and Management. http://dx.doi.org/10.1016/B978-0-12-810523-8.00005-7

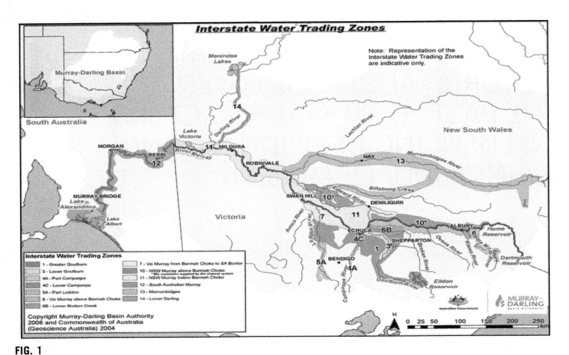

FIG. 1

Map showing the Southern Connected System of the Murray-Darling Basin.

INTRODUCTION

This chapter is primarily concerned with changes in irrigated agriculture in northern Victoria (Australia), although this story cannot be told in isolation from the wider Southern Connected Murray-Darling Basin (colored section in Fig. 1) since water trading takes place across this wider area.

The southern connected part of the Murray-Darling Basin system is one of the most important irrigation regions within Australia. It includes a number of irrigation areas, including Murrumbidgee Irrigation District around Griffith, the Riverland Irrigation District around Loxton, the Murray Irrigation District around Deniliquin, the Goulburn-Murray Irrigation District (GMID) around Shepparton and Kerang, and the Sunraysia Irrigation District around Mildura (Fig. 1).

Geographically, northern Victoria consists of three broad zones: the uplands, the slopes, and the northern plains. It is within the northern plains that irrigation has been practiced for over 125 years and is the driver of the socioeconomic aspects of the area. The northern Victorian region is dominated by a publicly owned irrigation system called the GMID. This comprises six large gravity-fed irrigation districts and three much smaller pumped districts. The GMID is set within the much larger Goulburn-Murray Water Region, which includes major storages and private diverters, both groundwater pumpers (around 100 GL/y) and direct from rivers/creeks (up to 100 GL/y). However, the large majority of the irrigation (1200–1700 GL/y) is delivered through the infrastructure of the GMID (Fig. 2).

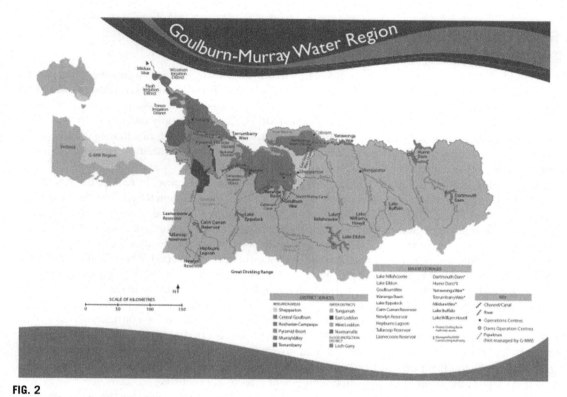

FIG. 2

Map showing the Goulburn-Murray Irrigation District.

The population in the GMID grew to around 100,000 by 1970 and continued to grow to 135,000 in 2015, a 31% increase. By contrast, the population of Australia grew by 83% over the same time period, although dryland farming areas of northwestern Victoria declined by 18% (ABS, 2016). Around 57,000 people are employed within the GMID, of which 7500 are in direct employment in agriculture, with a further 4000 in food manufacturing. Agriculture dominates the economic activity of the region, generating $2 billion farm gate value annually, of which 70% comes from irrigation. The GMID covers 500,000 ha with a mix of irrigation and dryland agriculture, but two-thirds of farm employees and 80% of those in food manufacturing are involved in irrigation-related activities (ABS, 2011).

This chapter covers the history of the growth of irrigated agriculture in northern Victoria from just after World War II to the current time, and summarizes the changes in water use over this period. We then analyze the major drivers of change in decision making in the various irrigated agricultural enterprises over time, including climate variability (particularly the Millennium drought), government policies, changes in agricultural practices, and the role of the water market.

GROWTH OF IRRIGATION IN NORTHERN VICTORIA
THE ROB RENDELL STORY

I live and have worked primarily within northern Victoria all my life, so this is a personal story as well as the story of irrigation in northern Victoria. My story begins with my 16 great-grandparents who were all (yes, every one of them!) of British/Scottish origins, and who came to Australia in the 1850s for a new beginning and became "selectors." The original "Rendell" was the only one of the 16 who was a convict; the others came seeking new opportunities. The history of agriculture and irrigation in northern Victoria is a story of farmers overcoming hardship and seizing opportunities. However, this was only possible at the expense of the indigenous Aboriginal inhabitants.

This section is about the post "selector" evolution of irrigation, because, in reality, the Aboriginal peoples have been denied their rightful part. This denial has continued even in the recent Murray-Darling Basin Plan (Aust Govt, 2012), where although there have been many cultural surveys, consideration of the environment and agriculture is still primarily "white fella" business. I hope we may combine the two stories 1 day, but at the moment the history is primarily about white people after the selectors.

Irrigation came to my family's part of northern Victoria after the second World War (WWII) and thus the family was the beneficiary of irrigation, which commenced in the 1950s. My parents owned a classic dryland sheep and wheat farm at "Baulkamaugh North" that was able to intensify production as a result of irrigation.

It seems incongruous in today's world, where water entitlements have considerable economic value, that entitlements were handed out to my family for no cost (today water entitlements represent around 40% of a typical dairy farm's total capital investment). Landholders were required to undertake irrigation works to utilize the entitlements and this involved major farm changes, but they received a valuable asset in the form of an irrigation entitlement effectively cost free.

Initially, upon receiving an irrigation supply, my parents tried dairy farming, but this was difficult in the early 1950s. My parents then used irrigation to increase our farm's sheep/wheat production. The advent of wheat quotas in 1969 challenged the status quo, and suddenly new crops (e.g., sunflowers) were tried and grown successfully and our farm income increased. In the late 1960s and 1970s, my five siblings and I all received university education funded by irrigation of new crops, building on the hard work of my "selector" family.

Following university I returned to the farm for a short period and I was fortunate to be part of a growth era in the irrigation industry in northern Victoria. The 1970s growth era saw increased areas of irrigation, increased water use, the search for groundwater to supplement surface-water irrigation supplies, and land forming to decrease labor requirements. These were heady days for irrigation in northern Victoria.

Like many farming industries, northern Victoria's irrigation industry has changed from many small individual, self-managed farms to today's larger farms, which often have several employees and seek specialist technical support. Traditionally, "extension" or technical farm assistance was provided in a patriarchal-type way, via government services (usually Departments of Agriculture) to these small owner-operator farms. In more recent times, the technical advice has been privatized and there are now more "off farm" specialists. I became one of those specialists, providing technical advice, employed initially by a water authority and more recently in my own business. I am now heavily involved in the modernization of the irrigation distribution system.

Our farm evolved over time, like the region. Initially the property was a dryland sheep and wheat operation, which, following the advent of irrigation in the 1950s, increased its sheep production tenfold before converting to higher-value irrigation crops in the 1970s. When my family sold in the early 1980s to a recent English immigrant farmer, the new farmer continued the farm much the same as we had done, i.e., irrigated cropping. Then, like many mixed farms, they felt compelled to convert to dairying in the 1990s in order to increase the farm's income. The farm was sold again in 2008 to a dairy farmer who wanted a bigger property, and it is now a large 700-cow dairy farm.

The evolution of the irrigation system on my family's farm has been replicated many times across the region. My father started with small "border-check" bays and small channels with dirt-controlled outlets, which remained until my time in the 1970s, when the irrigation layout became a "landformed" border-check system using large open channels with pipe outlets. Like much of the region, the farm suffered from rising water tables in the 1970s following several extremely wet years and the rapid expansion of irrigation. Unlike the areas to the west of the GMID region, where salinity was a big problem, the groundwater under our property was relatively fresh (1600 EC) and I was able to utilize this as a resource by installing a shallow (9 m) "spear point" pumping system. The layout of my day has now been superseded by an automated "pipe and riser" distribution supplying fast flows to large laser-controlled, graded-border check flows with a fully integrated recycle system.

Many of the recent improvements in farm irrigation practices on the old family farm have been financed by farm efficiency grants as part of the Murray-Darling Basin Plan (GBCMA, 2016). The farm has also benefited from the modernization of the public distribution system, where the smaller public channels have been replaced by an enhanced farm supply system. In fact, the channel that my father built as a contractor with horse and scoop for the State Rivers and Water Supply Commission following WWII is the same channel that has been rationalized by Goulburn-Murray Water, the current operator of the delivery system (GMW, 2016). My father at 94 was able recently to watch it being replaced by a pipe and riser on-farm delivery system.

The original Rendell family farm at Yalca "selected" by my convict great-great-grandfather is still being farmed by a descendant, but it is now four times larger and focused on dairying.

Only two descendants of my father's family of 50 cousins still own viable irrigation farms; another one is employed on an irrigated dairy farm but is also a small part-time irrigation farmer, and I am in the irrigation support industry. A number of other descendants have continued in agriculture but not in the region nor in irrigated agriculture. This is a typical story of the region where the percentage of farming families continuing in agriculture is reducing continually.

OTHER PEOPLES' STORIES

My story provides a typical story for the selector families and irrigation farms of northern Victoria, but there are some other typical stories of the region, which also help describe how the region has evolved over time. These include: the soldier settlers, the post WWII migrants on "fruit blocks," the "sharefarmers who made good," the outsider who buys a dairy block, the recent migration of New Zealand dairy farmers, the limited corporate story, the wine industry expansion, the "exotic industry" boom and bust, and the "managed investment schemes." These are discussed in the following sections.

Soldier settlement

Soldier settlement following WWII had a large impact on the region's irrigated agriculture, where typically 640 acre (ca. 260 ha) dryland sheep-wheat farms were provided with irrigation. These "selector" properties were compulsorily acquired and broken down into six dairy farms designed to milk 50 cows (on 40 ha each) or 15 orchards (on 15 ha each) producing fruit for the canning industry. This was social engineering at a large scale and utilized an estimated 20% of the region's water supply. Many of these settlers had neither experience in agriculture nor capital, but through sheer determination and the availability of low-interest loans were able to develop financially viable properties. Of course, not all were successful and over the years the more-successful properties expanded by purchasing the less-successful operations.

Very few of the soldier settlers are still living and while a number of their children became the dairy farmers in the 1990s, not many of their descendants remain in agriculture.

Fruit blocks—mainly a migrant story

Fruit blocks in the northern Victorian region have traditionally been limited to small areas on the most fertile and permeable soils around Cobram, Shepparton-Mooroopna and Girgarre. The fruit blocks have had a checkered history, starting with soldier settlements and ex-soldiers, who often struggled to make them successful. Properties changed hands and were bought by European immigrants with a farming history and immigrant attitudes to hard physical work and the desire to make good in a new country. The region suffered extensive floods in the mid-1970s, causing havoc for the permanent plantings on the fruit blocks and many areas had trees removed under a government-sponsored tree removal scheme. The rise of groundwater after the floods meant that groundwater pumps to control groundwater were installed in the orchard areas and flood irrigation practices were replaced by sprinkler methods. However, not all could survive the changes and many orchardists sold properties. The advent of the European Common Market limited access to overseas markets and Australia's progressive tariff reductions since the mid-1970s led to increased imports—all leading to a reduction in the processing fruit industry. The surviving orchardists have generally become much larger and have moved into the higher-value fresh fruit markets. The current horticulture business is dominated by a few large family corporate-style operations, mostly descendants from the migrant families of the 1960s and 1970s.

Dairy farming as a cheap way into agriculture

Dairy farming was traditionally seen as a way to get into farming because it was relatively low capital and had a high labor requirement suited to an owner-operator. Dairying also utilized a system of share farming (often the only way to get people to work long hours, 7 days a week), which enabled people with no or limited capital to enter the industry. Sharefarmers were often able to accumulate some capital (perhaps started by owning cows and then some equipment) and gradually were able to purchase their own farm. The long continuous hours also meant that dairy farmers often worked for only 10–15 years in the industry and then sold, meaning a constant turnover in dairy farms.

In more recent times, as farms have become bigger, the ability to enter the industry from outside agriculture with limited capital has become harder. The turnover of dairy farmers has entered a new stage where New Zealand dairy farmers have migrated to northern Victoria in this century as land values forced them to expand away from New Zealand.

Exotic and new industries—saviors of the economy!

Exotic industries that were going to "save" the economy of a region have often been promoted, and northern Victoria has had its share of these stories, with emu, ostrich, buffalo, and deer farming all having been tried and generally failing.

There have also been various attempts at high-value horticulture, mostly with external investment, including managed investment schemes. The wine industry expanded enormously in Australia in the

1990s and early part of the 2000s as a result of favorable overseas markets, changing customer demand and spurred on by mechanical harvesting and pruning, enabling efficiencies. As supply exceeded demand, prices crashed and money was lost (WGCSA, 2015). A managed investment scheme for olives was developed on wildly optimistic yield and price forecasts, which, when unmet, resulted in investors losing money. The development has been maintained, but with new capital and more modest expectations (News Weekly, 2009).

Niche industries—irrigation has provided other opportunities

Some irrigation farmers have ventured into niche products, such as organic farming, farm stays, and homemade produce (green tomatoes or locally made cheese) to be sold at farmer markets. Some have expanded a sideline of feeding a few pigs into large intensive piggery operations. Some service the horse industry in various ways. All these activities bring a new group of people to the region.

Remaining mixed farmers—most are part time

A major change in the region has been the decline of mixed irrigation farming—that is, those who used irrigation to increase production, but really have not changed the farming system from the selectors days. A very small number of these traditional mixed farmers have managed to survive by:

- Expanding to a scale that remains profitable (i.e., using ca. 3000 ML/y);
- Adopting some higher-value crops such as maize or lucerne;
- Selling water in years of scarcity when prices for temporary water are very high (>$200/ML);
- Combining irrigation with large areas of dryland farming.

Most of the remaining mixed farmers in the region have survived by becoming part-time farmers and selling part of their water entitlement. These part-time farmers are large in number and exceed the number of viable irrigation farm businesses.

Last-generation farmers—part of ongoing restructuring

The number of farmers is declining continuously. For example, the number of dairy farmers halves every 18–20 years (DOED, 2014; HMC, 2010). The evolution of dairy farms over the years, described in the following, illustrates how flexible farmers need to be to survive:

- *1950s postwar*—25-cow farm, walk-through dairy and 6 plant milking machines with hand separator.
- *1960s settler farm*—100-cow farm, one man with support of wife and/or children in a walk-through dairy.
- *1970s father and son farm*—180-cow farm, supporting two generations, more automated herringbone dairy.
- *1980s larger farm with milkers employed*—250-cow farm, purpose-built larger herringbone with bale-fed supplements.
- *1990s advent of rotaries and several employees*—500-cow farm, rotary milker with temporary feed pads.
- *2000s large rotaries*—1000-cow herds, rotary dairy, 8–10 employees and covered feed pads.

During this evolution of the dairy industry, the number of farms has declined in northern Victoria from 5000 to 1100, yet production has increased from 1500 L/cow to over 6000 L/cow (Ashton & Oliver, 2015; Harris, 2011).

Similarly, the processing tomato industry is producing the same number of tomatoes today with 15 growers as it did with 1400 growers in the 1960s, as the advent of mechanical harvesting, hybrid varieties, and drip irrigation meant yields increased from 25 to 130 tons/ha (Mann, 2015).

However, there are always farmers who fail to make the transition. They typically have minimal debt, no children at home and modest living requirements. Therefore, they are able to continue farming for a number of years with little capital investment until retirement or ill health forces the sale of the business. Then, typically, the business dissolves, the land and water are absorbed into other businesses and most of the farm infrastructure is "bulldozed" as of little value. So there are large numbers of "last-generation farmers" within the region who are often more than 60 years old.

Lifestyle—irrigation properties can be effective large house blocks

Large numbers of people are employed in the community in retail, education and other businesses. Many of these people now reside on the land on small farms that are no longer viable. Often these people have a few cows, horses, sheep or fruit trees, and utilize the regional irrigation infrastructure.

Growing viable farm businesses—the new era

The number of farm businesses within the GMID is declining. There are now fewer than 2000 reasonably sized businesses using 1200–1700 GL/y irrigation water collectively, with only 500 of these businesses producing the majority of the agricultural production (Dairy Australia, 2015). This is a long way from the 12,000 irrigation properties of the 1960s in the GMID (ABS, 2004).

These viable agricultural businesses are focused mainly around fresh fruit production and large dairy farms. A typical large horticulture business uses 100–500 ML/y on 20–100 ha of land, a dairy milking 500–1000 cows uses 1500–3000 ML/y, and the few remaining mixed farms use 1500–3000 ML/y, but also manage large dryland properties. These businesses often employ 5–50 people (ABS, 2011).

Thus, the GMID now has a mosaic of large properties in among part-time, last-generation and lifestyle properties, a very diverse mix of activity and land ownership.

THE STORY OF WATER USE—THE PAST AND THE FUTURE

This section discusses the various stages of water use that the GMID region has undergone.

PREIRRIGATION

Indigenous era

The northern Victoria area was inhabited by Australia's first people prior to irrigation development. They were mostly hunter-gatherers and had a strong affinity to the land and the river systems. At this time, the region was a mixture of open plains, sparse woodlands and forests, and the Aboriginal peoples managed the land with the use of fire (Barr & Cary, 1992). With European settlement in the 1840s, the area was "taken up" by squatters, and the indigenous population declined rapidly as a result of European diseases, direct conflict and policies to shift the remaining Aboriginal peoples to settlements like Cummaragunga at Barmah on the NSW side of the River Murray (ABC, 2012).

Squatter-selector era

The northern Victoria region was explored by Europeans in the 1850s. Following the explorers were the squatters who set up large runs within the region. At this time only about 30–40 squatters occupied what is now the whole of the northern Victorian irrigation region (Barr & Cary, 1992). The advent of sheep and the reduction in burning by the indigenous peoples resulted in major changes to the landscape.

Between the 1850s and 1880s the area was "opened up" to selectors, who were allowed 320 acres of land per family, and had to fence it, live on it and pay a certain amount to the government for the privilege. In order to survive, the settlers set about clearing the land, removing as many trees as possible, and started grazing and cropping the land.

The concept of allocating resources based on a "living area" (an area sufficient for a family to live on) underpinned the selection period. This concept was fundamental to the way agriculture developed in the region and continued right up until the mid-1960s, when irrigation entitlements were allocated to provide a "living area" to landholders, i.e., water entitlements were granted based on a formula that provided "enough water to generate a family income." The irrigation areas of Griffith in New South Wales, the Riverland in South Australia and the Sunraysia district around Mildura all followed this concept. The farm ownership structure in the southern connected basin today has remnants of that philosophy, which underpins many of the community attitudes to policies such as water trade and corporate investment in the region. The concept of living area was not applied to irrigation development in northern NSW or southern Queensland, and is part of the reason for the different attitudes and polices that have evolved in those regions.

Federation

Water policy has evolved with the evolution of politics and the existence of state boundaries, but also exists at a federal level. Federation occurred during one of the worst droughts in Australian history, which became a major driver for the states to work together in relation to the use of water resources (see Chapter 1).

At the same time each state has developed its own water policies and systems to reflect the differences between them. The Murray-Darling Basin Commission was founded in 1988 and evolved a consensus arrangement to manage the system. The current Basin Plan reflects this history (see Chapter 12).

The period from the 1950s saw a significant expansion in the scale of water extraction for irrigation, with the construction of dams and growth of irrigation districts. This period of growth reached a peak in 1995 with the imposition of a Cap (volumetric restriction) on total diversions across the Basin. Since then the focus of government policy has been to restore flows to the environment to establish a more sustainable diversions regime. This was seen in the National Water Initiative (NWC, 2004) and the Living Murray initiative (MDBA, 2011), and culminated in the signing of the Basin Plan, which now sets formal limits on the volumes that may be extracted for consumptive use (Chapter 12; Fig. 3).

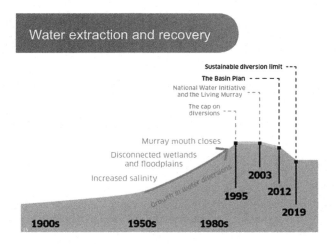

FIG. 3

Schematic showing the major water reforms leading up to the Basin Plan.

IRRIGATION ACROSS THE SOUTHERN CONNECTED SYSTEM FROM THE 1970S TO 2016

The Southern Connected System comprises a number of major river systems that link the main irrigation districts in NSW, Victoria and South Australia (Fig. 1). It includes the River Murray in each of the three states, as well as the Murrumbidgee River in NSW and the Goulburn River in Victoria. These irrigation areas now operate increasingly as a single integrated water market, as decisions by rice growers in southern NSW impact on dairy farmers in the GMID, so any discussion of water use needs to be seen in this wider context.

This section looks specifically at the period post-1970, when the region was in the middle of an era of dam construction and growth. The history of irrigated agriculture across the Southern Connected System can usefully be seen as falling into four broad time periods (Fig. 4; see also Chapter 2):

- *1970–85*: *Dams and growth*—This was an early development phase, with growth in supply and use supported by the construction of storages and the issuing of new entitlements. There was unconstrained growth in all sectors, with little effective competition for that water.
- *1986–2001*: *Cap and trade*—This period saw the introduction of water trading and the imposition of the Cap (upper limit)[1] on total diversions. These factors introduced competition and drove water to higher value uses. Water use by the mixed grazing industry declined dramatically, but rice, dairy and horticulture continued to grow.
- *2003–10*: *Drought and buyback*—This was an 8-year period of shocks with severe drought, combined with the federal government's buyback of entitlements and the impact of adverse commodity price cycles (RMCG, 2016). For the first time, the rice industry and the dairy industry were severely constrained. The wine industry contracted due to oversupply and lower

---

[1]http//:www.mdba.gov.au/what-we-do/managing-rivers/the-cap.

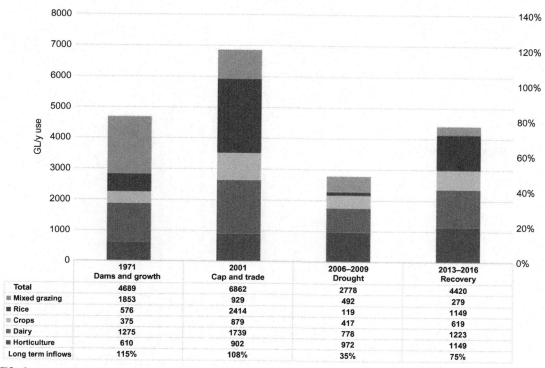

	1971 Dams and growth	2001 Cap and trade	2006–2009 Drought	2013–2016 Recovery
Total	4689	6862	2778	4420
Mixed grazing	1853	929	492	279
Rice	576	2414	119	1149
Crops	375	879	417	619
Dairy	1275	1739	778	1223
Horticulture	610	902	972	1149
Long term inflows	115%	108%	35%	75%

FIG. 4

Water use by sector over time in the Southern Connected System.

RMCG (2016). Basin Plan—GMID socio-economic impact assessment. *Report prepared for the GMID Water Leadership Forum by RMCG, Bendigo.*

prices for wine. On the other hand, horticulture, particularly the almond industry, continued to expand despite the drought (Cummins & Associates, 2016).

- *2013–16: Recovery*—This period has seen a recovery from the drought years with restoration of production volumes, but to a lower overall level than previously, due to lower inflows and a reduction in the size of the consumptive water pool due to government buyback of water for the environment.

Rice industry

The rice sector and other annual crops vary the area planted in any season in response to the level of early season announced water allocations, and the price of water in the temporary market. An analysis of the level of water use and rice production over the last 20 years shows a number of clear phases (Fig. 5; RMCG, 2016):

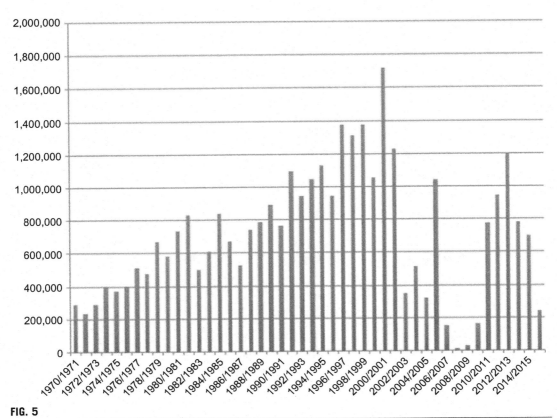

FIG. 5

Annual rice production (tons) in the NSW Riverina between 1970/71 and 2014/15.

Ricegrowers Association of Australia 2015.

- A period of steady overall growth from 1970 to 2001, with total production rising from 200,000 to 1,800,000 tons, but with significant annual variation to reflect the level of allocation in any season;
- A profound collapse in the level of production from 2002–03 to 2009–10 due to the Millennium drought when water availability and use was significantly reduced and water prices rose in the market;
- Recovery over the last 4 years (2011–12 to 2015–16), to a lower level of production than before the drought, and considerable variation in rice production depending on the level of allocation.

Dairy production

The dairy sector seeks to maintain the level of production between seasons. It has some flexibility to respond to variations in the level of water allocation between years by substituting bought-in feed for homegrown pasture, but this adds to the costs of the business. In more extreme climate scenarios, dairy farmers can sell off stock or pay for agistment elsewhere, although this reduces income as well as

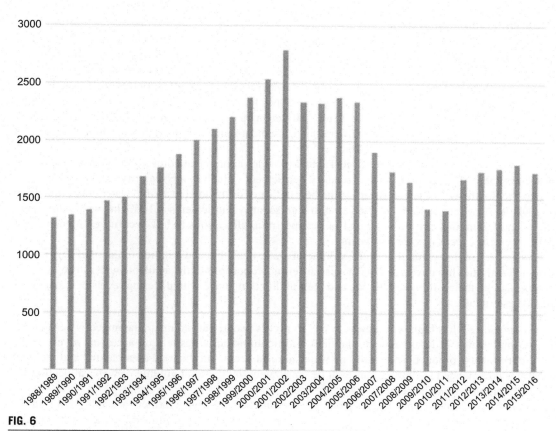

FIG. 6

Annual dairy production (million liters) in the GMID between 1988/89 and 2015/16.

Dairy Australia 2016.

increasing costs. In wet seasons, the sector can expand by bringing additional pasture into production to save costs of bought-in feed.

Fig. 6 provides a comparison of two periods (after the 2002–03 drought and then after the 2007–10 drought), which shows a reduction in the level of annual average milk production of around 600 ML/y. Between the two periods the water allocation levels were similar, but there was around 21% less water available due primarily to the federal government's buyback (see Chapter 12). RMCG (2016) proposed that the change in production over these two periods is a good indicator of the net effect of buyback.

THE DYNAMIC EQUILIBRIUM OF IRRIGATION ENTERPRISES

This section covers the changes in the decision making by farmers in the Southern Connected Systems and the GMID as a result of the influence of annual climate variability on water allocations and price.

Southern connected system

The volume of water available across the southern connected basin varies widely from year to year depending on the level of rainfall and inflows to storages. Table 1 shows the maximum volume available for irrigation is close to 6000 GL in wet years, while, by contrast, in drought years this volume may not be much more than 2000 GL. Prices in the temporary water market then reflect this relative scarcity, with values of $50/ML during wet years, but rising to well over ten times this value at $500/ML during drought.

Note that the volumes recorded in Table 1 are estimated after federal government buybacks and include 500 GL of groundwater. The figures also assume no further reductions post Jun. 2016, as part of the implementation of the Basin Plan.

The heading "Level of allocation" in Table 1 provides a reference point for the general volume of water available. The level of general security entitlement (which is the majority of entitlement type across the Southern Connected Basin) is the best indicator of the total available water across the basin as a whole. As a general rule, the allocation to high-security entitlements is 100% in every year except for drought scenarios. At the other end of the spectrum, allocations to low-security entitlements are generally zero except in very wet years.

As a result of this variability in the volume of water available and its price in the market, a mix of irrigated activities has developed that access differing proportions of the available water in any season, determined by their willingness to pay and their ability to manage with differing levels of water security. These include:

- *Horticulture*—has high willingness to pay to access water, but requires guaranteed water in each season because of its permanent plantings. Horticultural activities rely on accessing a very high security water product. They will buy water in most years, but the total area planted and the total water commanded by horticulture will be constrained by the total volume of water allocation available in a drought year.
- *Dairy*—buys water in most years from lower-return sectors. Dairy farmers are able to manage with a medium-high security product, as they can substitute bought-in feed for water when prices rise. However, this flexibility is limited and the overall level of production will be constrained to the net volume that is available in "medium" and "medium-dry" years after the horticultural demand is satisfied. Only in drought years will the dairy sector sell to horticulture.

Table 1 Water Availability and Price by Season Type in the Southern Connected Basin				
Scenario	Level of Allocation	Frequency (Over 20 Year)	Total Water Availability (GL)	Price ($/ML)
Wet	Some low security	3	5800	50
Medium-wet	95% general security	5	5200	80
Average	55% general security	5	4500	130
Medium-dry	20% general security	5	3700	225
Drought	50% high security	2	2300	575
RMCG (2016). Basin Plan—GMID socio-economic impact assessment. *Report prepared for the GMID Water Leadership Forum by RMCG, Bendigo.*				

- *Annual crops*—annual crops (e.g., rice, cotton) are able to manage with a lower-reliability water product as they can vary the area planted each season to match the available water. They will maximize production in average or wet seasons, and in drier seasons they will sell their allocation rather than plant crops themselves.
- *Mixed agriculture*—mixed farms will sell their water allocations in most years, but use that water when allocations are high and prices low, to produce feed for sale or for their own use. The sector has low willingness to pay but high flexibility.
- *Carry over*—all sectors will carry over allocations in wet years to boost security in the following seasons.

Fig. 7 provides an indication of the distribution of the available water between sectors under different climate scenarios, driven by differences in willingness to pay and reliance on security of supply (RMCG, 2016).

The preceding discussion and data in Fig. 7 suggest that there are essentially three different water products available in the Southern Connected System:

- A *very high-security product*—this is available every year, but is limited to the total volume available in a drought scenario, with a market price in drought years of around $575/ML. This

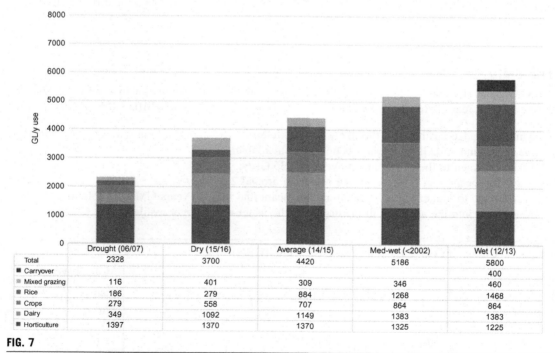

	Drought (06/07)	Dry (15/16)	Average (14/15)	Med-wet (<2002)	Wet (12/13)
Total	2328	3700	4420	5186	5800
■ Carryover					400
▨ Mixed grazing	116	401	309	346	460
■ Rice	186	279	884	1268	1468
■ Crops	279	558	707	864	864
▨ Dairy	349	1092	1149	1383	1383
■ Horticulture	1397	1370	1370	1325	1225

FIG. 7

Distribution of available water (GL/y) between farming sectors under a series of different climate scenarios.

RMCG (2016). Basin Plan—GMID socio-economic impact assessment. *Report prepared for the GMID Water Leadership Forum by RMCG, Bendigo.*

product is held predominantly by the horticultural sector. The volume of this product will define the size that the horticultural sector can grow to across the southern connected system. This is estimated at around 1400 GL/y once other sectors maintain a minimum presence.

- A *medium- to high-security product*—this product has a likely net total volume of around 1100 GL/y, after the very high-security product has been delivered, and a price of up to $225/ML. This product is used mainly by the dairy sector, with the volume available in a dry sequence defining the scale of the dairy sector.
- A *low-security product*—this product has a large volume of 2500 GL/y, with a price of less than $100/ML, but is only available in a third of years. This product is accessed by annual cropping sectors.

These different water products create an effective hierarchy of users, with lower-value sectors preferentially selling their allocations to other sectors with higher willingness to pay for water. However, this does not lead to a single irrigated monoculture across the southern connected basin, as the variability of inflows between years limits the growth of higher-value sectors to the volumes available in drier seasons, as they have a high reliance on the relative security of that water. This then provides an opportunity for other sectors to utilize the additional allocation available in wetter years.

These factors then create the observed dynamic equilibrium across the southern connected basin, with a mix of irrigation sectors that have differing ability to take advantage of different levels of security and price for their water requirements.

THE GMID

The GMID has also evolved over the same time period with a dramatic reduction in water use (Table 2). There are four reasons for the reduction in the level of water use in the GMID (RMCG, 2016):

- Climate change has reduced the likely inflows to around 75% of the long-term inflows, which means around 250 GL/y of water is no longer available;
- Water returned to the environment averages 335 GL/y;
- There has been a net trade out of the region of around 250 GL/y;
- Water trade to higher-value horticulture has meant that in most years GMID is able to trade the water back into the area, but in the dry years the water is no longer available.

Table 2 Change in GMID Water Use Over Time (GL/y)

Time Period	Drought	Average	Medium Wet
Pre-Millennium drought	1185	1930	2300
Millennium drought period	730	1640	1769
Post buyback predicted	330	1260	1480
Likely if "UpWater" implemented[a]	200	1000	1200

[a]*UpWater is a proposal to retrieve a further 450 GL/y from the consumptive pool for environmental flows.*
RMCG (2016). Basin Plan—GMID socio-economic impact assessment. Report prepared for the GMID Water Leadership Forum by RMCG, Bendigo.

UNDERLYING CHANGES

In this section we cover the major drivers that have influenced irrigated agriculture in the Southern Connected System. These include: general changes in agriculture, changes in government policies, the influence of water availability on its price, and the influence of water price on farming systems.

GENERAL CHANGES IN AGRICULTURE

Irrigation development is not undertaken in isolation of other changes happening in society and more specifically in agriculture and regional Australia.

Ongoing restructure

Farm businesses have been getting larger and fewer in total numbers, while output and productivity have increased (Aust Govt, 2016; Productivity Commission, 2005). Two examples of this can be seen in the processing tomato and dairy industries in the GMID.

In the *processing tomato industry*, there has been an increase in production from the 1970s when there were over 1400 producers producing around 150,000 tons of tomatoes, to the year 2000 when the number of producers had decreased to around 30 but they were producing over 300,000 tons of tomatoes, i.e., a doubling of production with one-fortieth the producers. In more recent times, the level of production has remained relatively constant but the number of producers has again halved to 14, although one producer (now vertically integrated with the one remaining factory) produces over half of the industry's production. At the same time, farm yields have gone from under 30 tons/ha to over 110 tons/ha, almost a four-fold increase. Similarly, the number of factories has decreased from 14 in the early 1990s to 2 today. The change in the tomato industry is an extreme example of industry rationalization triggered by a number of technologies such as landforming, laser-controlled grading, drip irrigation, hybrid varieties, mechanical harvesting, bulk handling of harvest, and mobile phones (Mann, 2015).

The *dairy industry* has seen a similar progression from very small farm operations—a 100-cow farm was the norm in the 1970s—to the 750-cow and larger herd sizes relevant today. The Australian dairy industry has the same number of cows today as it did in 1900, but the yield per cow is now over six times what it was then. Although the total herd size has remained constant, the number of dairy farmers has halved every 20 years since Federation (Ashton & Oliver, 2015; DOED, 2014; Harris, 2011; HMC, 2010). The dairy industry has benefited from better genetics, supplementary feeding of grain and fodder, rotary dairies, motor bikes, mobile phones, laser-controlled grading, and enhanced "fast flow irrigation."

This progression to larger and fewer businesses applies right across agriculture in Australia. The rate of change varies from industry to industry and is often triggered or accelerated by shock events. In northern Victoria, the two main shocks have been the Millennium drought, and the "buyback" of water entitlements for the environment. Both of these have resulted in water becoming more scarce.

Evolution of businesses

A critical factor in the evolution of irrigation businesses relates to the employment and management of labor. Employing labor is a substantial change in operations for farm businesses and many have resisted doing so other than through the use of family members. The rice industry still resists employing labor and has been able to increase production by using large-scale machinery. Many businesses function efficiently when the number employed is less than 10, as there is little need for documentation and systems, and instructions can be communicated verbally.

The dairy industry has moved through a series of "sweet spots" over the past 40 years, namely:

- The 1970s saw 100-cow farms become 200-cow farms as the dairy was modernized to a herringbone layout and two generations (father and son) operated a traditional gravity-fed irrigated pasture-based system.
- The 1980s saw the advent of laser grading and larger herringbone dairies with supplementary grain feeding in the dairy, enabling herds to reach 350 cows.
- The 1990s saw the advent of rotary dairies, and increased supplementary feeding, enabling up to 500 cow herds.
- The post-Millennium drought has seen an increase in a ration-fed diet, with more maize and fodder grown through fast-flow irrigation, allowing a larger area to be irrigated without an increase in labor requirements.

The dairy industry is about to enter a new "sweet spot" with herd sizes exceeding 1000 cows, which are housed, robot milked and fully ration-fed—that is, there will be no paddock grazing. These businesses will probably employ 20–30 people each and will involve staff supervisors and systems commensurate with such a larger-scale operation.

The horticulture industry in northern Victoria has increased its scale faster than has been observed in the dairy industry. This has come about partly through employment of seasonal-picking labor and partly through the creation of packing shed operations. These sheds add value to the primary product and enable direct contracts with wholesalers and supermarkets. As a result, some of the larger horticulture businesses are already employing up to 100 people (RMCG, 2013a).

The concept of removing labor and mechanizing

A key driver in the restructuring and evolution of farm businesses has been the need to reduce labor costs, in particular to reduce the need for the owner to do key tasks. Irrigation has traditionally been the most important task, which was generally undertaken by the owner, as any error was quickly translated into lost production. So, for many years farm size was dictated by how large an area the owner could irrigate by himself. For example, border check irrigation with dirt control structures limited dairy farms to 100 cows, processing tomatoes to 4000 tons, or horticulture to around 10 ha.

However, the advent of laser grading for dairy farmers, drip irrigation for processing tomatoes and fixed sprinklers for horticulture allowed expansion to occur. From 1970 through to the 1990s, the limiting resource on irrigated properties was labor (Harris, 2011). The shock of drought, water buyback and the increase in horticulture (particularly almonds) means that water is now the limiting resource, and so, while reducing labor costs is still a driver, it is counterbalanced by the need to increase water-use efficiency.

Change comes from adversity

One of the key paradoxes in the development of irrigated agriculture in northern Victoria is that adversity has brought change and improvements that would not have happened without the "shocks."

The Rendell family started growing sunflowers because wheat quotas were introduced, and the 1976 drought reduced water availability, so we investigated and found a groundwater supply as an alternative source of water. The advent of supplementary grain feeding followed the 1982 drought, and so on. There is no doubt that the recent Millennium drought, coupled with federal government buyback, has resulted in a significant change in the way water is now managed.

Water carryover was introduced in Victoria following the first year of the severe drought in 2006 (see also Chapter 2). Historically, any water that an irrigator did not use in one season was reallocated to the general pool to be distributed to all irrigators the next season. Changed accounting methods and policies have resulted in individuals being able to "carry over" any unused water from one year for use in the next season. The introduction of carryover has reduced the risk in a future drought year, as each irrigator will have had the opportunity to carry over water from previous seasons as an insurance policy. However, the greater take-up of carryover by farmers will see more water permanently stored in the dams, leading to an increased likelihood of spills and a reduction in the total yield from the entitlement.

The extension story—there is little publicly funded extension

Irrigation in Victoria has traditionally been supported by large numbers of government employees in extension and research. For example, in the GMID, there were research centers at Kyabram, Tatura, Kerang and Swan Hill, with extension officers in the major towns throughout the region. However, the last 20 years have seen the demise of these staff, with their transfer to supporting industry and regional benefit activities.

Different sectors have responded to this withdrawal in different ways:

- Mixed farming and grazing industries have declined and there is now almost no irrigated industry research or extension.
- Industry-based schemes are standard in irrigated cropping, rice, processing tomatoes and dairy, leading to extension officers in the dairy industry now being employed by Dairy Australia, and funded by producer levies and matched government funding.
- Larger businesses in horticulture, dairy and irrigated cropping all employ their own specialist advisors.

IRRIGATION POLICY IMPACTS

There have also been a number of government policy changes that have impacted on agriculture.

Federal government water buyback

A major policy that has impacted on irrigation in northern Victoria has been the implementation of the Murray Darling Basin Plan, which has included a water "buyback" scheme. Under this scheme, the federal government has purchased approximately 1100 GL/y of permanent water entitlements throughout the southern connected system, representing around 20% of the entitlements (DOE, 2016). Victoria sold around 450 GL/y of entitlements under the scheme during the period 2006–15 (Cummins & Associates, 2016).

The timing of the buyback followed one of the worst droughts in Australian history, at a time when many landholders were under financial stress. The commodity prices for the wine grape industry and the dairy industry were also well below long-term average trends. The very wet season of 2010–11 meant that there appeared to be plenty of water available and temporary water prices were very low (around $60/ML). The price offered for buyback was above long-term trend prices for permanent water entitlements. All these factors were instrumental in the way farmers responded, which saw several groups of farmers participate in the scheme:

- Mixed farmers who wished to retire;
- Dairy farmers who wished to retire;
- Dairy farmers who had suffered financial losses during the Millennium drought and used the system to effectively fund their losses by replacing the water sold by purchasing cheaper temporary water;
- Wine grape growers taking the opportunity to recoup some capital.

So buyback provided a very effective "drought relief" policy and limited the financial impact of the drought. However, longer term, the impact of buyback has been to reduce the level of production in the rice industry and dairy industry by around 20% (RMCG, 2016).

Farm efficiency grants

Part of the federal government's water recovery program within the Murray-Darling Basin Plan has involved on-farm water use efficiency grants. With these grants, approximately half the water saved as a result of any works is then transferred for use by the environment. This project is promoted as a win-win for both the landholder and the environment (GBCMA, 2016).

However, it is debatable how much the program changed practices and really saved water, or whether the program merely brought forward the timing of activities (e.g., upgrades) that would have happened anyway. Under this second scenario, the water saved would have been used by the landholder to expand their business or maintain the business's ongoing improvements in productivity.

In many cases, the landholder has gone to the water market and purchased water to replace the volume that was transferred to the environment, to enable ongoing expansion. However, these activities have taken water away from production elsewhere, and the regional overall production has reduced, but the landholder himself has benefited financially.

Proponents of the scheme argue that the scheme has facilitated improved productivity and "saved water." In practice, the program has effectively meant that the water recovered for the environment has been more expensive than if obtained by buyback and was water that was always going to be recovered and used within the region.

As always the truth is somewhere in the middle of the two arguments.

WATER SCARCITY DRIVES PRICES

Water has gone from being relatively unlimited in the "dams and growth" period of the 1970s to being quite scarce in the "drought and buyback" era (2000–15). It is likely that in the future this new dynamic equilibrium will see water becoming even scarcer, and that scarcity will drive higher prices in the water markets.

The close inverse correlation between these two variables (water availability and price) is demonstrated in Fig. 8, which shows that water prices on the Murray Irrigation Ltd. water exchange spiked to $700/ML in the drought, but slumped to $10/ML during the following floods.

Fig. 9 then shows the close correlation between the total volume of water available across the Southern Connected System and the average water price on the Murray Irrigation Ltd. water exchange (corrected to 2016 prices).

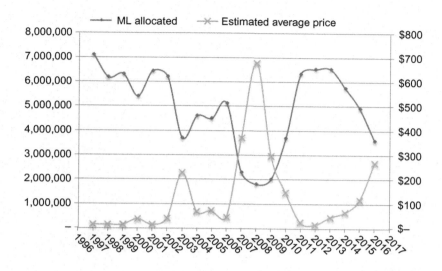

FIG. 8

Annual combined water volume (ML) allocated in the Southern Connected System and average recorded trade price ($/ML) over the period 1996/97 to 2015/16.

Murray Irrigation Ltd. 2016.

This graph demonstrates the clear relationship between the level of water availability and the temporary water price, which has stood the test of time despite all the underlying changes that have occurred over time. Although demand has increased, the water available for trade has also increased, as many more individuals participate in the market as both buyers and sellers, depending upon the particular season.

The allocation water price, when weighted by probability and different seasonal prices, represents on average about a 6% return, which is higher than lease rates for land but probably reflects the higher risk of water.

Fig. 10 then shows the trend in the price of permanent entitlement water, which has been rising at 8% per annum compound on average over the last 20 years. This is considerably higher than land values in agriculture, which have increased at closer to 3%–4% per annum.

WATER PRICES DRIVING FARMING SYSTEMS

In the 1980s, as a young extension officer in northern Victoria, I was involved in the adoption of laser-grading for border check irrigation on the basis of improved irrigation application efficiency. However, with time it became apparent that the real benefit of the technology was that it increased labor efficiency. It was the ability of a farmer to increase total production that drove the change. In those days water was relatively abundant and cheap.

While this emphasis on total farm production and farm business profit is still true today, the increase in water price has changed the approach and now production per megaliter is the critical factor.

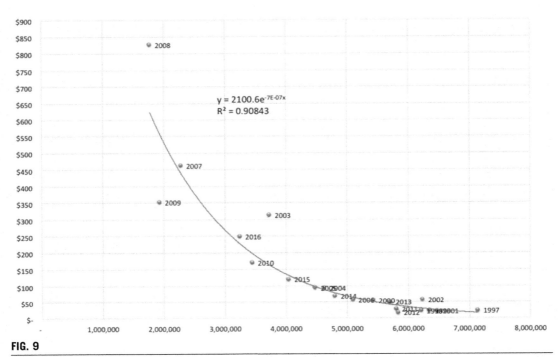

FIG. 9

Correlation between total water volume available (ML/y) and price in the temporary water market ($/ML) in the Southern Connected System.

RMCG (2016). Basin Plan—GMID socio-economic impact assessment. Report prepared for the GMID Water Leadership Forum by RMCG, Bendigo.

Up until 2005, the standard benchmark for production of pasture (dry matter) and crops (seed) was 1 ton for every megaliter of irrigation water applied (1 ton/ML), which translated into a gross income of $250/ML where:

- Mixed farming developed low-cost irrigation systems and typically sold water in dry years.
- Rice growers were able to generate large business profits by adopting large-scale systems.
- Irrigator croppers struggled to generate sufficient gross farm income because of difficulties with achieving large scale due to labor limits for irrigation (unlike rice).
- The dairy industry adopted a simple pasture-based system, based on direct grazing with pasture species that grew all year round and with minimal supplementary feeding.
- Horticulture took almost no notice of water efficiency but concentrated on quality (e.g., the wine industry reduced irrigation or the fruit industry made sure there was never water stress or water logging).

Now that water is worth around $150/ML on average and often up to $250/ML, a production of 1 ton/ML, or around $250 worth of production per ML, is no longer economic. Thus, we see considerable pressure to change farming systems as follows:

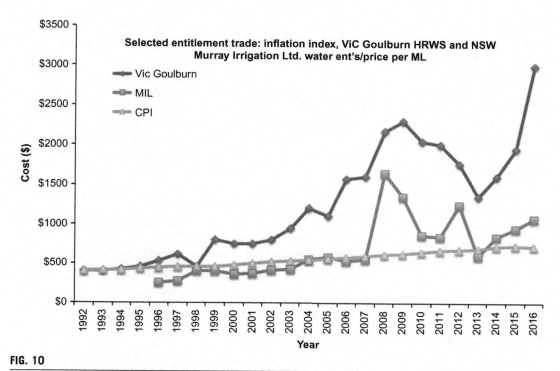

FIG. 10

Value of permanent water trades in major valleys in the Southern Connected System ($/ML).

RMCG (2016). Basin Plan—GMID socio-economic impact assessment. *Report prepared for the GMID Water Leadership Forum by RMCG, Bendigo.*

- *Mixed farm industry*—only survives in small numbers where existing infrastructure is used with the water sold in years the price exceeds $100/ML;
- *Rice industry*—is under considerable pressure to develop higher-yielding, shorter-season varieties that produce more tons per ML, for example a new rice variety was introduced in 2016 that increased yields to 1.2 tons/ha;
- *Irrigated cropping properties*—these properties are converting to higher-value crops like cotton and to more efficient crops like maize (production ca. 2 tons/ML). To do this requires fast-flow irrigation or pressurized irrigation systems like center pivots, which in turn requires a modernized delivery system to the farm;
- *Dairy industry*—can no longer sustain low cost, low water efficiency perennial pastures directly grazed by cattle. Rather it is converting to higher productivity annual pastures (closer to 1.5 tons/ML) and maize (3 tons/ML of dry matter), which is cut and fed as rations to cows as this increases feed conversion considerably. The sector is also switching to three milkings a day and full shedding of cows. However, this requires in excess of 1000 cows per business and possibly up to 3000 cows per business,
- *Horticulture*—the increased price of water only has a modest impact on farming systems in horticulture, as scale and product quality are still the main business drivers. Therefore, the irrigation

and farming systems are oriented around achieving those ends. However, if the almond price returns to the lower prices of 5 years ago, then this industry may start to adopt water-efficiency measures in order to reduce costs.

This change to higher productivity feed in the dairy sector is also driving changes in irrigation practice, with the adoption of "fast irrigation" practice combined with soil moisture monitoring technology. Here the irrigation water is applied at flows of 3–5 L/s per meter width of bay, which enables a bay to be irrigated in less than 2 h. This means that the soil is not waterlogged and the production of maize in particular is increased dramatically, up to 3 tons/ML of dry matter (compared to traditional pasture production of 1 ton/ML). However, to achieve this greater production requires a number of changes, including:

- Conversion from perennial pasture to maize cropping;
- Crop to be harvested and fed to dairy cows rather than grazed;
- Supply system to the farm has to be enhanced to 15–20 ML/day of constant supply compared to the old channels with Dethridge wheels supplying 8–10 ML/day of varying flow rates;
- The farm supply system requires larger channels and bay structures or can be replaced by a pipe and riser system.

To make these changes requires high levels of capital investment and amalgamation of properties to achieve the economies of scale required (RMCG, 2013b).

MODERNIZATION OF THE GMID DELIVERY SYSTEM

The modernization of the GMID delivery systems is the largest investment in irrigation infrastructure across Australia. The project is spending $2.2 billion to automate the control structures (including delivery channel regulators, farm offtake points and the associated meters), to remediate and line with plastic membrane the worst 10% performing sections of channel, and to rationalize and privatize spur channels to enable enhanced farm connection to the system. This will provide close to on-demand delivery at constant flows and supply head while reducing overall system losses from 834 to 310 GL/y, i.e. a saving of over 500 GL/y. This will also see a reduction in the length of public channels from 6000 km to around 4000 km in total.

Little had changed in the supply system

Despite farms restructuring and becoming substantially bigger, the GMID channel system was unchanged for 60 years. The original channel delivery system was built over a long period from the 1900s and mostly completed by the early 1950s. The channels were often built with horse and scoops and, in more recent times, bulldozers, but have been relatively unchanged since. The control structures were still manually operated "drop bars" changed twice a day. The Dethridge meter was the universal outlet structure.

Timing is the key to change

It became apparent to some local business leaders that:

- The system was badly in need of updating but the local irrigators could not afford to fund this;
- The level of service was inadequate for modern farm needs and so water was trading out of the GMID to areas where horticulture could develop with the appropriate supply system;

- The environment was demanding a bigger share of the water supply;
- There was a severe drought with a risk that Melbourne could run out of drinking water.

So a deal was hatched where Melbourne water customers would pay for the first stage of the modernization in return for the savings to be shared equally between the irrigators, the environment and Melbourne customers. A pipeline to transfer the savings to Melbourne was built in conjunction with the system modernization. The project provoked considerable community resistance but still proceeded. However, before the project was able to deliver the savings to Melbourne, the drought broke.

The second stage of the project is being funded by the federal government under the Murray-Darling Basin Plan to provide water savings to be transferred for use by the environment.

The jury is still out on this project as to the success of this GMID modernization. The project is around three-quarters finished and has just completed a "reset" in order to ensure satisfactory completion. It will be another 3 years before the final outcome will be able to be determined and evaluated.

However, what is known is that the automation of the "backbone" channels has been very successful, that the promised water savings have almost been made, and that the level of service to those who have been modernized (around 75% to date) has been greatly enhanced and exceeded expectations. Thus, the modernization story is yet to be told in its entirety.

CONCLUSIONS

Irrigated farming practice has changed beyond recognition since I first worked with my father planting sunflowers 40 years ago. Then, a 75-cow herd was the norm—now new dairies are being planned with over 2000 cows. Equally, 10 very large growers now dominate the apple, pear and stone fruit sector and have converted from canning fruit to fresh markets. These changes have been driven by a continuing production imperative to increase revenues and drive down costs.

Changes to irrigation practice have been just as dramatic. We used to rely on narrow border-check bays and small channels with dirt-controlled outlets. The need to control multiple small outlets limited the scale of irrigated enterprises and meant I had to get up through the night in order to open, close and monitor a succession of small bays. Now growers expect to be able to operate their high-flow gates and soil moisture monitors from a computer in the farm kitchen! That has allowed irrigation businesses to increase the level of production through farm amalgamation and achieve economies of scale.

Change has also been driven by a wide range of other factors including climate variability, access to export markets and changes in government policies. The water market has played a critical role in helping facilitate the transfer of water between locations and sectors. That has brought growth and change—not all of it always welcomed! Farmers used to operate at a small local scale with the water tied to the land. They now operate across the wider southern connected system, with water trading driving a new dynamic equilibrium between a mix of enterprises reflecting the highly variable rainfall between seasons.

ACKNOWLEDGEMENT

I wish to acknowledge the assistance provided by Matthew Toulmin in the preparation of this chapter.

REFERENCES

ABC. (2012). *Jimmy Little passes away but leaves a strong legacy.* http://www.abc.net.au/news/2012-04-02.

ABS. (2004). *The Australia dairy industry.* Canberra: Australian Bureau of Statistics.

ABS. (2011). *Employees by industry with division A and subdivision 01 agriculture.* Canberra: Population Census—Analysis Neil Clark and Associates, Australian Bureau of Statistics.

ABS. (2016). *Australian demographic statistics: Report #3101.0.* Canberra: Australian Bureau of Statistics.

Ashton, D., & Oliver, M. (2015). *Dairy farms in the Murray-Darling Basin: Research report 15.12.* Canberra: ABARES.

Aust Govt. (2012). *Water Act 2007—Basin Plan 2012, extract for the Federal Register of Legislative Instruments.* Canberra: Australian Government, Office of Parliamentary Counsel [245 pp.].

Aust Govt. (2016). *Agricultural competitiveness white paper.* Canberra: Department of Agriculture and Water Resources. http://agwhitepaper.agriculture.gov.au.

Barr, N., & Cary, C. (1992). *Greening a brown land: The Australian search for sustainable land use.* Melbourne: Macmillan.

Cummins & Associates. (2016). *Trends in northern Victorian water trade 2001–2015: Report prepared by Tim Cummins and Associates for Department of Environment, Land, Water and Planning, Melbourne.*

Dairy Australia. (2015). *Australian dairy industry in focus 2015.* Melbourne: Dairy Australia.

DoE. (2016). *Environmental water holdings.* Canberra: Department of the Environment. www.environment.gov.au/water/cewo/about/water-holdings.

DOED. (2014). *Dairy industry profile.* Canberra: Department of Economic Development, Jobs, Transport and Resources.

GBCMA. (2016). *Farm water program.* Shepparton: Goulburn Broken Catchment Management Authority. www.gbcma.vic.gov.au/sustainable_irrigation/farm_water.

GMW. (2016). *Connections project.* Tatura: Goulburn-Murray Water. http://www.connectionsproject.com.au.

Harris, D. (2011). *Victoria's dairy industry: An economic history of recent developments: Report prepared for DPI and Dairy Australia, Melbourne.*

HMC. (2010). *Changing land use in the GMID 2006–2010—Where have all the dairies gone?: Report prepared by HMC Property Group for Department of Primary Industries, Melbourne.*

Mann, L. (2015). *Annual industry survey 2015. Vol. 36.* Shepparton: Australian Processing Tomato Grower.

MDBA. (2011). *The living Murray story: Publication Number 157/11.* Canberra: Murray-Darling Basin Authority. (95 pp.) http://www.mdba.gov.au/media-pubs/publications/living-murray-story.

News Weekly. (2009). *Managed investment schemes: Behind the collapse of Timbercorp.* http://www.newsweekly.com.au/article.php?id=3969.

NWC. (2004). *Intergovernmental agreement on a national water initiative.* Canberra: National Water Commission [39 pp.].

Productivity Commission. (2005). *Trends in Australian agriculture.* Canberra: Productivity Commission.

RMCG. (2013a). *Goulburn Valley fruit growing industry roadmap.* Bendigo: RMCG.

RMCG. (2013b). *Cost benefit analysis of farm irrigation modernisation: Report prepared for Dairy Australia by RMCG, Bendigo. .*

RMCG. (2016). *Basin Plan—GMID socio-economic impact assessment: Report prepared for the GMID Water Leadership Forum by RMCG, Bendigo.*

WGCSA. (2015). *Market outlook for 2016: What the wineries say.* Adelaide: Wine Grape Council South Australia.

URBAN—MAJOR REFORMS IN URBAN WATER POLICY AND MANAGEMENT IN MAJOR AUSTRALIAN CITIES

5

J. Chong, S. White

University of Technology Sydney, Ultimo, NSW, Australia

CHAPTER OUTLINE

INTRODUCTION

The urban water sector in Australia is quite diverse. Service providers range from large integrated utilities, such as Sydney Water, Melbourne Water and the Water Corporation of Western Australia, through to disaggregated retail and bulk water utilities, and local government water authorities. Their responsibilities range from water supply only, to also include sewerage service provision and bulk stormwater management. Some utilities rely exclusively on local surface water, while sources used in other areas include large-scale intercatchment transfers, groundwater and desalination plants. The scales involved range from a few hundred connections to nearly two million users.

There have been major changes in the urban water sector over the last 20–30 years, as a result of deliberate economic and regulatory reforms, responses to the Millennium drought, and technological and social changes. This chapter seeks to describe some of these changes, and to provide insights into current and future changes in the Australian urban water sector.

BACKGROUND TO POLICY AND PLANNING REFORMS IN THE AUSTRALIAN URBAN WATER SECTOR

Urban water planning in Australia has a history of a strong public service ethic, reflecting the fact that many water supply institutions were created when water services were core functions of local government. These local governments emphasized the provision of safe and reliable water supplies and, where possible, the extension of sewerage networks to reduce the adverse impacts of sewage on public health. This emphasis was associated with a planning approach that primarily involved forecasting demand into the future, and planning for water supply infrastructure to meet that demand, with similar patterns occurring for wastewater service provision and stormwater services. Due to the high costs of providing wastewater and stormwater services and centralized sewerage systems, in some cases they had to wait for significant government intervention and financial support. For example, in the 1970s the provision of sewerage services to Western Sydney was funded by the Commonwealth government.

From the late 1980s through the 1990s, significant structural reforms in the urban water sector were promoted by the National Competition Policy (NCP) and were implemented from 1995 to 2005 by the Council of Australian Governments (COAG), Australia's peak intergovernmental forum. These reforms were based on the Hilmer Inquiry recommendations, which endorsed competitive markets as the basis for innovation and efficiency (Hilmer et al., 1993). The reforms saw the separation of the roles of regulator and operator of urban water utilities. In some states they involved the corporatization of state-owned utilities, a focus on commercial returns, the payment of dividends by water utilities, and clearer lines of regulatory control (Chong, 2014). Following these changes, the first independent economic regulators were established, including, in 1993, a government pricing tribunal in NSW, now called the Independent Pricing and Regulatory Tribunal (IPART). All states and territories now have some form of independent economic regulator, although their powers and functions vary widely from state to state.

The move to separate the regulatory and operational aspects of urban water supply has generated mixed results, due primarily to the historical weight and strength of the urban water utilities, particularly the major city–based utilities, which tend to dominate planning processes.

In terms of supply-and-demand planning, traditional planning methods have relied upon population projections with a simple multiplication of per capita demand, and in some cases projections of increased per capita demand based on the assumption that demand increases with affluence. While this was generally true from the postwar period up to the late 1980s, the opposite has been the case since the early 1990s, with declining per capita demand as a result of: pricing reform, changes in land use and lot size, improved efficiency of water-using appliances, changes in industry and commercial use, and increased community awareness of the need for water-efficient practices. The need remains for a more consistent and accurate demand forecasting approach, which disaggregates demand into its different forms and considers the impacts of technological advances and changes to social practices.

Other chapters in this book have highlighted the reforms achieved under the Intergovernmental Agreement on a National Water Initiative (NWI) and the history of water resource regulation, including water sharing arrangements and the move to competitive water markets in Australia. The impact on water use has been smaller in urban areas than in nonurban areas. This is because historically, urban water demand has been given a privileged position in water allocation systems, and because most of the capital city water systems in Australia are located outside of catchments with major irrigated agriculture. An obvious exception is Adelaide, which derives a large proportion of its water from the Murray-Darling system. Overall, the NWI reforms and the NWI approach to addressing environmental considerations predominantly targets nonurban water resource management (Chong, 2014). In its 2009 biennial assessment of progress against the NWI, the National Water Commission (NWC) found that the NWI did not sufficiently guide urban water reform, and that new challenges have arisen since the NWI was signed (NWC, 2011, p. 57). These challenges were particularly evident during the Millennium drought (1997–2009), and the responses to that drought are discussed in the next section.

THE MILLENNIUM DROUGHT AS A DRIVER OF CHANGE

The Millennium drought, which impacted Australian cities and agricultural areas from 1997 to 2009, tested urban water systems and the established ways of decision making (Horne, 2016).

Australian urban water systems have traditionally been designed to deal with highly variable rainfall and lengthy dry periods. For decades, the major cities were largely able to rely on dams and reservoirs, which were large enough to ensure sufficient supplies throughout droughts. However, across many of Australia's major cities and towns, the Millennium drought was more severe and persistent than previous droughts. Urban water systems, as well as approaches to decision making, were pushed to previously untested limits. For example, Melbourne storages dropped below 26% (Melbourne Water, 2017), and Southeast Queensland experienced a 1:100-year drought, broken by a 1:100-year flood (Seqwater, 2015).

Several cities already had drought response strategies in place, some prepared well before the Millennium drought. For example, in Melbourne, utilities had prepared drought response plans consistent with ministerial guidelines (DNRE, 1998) and a Drought Response Coordination Committee was established to coordinate the drought responses of utilities and the Victorian government (Rhodes, 2009). These plans included staged water restrictions that were to be progressively implemented as dam storage levels declined to specific levels. The "trigger levels" were calculated from estimates of likely water savings, and criteria were established to determine the acceptable frequency and duration of restrictions. However, in the city of Brisbane, water managers historically had to contend with flooding rather than droughts, and so planning for responses to the Millennium drought was done during the drought itself.

Across all locations, water restrictions were a core element of drought response strategies (Chong et al., 2009). These restrictions primarily targeted households' outdoor water use, for example by limiting garden watering to certain times of day and week, restricting the use of sprinklers, and banning or limiting the filling of swimming pools. These water restriction programs were built on decades of utility- and government-run water-wise public education campaigns, and during the drought they were also integrated with broader water savings campaigns (Lindsay et al., 2016). Restrictions rules commonly allowed efficient garden watering (e.g., by restricting watering to drippers or hoses with trigger nozzles, and by restricting watering to evenings and certain days of the week). Only in a very few locations did restrictions extend to total outdoor water bans.

There was strong community support for most levels of restrictions, with people understanding the need for all users to save water in an egalitarian manner (Chong et al., 2009; Lindsay et al., 2016).

Restrictions were highly valuable as a temporary means of saving water and slowing dam depletion, thus buying much-needed time before there was a need to commit to investing in the construction of large-scale infrastructure. However, as the drought progressed and restrictions remained in place, the situation became politicized, and there were calls for the "drought proofing" of cities—sometimes in a manner which would have completely removed the need for restrictions (Chong et al., 2009).

DEALING WITH HYDROLOGICAL UNCERTAINTY

The Millennium drought also helped drive the development of innovative ways to deal with uncertainty in major infrastructure planning. As the drought presented a real-time shift in the hydrological context of urban water systems, the industry was confronted by the need for new scenarios and ways of planning and decision making, beyond the 100-year rainfall records on which planning had conventionally been based (Horne, 2016). While in many locations long-term supply-demand plans had been established, with a range of options intended for staged implementation over 20- to 30-year time frames as population and demand grew, the drought drastically shortened the time frames for bringing forward options to supply or save water.

The Millennium drought brought with it an unprecedented level of uncertainty for water planners and managers (Head, 2014). The application of "real options" planning marked a significant conceptual and practical leap in how to make water management decisions against the backdrop of significant uncertainty. Real options approaches are based on the principles of readiness and adaptability, and involve staged, modular adaptive methods of planning and implementing high-cost capital works to allow maximum flexibility. A key example is "readiness to construct" strategies for desalination or independent potable reuse, should dam levels drop below predetermined trigger levels. Readiness strategies also involve establishing contractual and financial arrangements that allow for a staged approach to tendering, designing and implementing. It is only necessary to progress from one stage to the next if drought persists. For example, while requests for tenders for design and construction might be issued as dam levels fall, if it subsequently rains the preagreed cost of compensating tenderers for not progressing with the next stages will be far less than the cost of going ahead with construction (Giurco et al., 2014; White et al., 2006).

The Sydney Metropolitan Water Plan (White et al., 2006) saw the first application of real options planning for the desalination plant in Kurnell, which occurred over a 12-month period prior to the commitment to construct (Giurco et al., 2014). However, much of the recent infrastructure implementation in the Australian capital cities reflected a crisis-driven pursuit of "drought proofing" rather than a real options planning approach.

In several cities, crisis-driven decision making led to fast-tracked decisions to invest in large supply options that were at odds with plans, and with the analysis and evidence upon which those plans were based (White et al., 2008a). Infrastructure investments of billions of dollars were made without a full evaluation of the costs, benefits and risks involved (NWC, 2011; PC, 2011). The promise of a real options approach to desalination in Sydney did not eventuate, as ministerial direction was given to commence construction of a desalination plant before the trigger levels were reached. In South East Queensland, top-down, centralized investment decisions were accompanied by institutional changes that did not engender cooperation between state and local government entities in their responses to the drought (Head, 2014). Intercatchment transfers (such as the Victorian North-South pipeline) and dams (such as the Traveston Dam in Queensland) were proposed as ways of augmenting urban water supplies, but public confidence and support was eroded by poor community engagement in the affected regions, and by questionable assessments of the cost-effectiveness of these rainfall-dependent options (Chong & White, 2007).

In early 2010 the Millennium drought broke, with flooding rains in several cities. In the aftermath, large-scale infrastructure, ranging from desalination plants to large-scale recycling systems, were mothballed, some without ever being operated. The costs of these systems came under sharp scrutiny (PC, 2011; Vic Auditor-General, 2008). Postdrought analysis showed that short-term drought scarcity pricing, which had not seriously been contemplated, would likely have helped to reduce demand and may have avoided the need for new infrastructure (Grafton & Ward, 2008).

In the aftermath of the drought, utilities' demand management programs, which had been so successful in saving water prior to and during drought, were significantly scaled down, with associated declines in industry expertise. There are many reasons why several recycled water schemes funded during the drought were not financially viable in the long term (ISF, 2013). However, because restrictions are now largely off the table as a future drought response measure, access to "restriction-free" recycled water is no longer considered important, and this has resulted in sharply reduced private investment in recycled water schemes (Watson et al., 2013).

THE ROLE OF WATER EFFICIENCY IN SECURING URBAN WATER SYSTEMS

Building on decades of experience in demand management programs, utilities and governments across all drought-affected Australian states rolled out water efficiency initiatives at speed and scale during the drought. These programs were crucial to saving water and represented cost-effective drought response mechanisms (Turner et al., 2016). Programs covered end uses inside and outside the home and in a range of industry applications, and included targets, audits, retrofits, rebates, and water-saving plans for business and industry. Residential programs targeted rebates or retrofits of showerheads, toilets, washing machines, pool covers, and leakage reduction. These programs complemented programs implemented prior to the worst of the drought, including the Commonwealth government's Water Efficiency Labelling and Standards scheme for water-using fixtures and appliances (Aust Govt, 2017), and the NSW Government Building Sustainability Index (BASIX) performance-based rating scheme for new residential dwellings (NSW Govt, 2017).

One of the key benefits provided by these water efficiency programs was that, in combination with water restrictions, they rapidly reduced the rate of depletion of water storages. This meant that there was more time available for the introduction of new supply options, which had longer implementation times. In some cases, it meant that there was time for rainfall to replenish supplies.

While in most states there was strong institutional support for investment in water efficiency during the Millennium drought, there has been an inconsistent regulatory approach to water efficiency investment since then. For example, many regulators and utilities do not recognize that the issue of "foregone revenue" can be solved by allowing price pass-through of the costs of investment in water efficiency, including through a revenue cap rather than a price cap (White et al., 2008a). Similarly, although structural water-saving measures that permanently reduce water demand year on year through improved appliance or fixture efficiency can enable reduced capital requirements, they are usually classified as operating expenditures rather than as capital offsetting expenditures. These and other regulatory and market failures mean that there remains a significant underinvestment in water efficiency measures, which generally have lower unit costs than supply-side measures.

CURRENT AND FUTURE AUSTRALIAN URBAN WATER SECTOR

A number of emerging trends have the potential to make significant changes to the Australian urban water sector. Some of these have been triggered or accelerated by the Millennium drought, or by technological changes, while others were apparent for a number of years before the drought. These emerging trends are discussed in the following subsections.

A DIVERSITY OF WATER SERVICING OPTIONS

During the Millennium drought the range of water servicing options expanded. The need to bring forward supply options created the opportunity for innovation in implementing diverse water service options, some of them on a large scale. For example, the Western Australian Water Corporation has invested in managed aquifer recharge in the Perth region, and has developed a great deal of experience in community engagement in this form of indirect potable reuse (Moscovis, 2013).

The Millennium drought also saw the rapid expansion of a number of recycling schemes in urban centers, often with support from state and federal government grants (Horne, 2016). These supplied residential, industrial and irrigation water. Recycling schemes were often faced with operational challenges at commencement, and costs were many times higher than expected (ISF, 2013). However, many of these pioneering schemes were valuable as they enabled the industry to learn how to identify and deal with operational, institutional and regulatory risks. Regulatory innovations were also introduced. For example, the NSW Government's *Water Industry Competition Act* established a framework for licensing and regulating private suppliers of recycled water in that state (Chong, 2014).

White (2010) undertook an historical examination of the development of urban water systems in cities and towns in cities of the developed world (including Australia) since the 1800s and identified four epochs (or generations) in terms of objectives, planning and structure of these urban systems.

The *first-generation* water systems involved the unplanned, unmanaged and decentralized sourcing of water, and the unplanned, unmanaged and decentralized disposal of wastewater and stormwater.

However, as cities grew in size, these uncoordinated systems gave way to centralized systems. These *second-generation* systems provided water under pressure to enable firefighting, wastewater carried away in sewers to reduce cross-contamination, and stormwater systems that improved flood mitigation. These systems set the standard for over 100 years, and they still provide the model for many cities in emerging economies.

The *third-generation*, or transition, systems arose from the 1970s onwards in response to the growing impact for new water sources and more efficient effluent discharge. Community environmental consciousness, clean water regulations and awareness of the constraints on excessive water extraction resulted in significant expenditure on: "end of pipe" treatment systems, large-scale intercatchment water transfers, and (in Australia during the drought) the commitment of large sums to develop desalination capacity.

The emerging *fourth-generation* urban water systems reflect a new paradigm in urban water service provision. This has developed in response to market and technology changes, community demand and an emerging systems view of urban water. It is characterized by the following (White, 2010):

- Best practice efficiency in water use, system losses and pressure management.
- Source control of pollutants and nutrients.

- Maximized recovery of resources through processes which include wastewater and stormwater recycling, energy recovery, and nutrient and carbon recovery.
- A distributed supply of water and the treatment and reuse of wastewater.
- The integration of the three elements of the urban water cycle: water, wastewater, and stormwater.

These characteristics are reflected in a changing discourse in the urban water industry sectors, as evidenced by, for example, the International Water Association (IWA), which reflects the focus of water utilities, water industry firms and organizations, practitioners and researchers (IWA, 2016).

RESILIENCE AND LIVEABILITY

Significant practical learning outcomes have also been applied since the Millennium drought. In Melbourne, utilities have increased the sophistication of "drought scenarios" and they have used these scenarios to establish and communicate revised strategies for staging responses during droughts (Turner et al., 2016).

The experience of the drought also helped to firmly shift the goals for water systems towards resilience and liveability. This has included consideration of water in the urban landscape, health goals, and resilience and adaptability to climate change (Horne, 2016). Urban water planners are exploring innovations in more participatory, engaging ways to canvass community voices and perspectives on setting visions for how urban water systems are planned and managed (see e.g., Chong et al., 2017).

However, there are many challenges to reaching the goals of sustainability, liveability, and resilience. For example, the policy, institutional and regulatory settings can create a mutually reinforcing web of barriers that can limit innovation in distributed and resilient water planning and servicing (Chong, 2014; Watson et al., 2013). There is a significant need for further urban water reform, and if that reform is achieved, there will be opportunities for a substantial transformation of Australian urban water systems.

There is no question that the Millennium drought, and the national water reforms of the early 1990s, triggered major changes to the Australian water industry. Since then, a number of trends have emerged in the water sector and in other sectors that have the potential to impact on the urban water policy landscape.

In Australia's urban water systems, decision-making approaches have transformed markedly over the last few decades. There is now strong industry recognition of the need for "water services" to be more broadly defined, and for these services to play a pivotal role in meeting liveability and resilience goals for urban centers (Turner & White, 2017). As Australian cities grow, and as their population densities increase, there is a need for urban water systems to be managed to ensure: high-quality waterways; access to green spaces for recreation, amenity and public health; and increased density of tree canopies for urban cooling, especially during droughts. At the same time, solutions for supplying and saving water ideally encompass cost-effective, innovative and flexible technologies and systems.

Building resilience into the water system requires a multifaceted, multiscalar and multistakeholder approach. Urban water systems have the potential to play a pivotal role in increasing resilience, but strengthened coordination and partnerships are needed within the water industry, as well as between the water sector and other sectors, including other utility sectors (energy, waste, transport), planning agencies and local government.

There are many opportunities for building on the developments of the last 20 plus years to further transition urban water systems to more sustainable pathways. These include (IWA, 2016):

- Improved water pricing structures for local recycled water systems that reflect customer preferences and incorporate avoided costs.
- The coordination of water system planning and land use planning to move beyond the "business as usual" approach of extending centralized systems, towards considering and implementing innovative and sustainable water servicing solutions for future greenfield and infill developments.
- Integrated consideration of water in the landscape, energy use, recovery of organic material and nutrients, and urban food production to support a range of community goals for reducing greenhouse emissions, improving food security, increasing urban livability and moving towards a "circular economy," which maximizes resource efficiency and recovery (EMF, 2017).

COMMUNITY ENGAGEMENT

Community engagement has played a significant role in urban water policy, particularly through the strong reliance on water restrictions as a means of ensuring the sustainable yield of urban water supply systems. This has required strong community involvement in reducing water use in periods of drought. It has also meant that there is strong public support for restrictions, even during periods of severe drought when there have been quite onerous requirements over a long period of time on householders, especially in terms of outdoor water use. Of course, these restrictions will impact differently in different areas depending on climate, soil type and outdoor landscaping preferences.

However, there is another meaning of the term "community engagement" which has not historically been exercised in the Australian context, and which may represent an opportunity in the future. It relates to involving the community in determining preferences and in ascertaining values, in relation to water futures. For example, during the Millennium drought several jurisdictions changed the level of service for their water supplies without engaging with the community about the impacts and consequences of these changes (White et al., 2008b). These changes appear to have been driven by a prevailing view that restrictions were not an acceptable method to employ for water supply planning, with comparisons made with the electricity sector where loss of service or restrictions were not part of normal operating procedure. In this case, the increases in the long-run level of service, which effectively amounted to a reduction in the planned severity and duration of restrictions, had significant implications for the long-term yield of water supply systems. This influenced planning to the extent that additional supply options worth billions of dollars had to be incorporated into the planning process in order to attain the revised levels of service.

Innovative and robust methods of community engagement can and should be employed to determine community preferences regarding large-scale expenditure, environmental flows, reliability levels and levels of restriction. These methods should incorporate the three ideals of community engagement:

- *Representativeness*—a process involving the random selection of participants, and inviting those citizens who would not normally engage in these processes to have an opportunity to express their preferences in an informed way.
- *Deliberation*—a process in which engaged citizens have an opportunity to obtain information and increase their level of knowledge on issues by questioning stakeholders and experts, and by discussing and deliberating on the issues.

- *Influence*—involves the principle that the results of the engagement process be taken seriously by decision makers and are meaningful.

Many processes meet these criteria, including citizens' assemblies, citizens' juries, deliberative polls and similar processes (White, 2008).

It is likely that the future will see an increased demand for citizen engagement on these and other important matters. Certainly, in Australia major water supply decisions have strongly influenced state government election campaigns both before and during the most recent drought (Ferguson, 2014; Morgan, 2015). This, combined with the general alienation of the public from the political process, means that it would be highly beneficial to increase the level of engagement of citizens in the decision-making process in regard to water policy and other issues.

DIGITAL WATER FUTURES

In the last decade there has been a dramatic and rapid increase in the availability of new technologies for a range of applications in urban water systems. One example is the widespread use and availability of automatic reading systems for water meters, and the availability of digital and smart meters (Liu et al., 2013). The use of smart meters is much greater in the electricity sector, and has at least in the Australian context not been without controversy. However, in the urban water context, it has taken longer to secure the widespread adoption of smart meters. Several large-scale trials have taken place, including in some entire towns such as Hervey Bay in Queensland (Turner et al., 2010). Many of these trials and pilots are pioneering efforts in the sense that they have used technology, which has subsequently been improved or superseded.

The reasons for installing smart meters or automatic meter reading systems have varied, ranging from a desire to reduce the cost of meter reading to improving accuracy and increasing the level of service to customers by providing data on a range of areas including overnight leakage. There are indications that the ability to provide good feedback to customers, while not limited to digital or smart meters, does have some advantages in terms of water conservation and improved customer service. There have also been trials in the provision of detailed feedback from smart meters to commercial customers, in which the frequent provision of information on water use can support improved water efficiency.

One of the main reasons why water utilities have been hesitant about introducing smart metering has been a lack of certainty regarding their costs and benefits. In a review undertaken of the Hervey Bay trial, Turner et al. (2010) considered the full range of benefits, including the opportunity that is provided to retrofit with meters with improved accuracy, the full range of avoided costs associated with demand reduction, including reduced losses and peak demand reduction. This research found that when all these factors were considered, the benefits associated with smart meters could be increased and they can exceed the costs (Turner et al., 2010).

One of the key issues regarding the cost and functionality of automatic meter reading systems has been communications from the meter to repeater stations or base stations. The development of low-power WAN systems, which enable long-range, low-power, low-bandwidth communications to gateways that may be up to 10 or 15 km apart, may offer a significant opportunity for smart meter connectivity, and may enable other services to be provided consistent with the development of the "Internet of Things" (LoRa Alliance, 2017).

It is to be expected that future water systems will be characterized by the coupling of meters and other sensors, including soil moisture sensors and irrigation controllers, and by the use of monitoring and control systems to improve urban water systems generally. It is also likely that these systems will be used to control decentralized water and wastewater systems.

CONCLUSIONS

The urban water sector has changed considerably over the last 20–30 years, and even greater changes are likely in the next 20–30 years. These changes will be driven by: population increases (particularly in the major capital cities); an aspiration to increase resilience and adapt to a changing climate; the need to contribute to greenhouse gas emissions reduction; the goal of creating a "circular economy" through increased resource recovery; community demands for more livable cities; affordability; and a greater level of engagement by citizens and water consumers. There are opportunities to be grasped, but they will require further policy reform to facilitate the changes needed. These policy reforms include assessing proposed water supply options and water efficiency investments on an equal basis, and enabling utilities to generate an appropriate return on investments in water efficiency. It also includes reforms that enable new entrants into the water, sanitation and stormwater markets, integrated regulation of these different parts of the water cycle, and the removal of barriers to energy and resource recovery.

REFERENCES

Aust Govt. (2017). *Water efficiency labelling and standards scheme*. Canberra: Australian Government. http://www.waterrating.gov.au/.

Chong, J. (2014). Climate-readiness, competition and sustainability: An analysis of the legal and regulatory frameworks for providing water services in Sydney. *Water Policy, 16*(1), 1.

Chong, J., & White, S. (2007). Re-inventing sustainability: A climate for change. In: *Proceedings of Australia New Zealand Society for Ecological Economics conference, Noosa, Queensland*.

Chong, J., White, S., Herriman, J., & Campbell, D. (2009). *Review of water restrictions*. Report by the Institute for Sustainable Futures. Canberra: University of Technology Sydney for the National Water Commission.

Chong, J., Winterford, K., & Lederwasch, A. (2017). Community engagement on water futures: Using creative processes, appreciative inquiry and art to bring communities' views to life. In: *Ozwater conference, Sydney*.

DNRE. (1998). *Ministerial guidelines for developing and implementing a drought response plan—August 1998 (made under Division 1A of the Water Industry Act 1994)*. Melbourne: Department of Natural Resources and Environment.

EMF. (2017). *Circular economy*. Cowes: Ellen Macarthur Foundation. https://www.ellenmacarthurfoundation.org/circular-economy.

Ferguson, J. (2014). Billions in desalination costs for not a drop of water. *The Australian Newspaper,*(16 October) www.theaustralian.com.au/national-affairs/billions-in-desalination-costs-for-not-a-drop-of-water/news-story/1dc59f33b788c609e67c001786048fca.

Giurco, D. P., Turner, A., Fane, S., & White, S. B. (2014). Desalination for urban water: Changing perceptions and future scenarios in Australia. *Chemical Engineering Transactions, 42*, 13–18.

Grafton, R. Q., & Ward, M. B. (2008). Prices versus rationing: Marshallian surplus and mandatory water restrictions. *Economic Record, 84*, S57–S65.

Head, B. W. (2014). Managing urban water crises: Adaptive policy responses to drought and flood in Southeast Queensland. *Ecology and Society, 19*(2), 33. http://dx.doi.org/10.5751/ES-06414-190233.

Hilmer, F. G., Rayner, M. R., & Taperell, G. Q. (1993). *National Competition Policy*. Report of the National Competition Policy Review. Canberra: Commonwealth of Australia.

Horne, J. (2016). Resilience in major Australian cities: Assessing capacity and preparedness to respond to extreme weather events. *International Journal of Water Resources Development, 32*, 1–20. http://dx.doi.org/10.1080/07900627.2016.1244049.

ISF. (2013). *Making better recycled water investment decisions: Shifts happen*. Report by the Institute for Sustainable Futures. Sydney: University of Technology Sydney for the Australian Water Recycling Centre of Excellence. http://waterrecyclinginvestment.com.

IWA. (2016). *The IWA principles for water wise cities*. The Hague: International Water Association. http://www.iwa-network.org/wp-content/uploads/2016/08/IWA_Principles_Water_Wise_Cities.pdf.

Lindsay, J., Dean, A. J., & Supski, S. (2016). Responding to the Millennium drought: Comparing domestic water cultures in three Australian cities. *Regional Environmental Change, 17*(2), 565–577. http://dx.doi.org/10.1007/s10113-016-1048-6.

Liu, A., Giurco, D., Mukheibir, P., & Watkin, G. (2013). Smart metering and billing: Information to guide household water consumption. *Water (Australia), 40*(5), 73–77.

LoRa Alliance. (2017). *Wide area networks for internet of things*. LoRa Alliance. https://www.lora-alliance.org/.

Melbourne Water. (2017). *Storages over the years*. Melbourne: Melbourne Water. www.melbournewater.com.au/waterdata/waterstorages/Pages/Storages-over-the-years.aspx.

Morgan, R. (2015). Drought-proofing Perth: The long view of Western Australian Water. *The Conversation*, (16 February). https://theconversation.com/drought-proofing-perth-the-long-view-of-western-australian-water-36349.

Moscovis, V. (2013). *Groundwater replenishment trial: Final report*. Perth: Water Corporation of Western Australia. www.watercorporation.com.au/~/media/files/residential/water%20supply%20and%20services/gwrt/gwrt-final-report.pdf.

NSW Govt. (2017). *Planning tools: BASIX*. Sydney: NSW Government. https://www.planningportal.nsw.gov.au/planning-tools/basix.

NWC. (2011). *Urban water in Australia: Future directions*. Canberra: National Water Commission.

PC. (2011). *Australia's Urban Water Sector*. Final inquiry report no 55. Canberra: Productivity Commission.

Rhodes, B. (2009). Responding to a changing climate and drought in Melbourne, Australia. In: *Proceedings of efficient 2009, Conference of the International Water Association Efficient Urban Water Management Group, Sydney, October*.

Seqwater. (2015). *Water for life: South East Queensland's water security program 2015–2045*. Ipswich: Southeast Queensland Water. www.seqwater.com.au/waterforlife.

Turner, A., Retamal, M., White, S., Palfreeman, L., & Panikkar, A. (2010). *Third party evaluation of wide bay water smart metering and sustainable water pricing initiative project*. Report prepared by SMEC and the Institute for Sustainable Futures, University of Technology Sydney for the Department of the Environment, Water, Heritage and the Arts, Canberra. www.environment.gov.au/system/files/pages/514be549-1db1-4cb5-b620-1f51b6a36a3c/files/evaluation-feb2010.pdf.

Turner, A., & White, S. (2017). *Urban water futures: Trends and potential disruptions*. Report prepared for the Water Services Association of Australia by the Institute for Sustainable Futures, University of Technology Sydney, Sydney.

Turner, A., White, S., Chong, J., Dickinson, M.A., Cooley, H., & Donnelly, K. (2016). *Managing drought: Learning from Australia*. Report by the Alliance for Water Efficiency, Institute for Sustainable Futures, University of Technology, Sydney.

Vic Auditor-General. (2008). *Planning for water infrastructure in Victoria*. Melbourne: Victorian Auditor-General.

Watson, R., Mitchell, C., & Fane, S. (2013). *Distributed recycled water decisions-ensuring continued private investment*. Proceedings of OzWater13, Perth, Australia. http://cfsites1.uts.edu.au/find/isf/publications/watsonetal2013distributedrecycledwaterdecisionsabstract.pdf.

White, S. (2008). *Pathways to deliberative decision-making: Urban infrastructure and democracy*. In: *Proceedings of Conference on Environmental Governance and Democracy, Yale University, New Haven, CT*, pp. 1–8. www.academia.edu/5251277/Pathways_to_Deliberative_Decision-Making_Urban_Infrastructure_and_Democracy.

White, S. (2010). *Towards fourth generation water infrastructure*. Keynote address to the Urban Water Security Research Alliance, Brisbane, 28 September. www.urbanwateralliance.org.au/publications/forum-2nd-2010/Keynote-Stuart-White.pdf.

White, S., Campbell, D., Giurco, D., Snelling, C., Kazaglis, A., & Fane, S. (2006). *Review of the metropolitan water plan: Final report*. Report by Institute for Sustainable Futures at the University of Technology Sydney, ACIL Tasman and SMEC Australia, Sydney.

White, S., Fane, S. A., Giurco, D., & Turner, A. J. (2008a). Putting the economics in its place: Decision-making in an uncertain environment. In C. Zografos & R. Howarth (Eds.), *Deliberative ecological economics* (pp. 80–106). New Delhi: Oxford University Press.

White, S., Noble, K., & Chong, J. (2008b). Reform, risk and reality: Challenges and opportunities for Australian urban water management. *The Australian Economic Review, 41*, 428–434.

ENVIRONMENTAL WATER REFORM

6

S.E. Bunn

Griffith University, Brisbane, QLD, Australia

CHAPTER OUTLINE

INTRODUCTION

Consideration of the environment as a valued user of water is a relatively new chapter in the European history of water-resource management in Australia. Early settlers primarily looked to the river systems for water supply and as a means of navigation (Walker, 1992), and exploited them for their native fisheries (Cadwallader, 1978; Cadwallader & Lawrence, 1990). Development of water infrastructure for irrigated agriculture commenced in the late 1880s, primarily in the Murray-Darling Basin, and large-scale production was enabled with the construction of major storages in the 1950s and 1960s (Eastburn, 1990; Walker, 1992). A prevailing view throughout this period was that rivers should be "harnessed" to meet the nation's growing demands for food and water supply (Powell, 1989, 1991). Some reservoirs, such as the Canning Dam supplying water to the city of Perth, were designed to impound all in-flows with no compensation releases (Storey et al., 1991). The Snowy Mountains Hydro-Electric Scheme, Australia's largest engineering project, commenced during this time and involved the construction of major dams and interbasin transfers of water (Dann, 1969; Domicelj, 1981). Dam building for urban and agricultural water supply continued to grow through the 1970s and 1980s, including major structures on the Ord and Burdekin rivers in Australia's tropical north. As in many parts of the world at this time, the environmental consequences of these major water developments were scarcely considered.

Decision Making in Water Resources Policy and Management. http://dx.doi.org/10.1016/B978-0-12-810523-8.00007-0

However, as community attitudes toward the environment began to shift during this period, there was accumulating evidence of the ecological consequences of dams and flow regulation (e.g., Dexter, 1978; Doeg, 1984; Harris, 1984; Sheldon & Walker, 1993; Walker, 1979, 1985). There was also growing recognition from the scientific community and management agencies of the need to address these problems. Some of the first assessments of environmental flow needs for freshwater fishes in Australian rivers commenced in 1976 (Tunbridge, 1997, cited in Pusey, 1998), and considerable progress was made in developing environmental flow methodologies in subsequent decades (Arthington et al., 1992; Tharme, 2003).

CATALYST FOR CHANGE

Despite these early efforts, concerns about declining water quality and ecological health continued to grow, with the first national State of the Environment Report (DEST, 1996) highlighting the impact of land clearance, river regulation and introduced species on river and floodplain environments (see also Kingsford, 2000; Kingsford & Thomas, 1995). An audit of water use in the Murray-Darling Basin revealed that, under 1994 development conditions, drought-like conditions (i.e., a 1-in-20-year event) would be experienced in 61% of years (MDBMC, 1995). These changes to flow volume and flow pattern had resulted in an increase in the salinity of the River Murray, a reduction in the frequency of flooding of the floodplain wetlands, river conditions that were more suitable for the growth of blue-green algae, and a significant decline in native fish populations (MDBMC, 1995). A snapshot survey of river conditions in the Murray-Darling Basin (Norris et al., 2001) confirmed that the "current general state of ecological health in the Murray-Darling Basin is less than what is required for ecological sustainability."

Rising salinity in the River Murray had already become a serious issue for governments, not only threatening the environment but also water security and infrastructure (PMSEIC, 1998). Sufficient flushing flow was needed to move the salt leaching from the Basin's old soils and rocks (two million tonnes in an average year), and mobilized by the removal of tree cover and irrigation (MDBA, 2014). However, the 1000 km-long toxic algal bloom in the Darling River in 1991 provided a major catalyst for change, triggering the New South Wales Government to declare a state of emergency (Bowling & Baker, 1996; MDBMC, 1994). A subsequent report to the Prime Minister's Science and Engineering Council argued that "improved water resource management is imperative—a continuation of past policies on in-stream and off-stream values associated with the water resources of Australia and past practices, will severely and adversely affect every aspect of contemporary life" (PMSEIC, 1996).

In an initial attempt to halt the environmental decline in the Basin, a Cap on diversions was introduced by the Murray-Darling Basin Ministerial Council in 1995, limiting extractions to 1993–94 levels of development. This was a political "line in the sand," drawn without any specific reference to ecological needs (Hamstead, 2009). However, Queensland never signed up for the agreement and New South Wales regularly flouted its commitment (Cullen, 2011).

Some early efforts were made to provide environmental water allocations to key ecological assets within the Murray-Darling Basin. Burrendong Dam, constructed on the upper Macquarie River in 1967, included a formal environmental water allocation (18,500 ML/y) for the Macquarie Marshes with its first "experimental" release in 1980 (NWC, 2012a). The subsequent water management plan for the

Marshes included the delivery of an increased 50,000 ML/y environmental water allocation (DWR-NPWS, 1986). An environmental water allocation of 100 GL/y for the Barmah-Millewa Forest was also approved by the Murray-Darling Basin Commission in 1993 (MDBA, 2011), with the first allocation occurring in 1998.

FORMAL RECOGNITION OF THE ENVIRONMENT AS A USER OF WATER

In 1994, the Council of Australian Governments (COAG) agreed to implement a strategic water-reform framework that included provisions for environmental water requirements to maintain the health of river systems and groundwater basins (NWC, 2014b). To guide this, the Agriculture and Resource Management Council of Australia and New Zealand (ARMCANZ) and the Australian and New Zealand Environment and Conservation Council (ANZECC) developed a set of National Principles to provide policy direction on dealing with environmental water (Table 1; ARMCANZ & ANZECC, 1996). The stated goal for providing water for the environment was "To sustain and where necessary restore ecological processes and biodiversity of water dependent ecosystems." Environmental allocations were defined as "… descriptions of the water regimes needed to sustain the ecological values of water-dependent ecosystems at a low level of risk."

Following the COAG agreement, most jurisdictions undertook a program of water reforms that included provisions for allocating water to the environment (NWC, 2014b). Water plans were developed and adopted in New South Wales, Queensland, and South Australia that gave statutory effect to environmental water provisions (NWC, 2012a). Queensland's Water Act 2000 (section 46), outlined that Water Resource plans must state: the outcomes, including … ecological outcomes; the water and natural ecosystem monitoring requirements to assist in assessing the effectiveness of the proposed management strategies; criteria for addressing degradation that has occurred in natural ecosystems; environmental flow objectives; and performance indicators for environmental flow objectives (Queensland Government, 2000). Western Australia developed a statewide policy for Environmental Water Provisions, to provide protection of water-dependent ecosystems (WRC, 2000). The Victorian Government implemented its Environmental Water Reserve, which included objectives to: maintain the environmental values of the water system and the other water services that depend on environmental condition; sustain biodiversity, ecological functioning and water quality; and have legal status and be held by the Crown (DSE, 2004). An intergovernmental agreement for the Lake Eyre Basin was also signed by the Commonwealth and the states of Queensland and South Australia in 2000, and the Northern Territory in 2004 (Anon, 2000), with guiding principles that included that the water requirements for ecological processes, biodiversity and ecologically significant areas within the Lake Eyre Basin Agreement Area should be maintained, especially by means of flow variability and seasonality.

Environmental water provisions were most commonly met through setting annual allocation limits and access rules in water plans to leave behind sufficient water to meet environmental objectives (NWC, 2014b). These planned environmental water provisions included cease-to-pump rules, flow sharing arrangements, passing-flow releases from water storages and environmental water allowances. In some jurisdictions, "held" environmental water provisions were made through the purchase or recovery of water licenses (see Chapter 13).

Table 1 National Principles for the Provision of Water for Ecosystems (ARMCANZ & ANZECC, 1996)	
Basic premise of principles	
Principle 1	River regulation and/or consumptive use should be recognized as potentially impacting on ecological values
Determining environmental water provisions	
Principle 2	Provision of water for ecosystems should be on the basis of the best scientific information available on the water regimes necessary to sustain the ecological values of water dependent ecosystems
Provision of water for ecosystems	
Principle 3	Environmental water provisions should be legally recognized
Principle 4	In systems where there are existing users, provision of water for ecosystems should go as far as possible to meet the water regime necessary to sustain the ecological values of aquatic ecosystems while recognizing the existing rights of other water users
Principle 5	Where environmental water requirements cannot be met due to existing uses, action (including reallocation) should be taken to meet environmental needs
Principle 6	Further allocation of water for any use should only be on the basis that natural ecological processes and biodiversity are sustained (i.e., ecological values are sustained)
Management of environmental water allocations	
Principle 7	Accountabilities in all aspects of management of environmental water provisions should be transparent and clearly defined
Principle 8	Environmental water provisions should be responsive to monitoring and improvements in understanding of environmental water requirements
Other uses	
Principle 9	All water uses should be managed in a manner which recognizes ecological values
Principle 10	Appropriate demand management and water-pricing strategies should be used to assist in sustaining ecological values of water resources
Further research	
Principle 11	Strategic and applied research to improve understanding of environmental water requirements is essential
Community involvement	
Principle 12	All relevant environmental, social and economic stakeholders will be involved in water-allocation planning and decision making on environmental water provisions

ADDRESSING OVERALLOCATION

In 2002, the Murray-Darling Basin Ministerial Council proposed the Living Murray river restoration program, reinvigorating discussion about restoring the health of the River Murray system (MDBA, 2011). An independent scientific advisory panel concluded that, for a "working river" like the Murray-Darling, there was a substantial risk it would not be in a healthy state when key attributes of the flow regime were reduced below two-thirds of their natural level, and recommended an average 1500 GL/y be returned to the River Murray, over and above volumes provided for by the Cap (Jones et al., 2003).

The establishment of the National Water Initiative (NWI) in 2004 was a significant milestone in the reforms to address overallocation in the Murray-Darling Basin, and the Living Murray Initiative commenced. Partner governments committed $500 million to recover 500 GL/y of water and $150 million for water management structures to facilitate its delivery to six icon sites: Barmah-Millewa Forest, Gunbower-Koondrook-Perricoota Forest, Hattah Lakes, Chowilla Floodplain and Lindsay-Wallpolla Islands, the Lower Lakes, Coorong and Murray Mouth (in South Australia), and the River Murray Channel (running from near Albury to the sea) (MDBA, 2011). The NSW Government established a RiverBank environmental fund in 2005 to purchase water for the state's most stressed and valued inland rivers and, with additional funding from the federal government, supported water purchases for five of the most important wetlands in the NSW Murray-Darling Basin (NWC, 2014b). The NSW and Australian governments also jointly funded a Wetland Recovery Program in 2006, to support further research and water purchases for the Gwydir wetlands and Macquarie Marshes.

The Commissioner, Snowy Water Inquiry (1998) was initiated to investigate "the environmental issues arising from the pattern of water flows in rivers and streams … caused by the operation of the Snowy Mountains Hydro-electric Scheme and to report on options for dealing with the issues and the environmental, economic, agricultural and other impacts of those options." The New South Wales, Victorian and Commonwealth Governments agreed in 2002 to implement the outcomes arising from the inquiry through a legally binding intergovernmental agreement, with a commitment to return 212 GL/y or 21% of the mean natural flow by 2012 and ultimately 28%; close to the level recommended by the independent expert panel in 1996 (NSW Office of Water, 2010; Rose & Erskine, 2011).

Addressing overallocation and overuse was a central goal of the 2004 NWI, which sought to "implement firm pathways and open processes for returning previously over-allocated and/or overdrawn surface and groundwater systems to *environmentally sustainable levels of extraction*" (Clause 25). The states and territories also agreed that their water-access entitlements and planning frameworks would "provide a statutory basis for environmental and other public benefit outcomes in surface and groundwater systems to protect water sources and their dependent ecosystems," and be "characterised by planning processes in which there is adequate opportunity for productive, environmental and other public benefit considerations to be identified and considered in an open and transparent way" (Clause 25).

Importantly, the NWI also sought to ensure that water provided by the states and territories to meet agreed environmental and other public benefit outcomes was to "be given statutory recognition and have at least the same level of security as water access entitlements for consumptive use and be fully accounted for" (Clause 35). A desired outcome of the NWI was to identify the environmental and other public benefit outcomes with as much specificity as possible within water resource plans, and establish and equip accountable environmental water managers with the necessary authority and resources to provide sufficient water at the right times and places to achieve these outcomes (Clause 78). Another key outcome of the NWI was to ensure that adequate measurement, monitoring and reporting systems were in place to support public and investor confidence in the amount of water being traded, extracted for consumptive use and recovered and managed for environmental and other public benefit outcomes (Clause 80).

DEFINITIONAL IMPEDIMENTS TO ENVIRONMENTAL WATER REFORM

Meeting the environmental commitments of the NWI would prove to be an extremely difficult challenge, not just because of the unfolding Millenium drought (van Dijk et al., 2013), but also because of the lack of clarity about key terms and definitions. In its first biennial assessment, the National Water

Commission regarded the lack of specificity of environmental outcomes being sought as a significant risk to achieving NWI outcomes (NWC, 2007). This was attributed to shortcomings in water planning and, in particular, the adequacy of the science underpinning the water plans. The concern was that these inadequacies would lead to broad-brush management regimes, difficulty in monitoring outcomes and consequent low levels of accountability (NWC, 2007).

Jurisdictions found it difficult to agree if a water system was overallocated or overused because that required a comparison of the actual level of extraction to what would be allowed under an "environmentally sustainable level of extraction" (Hamstead, 2009). The latter was defined as "the level of water extraction from a particular system which, if exceeded would compromise key environmental assets, or ecosystem functions and the productive base of the resource." However, it was often not clear what was actually meant by "environmental assets, or ecosystem functions and the productive base of the resource," or how to tell which were "key" and which were not, and what it means to "compromise" them (Hamstead, 2009). There was also disagreement as to whether it was a legitimate outcome under the NWI to trade off environmental values to support consumptive requirements within water plans.

In an effort to address the "disappointingly slow progress" in returning overallocated or overused systems to environmentally sustainable levels of extraction, the National Water Commission decided to undertake an assessment in 2009 of the extent to which surface and groundwater systems have been hydrologically altered because of water extraction, regulation or flow alteration. Some jurisdictions were strongly opposed to this review, arguing that overallocation and overuse were judgements that could only be made in their water plans, and as a consequence the final review focused on evidence of "water stress" (NWC, 2012b). A key finding was that water stress was not confined to the impacts of irrigated agriculture, and some rivers and aquifers in and near capital cities were also hydrologically stressed.

These and other studies highlighted that, while water stress is most often associated with reductions in flow magnitude or volume due to extraction, this is not the only cause, and other changes to the flow regime (e.g., seasonality, predictability) were also important to consider (Bunn & Arthington, 2002). The assumption that a "sustainable level of extraction" could be simply characterized as a single number (e.g., the "two-thirds rule" proposed by Jones et al., 2003) was clearly a problem (Hamstead, 2009), and there was concern from the scientific community that simplistic "rule of thumb" provisions such as these would undermine a decade of progress in environmental flow science and management (see Arthington et al., 2006). Although a high ratio of extraction based on annual flows is likely to be associated with ecosystem degradation, low ratios do not necessarily imply there is little risk, especially when development affects important low flow attributes (see Chan et al., 2012).

THE ENVIRONMENT WAS THE BIG LOSER DURING THE DROUGHT

The Millenium drought (van Dijk et al., 2013) provided a harsh test of the progress made in environmental water reform in Australia. Audits of river health in the Murray-Darling Basin at this time showed that, of 62 zones in 23 valleys, only 2 were rated in Good ecological health, 11 were in Moderate health and the remaining 49 were in Poor (19 zones), Very poor (27 zones) or Extremely poor health (3 zones) (Davies et al., 2008, 2010). At the end of the Millenium drought, only the Paroo valley was found to be in Good ecosystem health and only the Warrego was rated in Moderate health, with most valleys rated as Poor or Very Poor (MDBA, 2012a).

The National Water Commission's 2009 biennial assessment concluded that progress with Australia's environmental water reform was slow. While many water-resource plans were consistent with NWI objectives with respect to environmental sustainability, some were not (NWC, 2009). Furthermore, many that were consistent in principle failed in practice to deliver desired environmental outcomes, and trade-offs between the environment and consumptive were not transparent. Some water plans were suspended by state governments during the drought, showing that the security of environmental flow entitlements was clearly less than that of consumptive users (NWC, 2009).

The National Water Commission also found there was no transparent, accessible, and accountable mechanism for registration of environmental water and, as a consequence, it was not possible to assess the level of compliance with environmental entitlements and rules and the risks associated with non-compliance (NWC, 2009). The Commission recommended the development of nationally consistent registration of environmental water across Australia, as well as annual public reporting of the existence, delivery and outcomes of environmental water. A series of comprehensive environmental water assessments was subsequently produced (NWC, 2010, 2012a, 2014b).

By 2011, there had been noticeable improvements in determining environmental water objectives and environmental water provisions through water plans (NWC, 2011). The newer plans were found to be more comprehensive and used better science to determine environmental water requirements, and many showed improvement in the clarity of trade-off processes. However, monitoring and reporting of water-plan provisions were not being done systematically or comprehensively and it was difficult to tell if water plans were achieving environmental objectives (NWC, 2011, 2012a). These reports also noted that it was often difficult to determine the extent to which rules-based commitments in water plans resulted in environmental water being made available. Water ecosystem outcomes and targeted environmental water regimes needed to be made more specific and clear, and a more systematic approach was required for identifying risks associated with different water regime scenarios and their associated consequent impacts to aquatic ecosystems (NWC, 2012a).

RECOVERY OF ENVIRONMENTAL WATER CONTINUES

The making of the Murray-Darling Basin Plan in November 2012 represented another significant milestone in the environmental water reform journey (Chapter 13; MDBA, 2012b). The Plan established long-term average sustainable diversion limits (SDLs) for all surface and groundwater resources within the Murray-Darling Basin that will be progressively phased in from Jul. 2019 (NWC, 2013a). This has resulted in an agreed reduction in average surface-water diversions of 2750 GL/y to meet the SDLs, in addition to water that was previously recovered (e.g., through the Living Murray initiative) (NWC, 2013b).

The Australian Government began to recover water for the environment, primarily in the Murray-Darling Basin, as part of a $12.9 billion national program, which includes investment in water-saving infrastructure and ~$3.1 billion of water purchases (NWC, 2013b). Water recovered for the environment is administered by the Commonwealth Department of Agriculture and Water Resources. The Commonwealth Environmental Water Holder (CEWH), established under the Water Act 2007, has the responsibility of managing the Commonwealth's environmental water holdings (Docker & Robinson, 2014). The CEWH, together with the Victorian Environmental Water Holder (established in July 2011), the Murray-Darling Basin Authority and the New South Wales and South Australian governments are now in a position to deliver significant volumes of water for the environment (NWC, 2014b). As on Aug. 31, 2016, the Commonwealth environmental water holdings totaled 2457 GL of registered entitlements with a long-term average annual yield of 1709 GL.[1] Although this is a significant volume of water to be reallocated from the consumptive pool, it should be noted that most environmental water in the Basin will continue to be provided via rules in water-sharing plans (MDBA, 2012b). A Basin-wide environmental watering strategy has been developed to help environmental water holders, Basin state governments and waterway managers plan and manage environmental watering at a Basin scale and over the long term to meet the agreed environmental objectives (MDBA, 2014; Swirepik et al., 2015).

The CEWH has established long-term arrangements for intensive monitoring at seven indicator sites across the Basin, where around 90% of Commonwealth environmental water is held (Gawne et al., 2013). The Long Term Intervention Monitoring Project will be essential to evaluate the contribution of Commonwealth environmental water in meeting environmental objectives within the Basin Plan (NWC, 2014b). In Victoria, monitoring of environmental flow outcomes in eight major regulated rivers is undertaken through a standard approach referred to as the Victorian Environmental Flows and Assessment Program (VEFMAP) (NWC, 2014b).

Environmental recovery of water has not only been limited to surface-water systems in the Murray-Darling Basin. Concerns about falling groundwater pressure over much of the Great Artesian Basin because of free-flowing bores and open channels were formally addressed with the adoption of the Great Artesian Basin Strategic Management Plan in 2000 by the Australian, Queensland, New South Wales, South Australian, and Northern Territory governments (GABCC, 2000). The plan featured an initial 15-year $230 million Great Artesian Basin Sustainability Initiative (GABSI), which included capping and piping of hundreds of bores, elimination of thousands of kilometers of open-bore drains and installation of piping (along with associated tanks and troughs) to deliver water to stock. The overall GABSI package had a dual purpose to recover water for the environment and to maintain aquifer pressure (NWC, 2013b). This ultimately benefited landholders and assisted in the protection of groundwater-dependent surface-water systems, like mound springs. By the end of June 2013, this initiative had saved approximately 200,000 ML/y, and it was extended in 2014 for another 3 years (GABCC, 2015).

WORKS AND MEASURES TO IMPROVE ENVIRONMENTAL OUTCOMES

There is a provision in the Murray-Darling Basin Plan to adjust the SDLs to enable improved environmental or socioeconomic outcomes (Hart, 2016a, 2016b; NWC, 2014b). These SDL adjustment mechanisms can include supply measures (i.e., mechanisms to reduce the volume of water to be recovered for the environment provided that equivalent environmental outcomes can be maintained), and/or efficiency measures (i.e., mechanisms to increase the volume of water available for the environment provided that social and economic outcomes are maintained or improved). Efficiency measures include improving the efficiency of on-farm irrigation and transferring the water savings for environmental

---

[1]https://www.environment.gov.au/water/cewo/about-commonwealth-environmental-water.

use. Under this SDL adjustment mechanism, up to 450 GL/y of additional water for the environment may be acquired through efficiency measures (NWC, 2014b).

Under the Living Murray initiative, around $280 million has been committed to environmental works and measures at icon sites (MDBA, 2011). These include: regulators (water-controlling structures that can be opened or closed to produce controlled flooding and drying of wetlands to reinstate more natural flood cycles); channels (excavated creeks or waterways, sometimes with containment banks, used to direct water in or out of an area); levee banks (earthen embankments that keep water inside a site that is being watered and enable placement of regulators for improved management); and fish-ways (site-specific structures that allow fish to pass through or around physical barriers such as dams, weirs, and road crossings) (MDBA, 2011).

An example is the environmental watering and works and measures on Hattah-Kulkyne Lakes, in Victoria (NWC, 2014b). In 2005 and 2006, temporary pumps were used to deliver environmental water to the lakes, some of which had been dry since 1996. A $32 million infrastructure package, including a permanent pump station, regulators and environmental levees, was opened in December 2013. These works were designed to deliver water to the lakes and achieve environmental benefits that would normally require a natural overbank flood, while the river is operating at normal flow levels. Primarily intended to support riparian vegetation and waterbirds, initial monitoring has shown evidence that some small-bodied native fish can survive the transfer via pumps and subsequently recruit (Vilizzi et al., 2013).

There is concern that, although water-efficient and capable of achieving some ecological targets, these engineering solutions do not mimic the natural flood pulse and extent and duration of lateral connectivity and, as a consequence, may not achieve important ecological outcomes in terms of biotic connectivity, river-floodplain productivity, and water quality (Bond et al., 2014; Pittock et al., 2013).

FUTURE CHALLENGES

For reasons evident in the previous discussions, much of the focus and investment in environmental water reform in Australia has been in the Murray-Darling Basin. There are few examples outside of the Basin where water entitlements have been acquired for environmental purposes and in some jurisdictions, entitlement frameworks do not currently facilitate the purchase of water for the environment (NWC, 2014b). For example, the longer-term security of environmental water provisions in Western Australia continues to be at risk given the nonstatutory nature of water allocation plans. Although the WA Government released its Securing Western Australia's Water Future policy paper in 2013 (DoW, 2013), which includes several reforms that will build on existing arrangements for environmental water, the drafting of the new Water Resources Management Bill is still in progress. Regular comprehensive reviews of environmental water management (e.g., NWC, 2010, 2012a, 2014a) ended when the National Water Commission was abolished in 2015, and there is a risk that public confidence in the reform process will be eroded.

As efforts continue to recover more environmental water in the Murray-Darling Basin under the SDL, there are growing calls from rural communities for the use of complementary measures, such as carp control, mitigation of cold-water pollution below dams, and feral animal control in wetlands, to offset the need for additional water and achieve improved river health. It is well recognized there remains a need to better coordinate land and water management planning to identify the non water-regime related actions needed to complement environmental water-management arrangements (Davis et al., 2015; NWC, 2011, 2014a). Under a drying climate, complementary measures such as these will be important to offset the effect of reduced availability of environmental water, especially if current water-sharing arrangements result in a greater risk to the environment than consumptive users. However, we currently lack a robust and transparent approach to inform how trade-offs could be made, and pursuing this agenda threatens to undermine efforts to determine the current SDL figures.

Our ability to measure and model low flows and the connections between groundwater and surface water remains a major technical challenge for environmental flow management (Barma & Varley, 2012a, 2012b). Stream gauging networks and the hydrological models used in water-resource planning do not perform well for many of the low-flow attributes known to be important for environmental outcomes (Marsh et al., 2012; Rolls et al., 2012). Better tools are required to improve measurement and modeling of the timing, frequency and duration of low (and no) flows.

Given the considerable investment in environmental water reform to date, and particularly in the recovery of water in the Murray-Darling Basin, it is imperative that Australia continues to invest in robust science and monitoring programs to determine whether desired long-term objectives are being met. Unfortunately, funding for these activities has dramatically declined in recent years, with the demise of several key institutions, including Land and Water Australia, the National Water Commission and several public-good Cooperative Research Centers. Transparent reporting of outcomes to the broader community will also be essential.

Many of the environmental water reforms described here coincided with one of the most extreme droughts in European history, and it is arguable that the observed ecological degradation could have been much worse had these reforms not been implemented at this difficult time. The next drought will be a true test of the effectiveness of Australia's water reforms and the willingness of governments to honor their commitments to the National Water Initiative.

REFERENCES

Anon.. (2000). *Lake Eyre basin intergovernmental agreement.* http://webarchive.nla.gov.au/gov/20130904072343/ http://www.environment.gov.au/water/publications/environmental/rivers/lake-eyre/agreement.html [Accessed 04.11.16].

ARMCANZ & ANZECC. (1996). *National principles for the provision of water for ecosystems.* Occasional paper SWR no. 3, Canberra: Agriculture and Resource Management Council of Australia and New Zealand and Australia and New Zealand Environment and Conservation Council. 14 p.

Arthington, A. H., Bunn, S. E., Poff, N. L., & Naiman, R. J. (2006). The challenge of providing environmental flow rules to sustain river ecosystems. *Ecological Applications, 16,* 1311–1318.

Arthington, A. H., King, J. M., O'Keeffe, J. H., Bunn, S. E., Day, J. A., Pusey, B. J., et al. (1992). Development of an holistic approach for assessing environmental flow requirements of riverine ecosystems. In *Proceedings of an international seminar and workshop on water allocation for the environment* (pp. 69–76). Armidale: Centre for Water Policy Research.

Barma, D., & Varley, I. (2012a). *Hydrological modelling practices for estimating low flows—Stocktake, review and case studies.* Canberra: National Water Commission.

Barma, D., & Varley, I. (2012b). *Hydrological modelling practices for estimating low flows—Guidelines.* Canberra: National Water Commission.

Bond, N. R., Costello, J., King, A., Warfe, D., Reich, P., & Balcombe, S. (2014). Risks and opportunities from artificial watering as a means of achieving environmental flow objectives. *Frontiers in Ecology and Environment, 12*, 386–394.

Bowling, L. C., & Baker, P. D. (1996). Major cyanobacterial bloom in the Barwon-Darling River, Australia, in 1991, and underlying limnological conditions. *Marine and Freshwater Research, 47*, 643–657.

Bunn, S. E., & Arthington, A. H. (2002). Basic principles and ecological consequences of altered flow regimes for aquatic biodiversity. *Environmental Management, 30*, 492–507.

Cadwallader, P. L. (1978). Some causes of decline in range and abundance of native fishes in the Murray-Darling River system. *The Proceedings of the Royal Society of Victoria, 90*, 211–224.

Cadwallader, P., & Lawrence, B. (1990). Fish. In N. Mackay & D. Eastburn (Eds.), *The Murray* (pp. 317–335). Canberra: Murray Darling Basin Commission.

Chan, T., Hart, B., Kennard, M. J., Pusey, B. J., Shenton, W., Douglas, M., et al. (2012). Bayesian network models for environmental flow decision-making in the Daly River, Northern Territory, Australia. *River Research and Applications, 28*, 283–301.

Commissioner, Snowy Water Inquiry. (1998). *Snowy water inquiry final report*. Sydney: Snowy Water Inquiry.

Cullen, P. (2011). Facing up to the water crisis in the Murray-Darling Basin. In H. C. Cullen (Ed.), *This land, our water: Water challenges for the 21st century* (pp. 147–158). Hindmarsh: ATF Press.

Dann, H. E. (1969). Two decades of engineering—The Snowy Mountains scheme. *Journal of the Institute of Engineers Australia, 41*, 23–30.

Davies, P. E., Harris, J. H., Hillman, T. J., & Walker, K. F. (2008). *SRA report 1: A report on the ecological health of rivers in the Murray–Darling Basin, 2004–2007*. Canberra: Prepared by the Independent Sustainable Rivers Audit Group for the Murray–Darling Basin Ministerial Council.

Davies, P. E., Harris, J. H., Hillman, T. J., & Walker, K. F. (2010). The Sustainable Rivers Audit: Assessing river ecosystem health in the Murray-Darling Basin, Australia. *Marine and Freshwater Research, 61*, 764–777.

Davis, J., O'Grady, A. P., Dale, A., Arthington, A. H., Gell, P. A., Driver, P. D., et al. (2015). When trends intersect: The challenge of protecting freshwater ecosystems under multiple land use and hydrological intensification scenarios. *Science of the Total Environment, 534*, 65–78.

DEST. (1996). *Australia state of the environment*. Department of Environment, Sport and Territories. Collingwood: CSIRO.

Dexter, B. D. (1978). Silviculture of the river red gum forest of the central Murray floodplain. *Proceedings of the Royal Society of Victoria, 93*, 175–191.

Docker, B., & Robinson, I. (2014). Environmental water management in Australia: Experience from the Murray-Darling Basin. *International Journal of Water Resources Development, 30*, 164–177.

Doeg, T. J. (1984). Response of the macroinvertebrate fauna of the Mitta Mitta River, Victoria, to the construction and operation of Dartmouth Dam. 2. Irrigation release. *Occasional Papers of the Museum of Victoria, 1*, 101–108.

Domicelj, S. (1981). The Australian Snowy Mountains scheme: National growth and regional development. *Habitat International, 5*, 601–616.

DoW. (2013). *Securing Western Australia's water future. Position paper—Reforming water resource management*. Perth: Department of Water.

DSE. (2004). *Securing our water future together: Victorian Government white paper*. Melbourne: Department of Sustainability and Environment.

DWR-NPWS. (1986). *Water management plan for the Macquarie Marshes*. Sydney: Department of Water Resources and National Parks and Wildlife Service.

Eastburn, D. (1990). *The River Murray: History at a glance*. Canberra: Murray-Darling Basin Commission.

GABCC. (2000). *Great Artesian Basin strategic management plan*. Canberra: Great Artesian Basin Consultative Council.

GABCC. (2015). *Great Artesian Basin Coordinating Committee annual report 2014–2015.* Canberra: Great Artesian Basin Consultative Council.

Gawne, B., Brooks, S., Butcher, R., Cottingham, P., Everingham, P., Hale, J., et al. (2013). *Long term intervention monitoring logic and rationale document.* Final report prepared for the Commonwealth Environmental Water Office by The Murray-Darling Freshwater Research Centre, MDFRC Publication, 01/2013, May, 109 p.

Hamstead, M. (2009). *Improving environmental sustainability in water planning.* Waterlines report. Canberra: National Water Commission.

Harris, J. H. (1984). Impoundment of coastal drainages of south-eastern Australia, and a review of its relevance to fish migrations. *Australian Zoologist, 21,* 235–249.

Hart, B. T. (2016a). The Australian Murray-Darling Basin Plan: Challenges in its implementation (part 1). *International Journal of Water Resources Development, 32,* 819–834. http://dx.doi.org/10.1080/07900627.2015.1083847.

Hart, B. T. (2016b). The Australian Murray-Darling Basin Plan: Challenges in its implementation (part 2). *International Journal of Water Resources Development, 32,* 835–852. http://dx.doi.org/10.1080/07900627.2015.1084494.

Jones, G., Hillman, T., Kingsford, R., McMahon, T., Walker, K., Arthington, A., et al. (2003). *Ecological assessment of environmental flow reference points for the River Murray system.* Interim report prepared for the Murray-Darling Basin Commission, Living Murray Initiative. Canberra: Murray-Darling Basin Commission.

Kingsford, R. T. (2000). Ecological impacts of dams, water diversions and river management on floodplain wetlands in Australia. *Austral Ecology, 25,* 109–127.

Kingsford, R. T., & Thomas, R. F. (1995). The Macquarie Marshes in arid Australia and their waterbirds: A 50 year history of decline. *Environmental Management, 19,* 867–878.

Marsh, N., Sheldon, F., Wettin, P., Taylor, C., & Barma, D. (2012). *Guidance on ecological responses and hydrological modelling for low-flow water planning.* Canberra: National Water Commission. ISBN 978-1-921853-55-5.

MDBA. (2011). *The Living Murray story—One of Australia's largest river restoration projects.* Canberra: Murray-Darling Basin Authority. Publication no. 157/11.

MDBA. (2012a). *Sustainable rivers audit 2: The ecological health of rivers in the Murray–Darling Basin at the end of the millennium drought (2008–2010). summary.* Canberra: Murray-Darling Basin Authority Publication no. 75/12.

MDBA. (2012b). *Basin plan.* Canberra: Murray-Darling Basin Authority. [Accessed 04.11.16]. http://www.mdba.gov.au/basin-plan.

MDBA. (2014). *Basin-wide environmental watering strategy.* Canberra: Murray-Darling Basin Authority. Publication no. 20/14.

MDBMC. (1994). *Algal management strategy for the Murray-Darling Basin.* Canberra: Murray-Darling Basin Commission.

MDBMC. (1995). *An audit of water use in the Murray-Darling Basin: Water use and healthy rivers—Working towards a balance.* Canberra: Murray-Darling Basin Commission.

Norris, R. H., Liston, P., Davies, N., Coysh, J., Dyer, F., Linke, S., et al. (2001). *Snapshot of the Murray-Darling Basin River condition.* Canberra: Murray Darling Basin Commission.

NSW Office of Water. (2010). *Returning environmental flows to the Snowy River. An overview of water recovery, management and delivery of increased flows.* Sydney: Department of Environment, Climate Change and Water.

NWC. (2007). *National Water Initiative: First biennial assessment of progress in implementation.* Canberra: National Water Commission.

NWC. (2009). *Australian Water Reform 2009: Second biennial assessment of progress in implementation of the National Water Initiative.* Canberra: National Water Commission.

NWC. (2010). *Australian environmental water management report 2010.* Canberra: National Water Commission.

NWC. (2011). *The National Water Initiative—Securing Australia's water future: 2011 assessment.* Canberra: National Water Commission.

NWC. (2012a). *Australian environmental water management: 2012 review.* Canberra: National Water Commission.

NWC. (2012b). *Assessing water stress in Australian catchments and aquifers.* Canberra: National Water Commission.

NWC. (2013a). *Murray–Darling Basin Plan implementation: Initial report.* Canberra: National Water Commission.

NWC. (2013b). *Water management and pathways to sustainable levels of extraction: Issues paper.* Canberra: National Water Commission.

NWC. (2014a). *Australia's water blueprint: National reform assessment 2014.* Canberra: National Water Commission.

NWC. (2014b). *Australian environmental water management: 2014 review.* Canberra: National Water Commission.

Pittock, J., Finlayson, C. M., & Howitt, J. (2013). Beguiling and risky: "Environmental works and measures" for wetland conservation under a changing climate. *Hydrobiologia, 708,* 111–131.

PMSEIC. (1996). Managing Australia's inland waters. Roles for science and technology. In *A paper prepared by an independent working group for consideration by the Prime Minister's Science, Engineering and Innovation Council at its fourteenth meeting, 13 September.* Canberra: Department of Industry, Science and Tourism.

PMSEIC. (1998). Dryland salinity and its impacts on rural industries and the landscape. Occasional paper number 1. In *A paper prepared by an independent working group for consideration by the Prime Minister's Science, Engineering and Innovation Council at its second meeting, 4 December.* Canberra: Department of Industry, Science and Resources.

Powell, J. M. (1989). *Watering the garden state: Water, land, and community in Victoria, 1834–1988.* Sydney: Allen and Unwin.

Powell, J. M. (1991). *Plains of promise, rivers of destiny: Water management and the development of Queensland 1824–1990.* Brisbane: Boolarong Publications.

Pusey, B. J. (1998). Methods addressing the flow requirements of fish. In A. H. Arthington & J. M. Zalucki (Eds.), *Comparative evaluation of environmental flow assessment techniques: Review of methods* (pp. 66–105). Canberra: Land and Water Resources Research and Development Corporation.

Queensland Government. (2000) *Water Act 2000.* https://www.legislation.qld.gov.au/LEGISLTN/CURRENT/W/WaterA00.pdf [Accessed 31.10.16].

Rolls, R. J., Leigh, C., & Sheldon, F. (2012). Mechanistic effects of low-flow hydrology on riverine ecosystems: Ecological principles and consequences of alteration. *Freshwater Science, 31,* 1163–1186.

Rose, R., & Erskine, W. (2011). Channel recovery processes, rates and pathways following environmental flow releases to the Snowy River, Australia. *River, coastal and estuarine morphodynamics: RCEM2011.* Beijing: Tsinghua University Press.

Sheldon, F., & Walker, K. F. (1993). Pipelines as a refuge for freshwater snails. *Regulated Rivers: Research and Management, 8,* 295–299.

Storey, A. W., Edward, D. H., & Gazey, P. (1991). Recovery of aquatic macroinvertebrate assemblages downstream of the Canning Dam, Western Australia. *Regulated Rivers: Research and Management, 6,* 213–224.

Swirepik, J. L., Burns, I. C., Dyer, F. J., Neave, I. A., O'Brien, M. G., Pryde, G. M., et al. (2015). Establishing environmental water requirements for the Murray-Darling Basin, Australia's largest developed river system. *River Research and Applications, 32,* 1153–1165.

Tharme, R. E. (2003). A global perspective on environmental flow assessment: Emerging trends in the development and application of environmental flow methodologies for rivers. *River Research and Applications, 19,* 397–441.

Tunbridge, B. R. (1997). *Environmental flows: review of the environmental flow methodology used by the freshwater ecology section—Including comparison with United States instream flow methodology*. Report for the Victorian Department of Natural Resources and Environment, Melbourne, 22 p. (cited in Pusey, 1998).

van Dijk, A. I. J. M., Beck, H. E., Crosbie, R. S., de Jeu, R. A. M., Liu, Y. Y., Podger, G. M., et al. (2013). The Millennium Drought in southeast Australia (2001–2009): Natural and human causes and implications for water resources, ecosystems, economy, and society. *Water Resources Research, 49*, 1040–1057.

Vilizzi, L., McCarthy, B. J., Scholz, O., Sharpe, C. P., & Wood, D. B. (2013). Managed and natural inundation: Benefits for conservation of native fish in a semi-arid wetland system. *Aquatic Conservation: Marine and Freshwater Ecosystems, 23*, 37–50.

Walker, K. F. (1979). Regulated streams in Australia: The Murray-Darling system. In J. V. Ward & J. A. Stanford (Eds.), *The ecology of regulated streams* (pp. 143–163). New York, NY: Plenum Press.

Walker, K. F. (1985). A review of the ecological effects of river regulation in Australia. *Hydrobiologia, 125*, 111–129.

Walker, K. F. (1992). The River Murray, Australia: A semiarid lowland river. In P. Calow & G. E. Petts (Eds.), *The rivers handbook: Vol. 1.* (pp. 472–492). London: Blackwell Scientific Publications.

WRC. (2000). *Environmental water provisions policy for Western Australia*. Statewide policy no. 5. Perth: Water and Rivers Commission.

INPUTS INTO THE WATER RESOURCES DECISION-MAKING PROCESS

ENVIRONMENTAL FLOWS AND ECO-HYDROLOGICAL ASSESSMENTS IN RIVERS

7

M.J. Stewardson, J.A. Webb, A. Horne
The University of Melbourne, Melbourne, VIC, Australia

CHAPTER OUTLINE

INTRODUCTION

The protection of environmental flows in rivers is a new practice developed over recent decades in response to the decline of ecological condition, function and biodiversity of rivers produced by human modification to river flow regimes (Horne et al., 2017b). Management of environmental flows includes the stages of planning, implementation, monitoring, and evaluation (NWC, 2014), and requires consideration of both technical and social elements for a successful outcome.

Decision Making in Water Resources Policy and Management. http://dx.doi.org/10.1016/B978-0-12-810523-8.00008-2

This chapter deals specifically with methods of environmental flows assessment during the planning stage of environmental water management. These assessment methods are used to establish: the level of human impact on river, estuary, groundwater-dependent, and wetland ecosystems; their environmental water requirements; and potential environmental risks if environmental water allocation scenarios fall short of the required environmental water. We constrain our attention to the technical aspects of the planning phase, acknowledging that these methods must be applied within a planning process that engages stakeholders in decision making. We also constrain our attention to environmental flows assessment in rivers, including both channels and floodplains.

Strong national water policy requiring governments to protect environmental water has accelerated efforts to plan and deliver environmental flows across Australia since the mid-1990s (see Chapter 6). State governments, with primary responsibility for environmental water management, have responded with innovative and practical approaches to integrate environmental flows within their water planning frameworks. The last two decades have seen an explosion of environmental flows assessment methods and case studies across Australia (Saintilan & Overton, 2010). Initial studies were focused on regulated rivers and in-channel flows. This quickly broadened to include flow requirements for floodplains, recognizing the important role floodplains have in river ecosystem function and the widespread impact of large irrigation and urban water supply dams on floodplain hydrology in Australia (Saintilan & Overton, 2010). More recently, environmental flows planners in Australia have also considered wetlands, including those that are groundwater dependent, and estuaries, but these assessments are beyond the scope of this chapter (e.g., Lloyd et al., 2012).

Environmental flows assessments in Australia have been applied at the scales of individual reaches, rivers, basins and regions. At the basin and regional scales, environmental flows assessments use hydrological methods, mostly to identify rivers that are potentially "flow stressed." These rivers may receive special management attention, including more detailed river-specific environmental flow studies. Hydrological methods rely on streamflow data alone and may use a variety of statistics characterizing flow regimes or their alteration to assess hydrological alterations and the environmental water requirements (Davies et al., 2012; Poff et al., 2017; Tharme, 2003).

Hydrological assessments have also been used to establish regional or basin-wide sustainable diversion limits including in Victoria (SKM, 2005) and the Murray-Darling Basin (MDBA, 2011; Swirepik et al., 2016). These hydrological approaches are based on an assessment of alteration since a predevelopment state. In this way they are best described as a top-down approach (e.g., Richter et al., 1997), where constraints are set on the degree of flow alteration. This chapter describes hydrological assessment methods used for basin-scale and regional environmental flows assessment in Australia.

At the reach or river scale, environmental flows assessments have generally adopted a holistic methodology. Holistic methods consider the role of multiple flow components including minimum flows and flow pulses in supporting multiple aquatic species life stages and potentially ecological functions (Poff et al., 2017; Tharme, 2003). Indeed, Australian researchers and practitioners have been among the leaders in developing these methods. Early examples include studies in Barkar-Barambah Creek, Queensland (Arthington et al., 1992) and Thomson River, Victoria (Gippel & Stewardson, 1995). In most cases, the regulation of flows for human use began decades prior to the commencement of environmental flow planning, and environmental flows assessments have revealed widespread over-allocation of water (i.e., environmental water requirements are not being met under the historic arrangements). In such circumstances, it is common practice in Australia for holistic methods to design environmental flow regimes using a "bottom-up" approach where critical components of the flow regime are protected. This is consistent with the concept of environmental flow design described by Acreman et al. (2014) and is quite different from the "top-down" approach where limits are imposed on hydrological alterations from the natural state (e.g., Richter et al., 1997).

An integral component of holistic environmental flows assessments is the prediction of ecological responses to changing flow regimes. It is a reasonable statement that the direct modeling of ecological responses has lagged well behind the modeling of hydrologic and hydraulic responses in environmental flows assessments. Rather, expert opinion has often been used to predict likely ecological responses to specific changes in discharge and/or hydraulic habitat. With increasing availability of large-scale monitoring data sets and an increasing range of analytical and software methods to model ecological responses, there is good potential for increased use of ecological models in future environmental flows assessments (refer to the section "Improving Quality of Ecological Predictions").

Environmental flows assessments must be integrated within the broader task of water resource allocation planning. Water resource models are critical to the planning processes and provide the means to evaluating consequences of alternate water allocation scenarios. With the evolution of environmental flow planning in Australia, water resource models have been adapted to represent environmental water allocation scenarios. Environmental water demands have several distinctive features compared to other human uses of water which can be a challenge for modelers. The Australian water resource modeling tools and the methods used to represent environmental water demands are addressed in Section Integrating Environmental Flows in Water Resource Modeling.

HYDROLOGICAL METHODS

The hydrological methods applied in Australia have been largely desktop procedures evaluating flow stress using streamflow data from multiple sites across a basin or region. These are used to prioritize rivers for detailed environmental flows assessment based on the severity of the flow stress, or to establish sustainable diversion limits beyond which further flow modifications will lead to an unacceptable environmental risk.

We describe the three stages of these assessments in this section: establishing a reference hydrological state prior to water resource developments; calculating measures of flow stress; and defining thresholds for these measures. Two examples are used to illustrate the approach used in Australia. The first is the assessment of hydrological alteration using the Flow Stress Ranking procedure (SKM, 2005) designed for setting sustainable diversion limits in Victoria and subsequently modified for evaluating hydrological condition in the Murray-Darling Basin for the Sustainable Rivers Audit (Davies et al., 2012). The other example is the method used for planning sustainable diversion limits in the Murray-Darling Basin for the Basin Plan (Swirepik et al., 2016).

PREDEVELOPMENT HYDROLOGICAL STATE

Even in rivers subject to a long history of river regulation, consideration of environmental flow requirements is dependent upon analysis of hydrological alterations from natural or predevelopment conditions. This is generally based on a comparison of streamflow series corresponding to *developed* or *impacted* conditions, and a streamflow series representing predevelopment conditions. Derivation of these two streamflow series can itself pose a challenge. Gauged streamflow data can include significant nonstationarity as a result of changes in catchment condition and water use over time, and long-term climate variability.

In catchments with significant levels of human activity during the period of gauged records, modeling techniques are required to derive an estimated target flow series (Blöschl et al., 2013). Three approaches are described here (Stewardson et al., 2017).

Rainfall runoff modeling—A rainfall runoff model may be calibrated to gauged data from the pre-impact time period. This model can then be used to estimate reference stream flows for the postimpact period. Abstractions/returns may be added to create any scenario of target regime. This approach requires that streamflow records commence sometime before the onset of hydrological alteration.

Streamflow transposition—If there is no record available for a site prior to the onset of hydrological alteration, then streamflow records from a nearby unimpacted catchment may be *transposed* to estimate the unimpacted stream flows at the site of interest using a transposition factor (Lowe & Nathan, 2006). There is a significant body of work around the transposition of information from paired or experimental catchments. This is particularly relevant when assessing the alterations due to land use changes (Brown et al., 2005). This method requires that there are streamflow records available for a nearby *hydrologically similar* catchment, which is unimpacted by human disturbances.

Reversing water use and regulation—Where there is information available about the level of human impact (including the timing and volume of water extractions and catchment changes, such as farm dam development), these effects can be removed from the gauged record for the catchment. In a river system managed for consumptive water supply, the water resource models can often be used to model stream flows that would have occurred in the absence of water resource developments.

In the Murray-Darling Basin, developed and reference flow series have been derived for the main-stem rivers (Davies et al., 2012). These have predominately used the third method of reversing water use and regulation. The most recent Sustainable Rivers Audit considered both farm dam impacts and land use change in the derivation of these flow series (Davies et al., 2012). While these impacts are considered within the Basin Plan and water sharing processes, they have not historically been considered when deriving the reference flow series to inform environmental flows assessment studies.

FLOW STRESS RANKING PROCEDURE

Developed for use in Victoria (southeast Australia), the Flow Stress Ranking procedure uses a set of metrics to characterize hydrological alteration (SKM, 2005). It was initially applied statewide based on metrics calculated for 551 sites. The ranking indicates the level of threat to river health based on the level of water extractions by rural, urban and industry users. The metrics characterize the degree of hydrologic stress under current management conditions relative to "predevelopment" flow conditions: that is, the flow regime that would occur if all anthropogenic extractions, water harvesting, and impoundments were removed.

The method was developed with expert ecological input advising the design of metrics but constrained by the availability of data for a statewide assessment. The development of metrics was iterative with expert advisers meeting to consider results of testing of preliminary metrics, making adjustments and then reviewing these when calculations were revised. Experts evaluated the metrics based on preliminary ranking of test sites, where the eco-hydrological state had previously been established using more detailed environmental flow studies. The expert team settled on 10 metrics they felt represented flow stress across a range of flow components.

The initial set of metrics was calculated using daily streamflow data, recognizing the importance of daily variations for stream ecosystems. However, statistical testing using a sample of 50 river stations showed that only five (largely independent) indices, calculated using monthly data instead of daily, were required to characterize variation across the 10 daily flow metrics proposed by the panel (SKM, 2005). The use of monthly flow indices dramatically simplifies compilation of streamflow data and modeling pre-development streamflows across the region. The five monthly indices used to measure flow alterations were:

- low flow magnitude,
- high flow magnitude,
- proportion of time that the flow is zero (or nearly so),
- variability in monthly streamflow,
- seasonal timing of when low and high flows occur.

Using bootstrap confidence intervals, it was shown that at least 15 years of data were required to reduce uncertainty to acceptable levels (SKM, 2005).

To illustrate the form of the metrics, the metric related to low flow magnitude is calculated using:

$$LF = \frac{LF_{91.7} + LF_{83.3}}{2}$$

$$LF_{91.7} = 1 - 2|P_{ile}(Q91.7_u) - P_{ile}(Q91.7_c)|$$

$$LF_{83.3} = 1 - 2|P_{ile}(Q83.3_u) - P_{ile}(Q83.3_c)|$$

where $Q91.7_x$ refers to the flow exceeded 91.7% of the time in the unimpacted ($x=u$) or current ($x=c$) streamflow series (similarly for $Q83.3_x$). The 91.7 and 83.3 percentile flows correspond to flows exceeded on average 11 and 10 months out of 12, respectively. $P_{ile}(Q91.7_x)$ is the proportion of years when $Q91.7_x$ is exceeded in the unimpacted streamflow series. The metric is based on change in percent of time low flows are exceeded rather than the actual magnitude of the flow. This procedure is described as producing a range-standardized metric, meaning that flows with a greater variability in the unimpacted flow series will produce a lower flow stress measure for the same change in discharge at the 91.7 percentile level.

The major strength of the Flow Stress Ranking procedure is that it is an authoritative ecologically based method that can be applied across a large region using available data. It has provided a critical screening tool for the state water management agency in Victoria to identify catchments where flow extractions are approaching or exceeded unsustainable levels. Water diversions in such catchments are capped (i.e., prevented from further expansion). More detailed environmental flows assessment (e.g., the holistic method described in the following) can be prioritized for these rivers to assess the extent of overallocation and inform future water planning with the possibility of recovering water for the environment.

UMBRELLA ENVIRONMENTAL ASSET APPROACH

A hydrological method, described initially as the Environmental Sustainable Level of Take (ESLT) method and later as the Umbrella Environmental Asset (UEA) approach, has been used to set sustainable diversion limits across the 1 million km^2 Murray-Darling Basin (MDBA, 2011; Swirepik et al., 2016).

This method was a core element of the Basin Plan, finalized in 2012, that sets sustainable diversion limits (SDLs) for the basin's subcatchment, to be implemented by state governments (Hart, 2015a, 2015b, 2016). The Basin Plan was the result of over 4 years of work, led by the Murray-Darling Basin Authority, and included consultation with a range of government jurisdictions and regional communities across the Basin. The Basin Plan requires a reduction on the previous diversion limit (or "Cap") of 2750 GL calculated based on average water withdrawals. This represents ∼20% of the total annual water withdrawals allowed on average previously (Hart, 2016).

A key challenge in environmental water planning across such a large and diverse river basin is patchy knowledge concerning environmental flow needs. The UEA approach addresses this challenge by assuming the accumulated environmental flow requirements at carefully selected "information-rich" sites is equivalent to, or exceeds, the environmental flow required to meet the needs of all assets throughout the basin (Swirepik et al., 2016). The UEA approach has two key stages described here: selecting the EUAs, and identifying key flow components required to achieve UEA environmental objectives.

The Murray-Darling Basin was represented as the accumulation of 18 main river valleys, but Sustainable Diversion Limits were not considered for seven of these because they were largely unmodified, held a small volume of the surface water resource, or there was already a detailed valley-wide environmental flows assessment that could be used instead of the UEA approach (Swirepik et al., 2016). A total of 24 UEAs were selected from within the remaining 11 valleys using five criteria: high ecological value; representative of water requirements; spatially representative; significant flow alteration; and availability of information. The selected assets were mostly large wetland systems at the downstream end of each valley and below the major flow diversion points. It is assumed that the large environmental water demands of these assets will exceed the environmental water requirements of upstream environmental assets, although this is untested.

The environmental flow requirements of multiple ecosystem values at each UEA were evaluated using available information from previous environmental flow studies at these sites (Swirepik et al., 2016). For these large wetland systems, typical water requirements included in-channel and overbank flow pulses specified using a threshold magnitude, duration, and frequency that should be exceeded. In most cases, these flow pulse requirements were also required to occur within a specified "season" within the year. Sustainable diversion limits for each valley were tested and incrementally adjusted to improve compliance with the environmental water requirements of the UEAs across the basin. Full compliance with UEAs was not a requirement of the analysis. Instead, a trade-off was applied to improve compliance with environmental water requirements, while attempting to reduce adverse outcomes for irrigators and rural communities resulting from reduced irrigation diversions.

The UEA approach to environmental water assessment is pragmatic in the face of significant data and knowledge gaps at the basin scale and the resulting sustainable diversion limits is a critical input to water planning. However, more detailed holistic environmental flows assessments are still required at the local scale to optimize water-sharing arrangements across multiple water users, within the constraints imposed by the sustainable diversion limits. These detailed assessments are also critical where environmental water allocation mechanisms include held environmental water entitlements that must be actively managed (Horne et al., 2017a). These detailed studies provide more specific advice and models to prioritize active decisions to deliver environmental water at times and in locations that give the best environmental outcome.

Methods of selecting UEAs, and the assumption that a set of UEAs can be identified that truly represents environmental water requirements of the full basin, are yet to be tested. Such tests will be difficult at the large basin scale but may be possible in an intensively studied basin. Testing of such an approach is important, but it is likely the largest source of uncertainty will be in environmental water requirements for the selected UEAs. Even with the application of holistic methods (described in the next section), there are major uncertainties and knowledge gaps that make these assessments uncertain.

HOLISTIC METHODS

Holistic methods for environmental flows assessment have emerged over the last 20 years, with Australian researchers being at the forefront of their development. Several holistic methods have been applied across the Murray-Darling Basin (Table 1), but all share the elements of: (a) considering the broad ecological requirements of the entire ecosystem, and (b) recommending specific flow components believed to fulfil specific ecological purposes.

FLOWS METHOD

Victoria has the most standardized holistic method for environmental flows assessment, with the FLOWS method (DEPI, 2013) being used for nearly all assessments across the state. The FLOWS method is well documented with comprehensive description of the data sources, tasks and outputs (DNRE, 2002). It has also been tested in applications to many rivers across Victoria with a broad range of water uses and water management infrastructure. The method draws on available information with additional hydraulic data and models where necessary to characterize important eco-hydraulic conditions and their response to discharge.

In Victoria, most FLOWS studies are focused on in-channel environmental flow requirements, with hydraulic data derived using a one-dimensional hydraulic modeling based on reach surveys of the stream channels. Water resource modeling provides streamflow scenarios, including at least the predevelopment and current conditions, but future scenarios are also required if water management is expected to change.

The key elements of the FLOWS method are: the use of a multidisciplinary expert panel; characterizing flow objectives using hydraulically defined flow components; and linking environmental flow objectives to environmental objectives. The expert panel includes the range of disciplinary expertise required to undertake the assessment, including freshwater ecology, fluvial geomorphology, aquatic biogeochemistry, eco-hydraulics and hydrology, depending on the specific needs of the study. The panel develops an issues paper based on available reports and data combined with an inspection of sites along the river. This preliminary report identifies environmental values and threats and their dependencies on streamflow and other human disturbances.

Drawing on existing policy documents, the panel defines environmental objectives to guide the design of an environmental flow regime. Importantly, these objectives are consistent with environmental objectives set through government-led catchment planning processes at the state and regional scale. The panel is responsible for recommending environmental flow components to achieve these environmental objectives. Flow components can be expressed as minimum flows, low flows (periods when

Table 1 Summary of the Different Holistic Environmental Flows Methods Used Across the Murray-Darling Basin

State	Method	Aim(s)	Steps
Victoria	FLOWS – a method for determining environmental water requirements in Victoria (DEPI, 2013)	• To describe key flow components as part of a recommendation for an environmental flow regime, rather than a minimum flow recommendation	1. Identify the current environmental assets 2. Identify assets expected to be associated with a "healthy" waterway 3. Development environmental objectives 4. Identify key flow-related events and flow components to meet each environmental objective 5. Develop flow objectives 6. Develop recommendations to meet each flow objective
South Australia	Scientific Panel Process/ Assessment (Arthington et al., 1998) – Murray River Lower Lakes and Coorong	• To provide recommendations for management actions to achieve environmental outcomes • To assess and integrate ecological, geomorphological and hydrological parameters of the river system using specialists	1. Rapid field assessment supported by previous information and the analysis of natural and current flow regimes 2. Provide broad management recommendations on ecosystem health principles for river reaches and important ecological areas 3. Identify short- and longer-term recommendations and a range of options
	Scientific Panel Habitat Assessment Method (SPHAM) (Favier et al., 2000) – Mid North Region – Eastern Mount Lofty Ranges, Marne River	• To determine the water requirements necessary to maintain the ecological health of the river and its major tributaries using rapid appraisal methods from a multidisciplinary team • To develop flow benchmarks that could be used for future monitoring programs	1. Preworkshop phase: assemble scientific panel, catchment tour, identify geomorphic zones, data collection, hydrological modeling 2. Scientific panel workshop: identify links for each flow zone, key flow levels, quantify water requirements, produce a summary, identify knowledge gaps and monitoring needs 3. Post-workshop: flow modeling to test results and future development scenarios, produce technical reports on links.

State	Method	Objectives	Steps
New South Wales	"Environmental Flow Rules" – Regulated section of the Murrumbidgee River below Burrinjuck Dam (Shields & Good, 2002)	• To identify and quantify the benefits to the environment of this limited additional volume of water • To determine the most effective use of water • To ensure natural flow and variability in the river are maintained as much as possible	1. Identify all native flora and fauna dependent on the waters of the river and minimum flow regime 2. Hydrographs developed from modeled minimum water requirements 3. Refine and amalgamated into four hydrographs to provide optimal river flow regimes for all competing environmental needs 4. Compare minimum environmental flow regime needs with the interim flow rules 5. Development of an "environmental spreadsheet matrix" and four flow rules
Queensland	Benchmarking methodology (Brizga, 2007)	• Top-down methodology. Examines the consequences of altering or removing various components of the flow regime • Recognizes the role of the government as a decision maker • Focuses on providing the best available information to inform its decisions	1. Uses natural or existing flow regimes as a starting point 2. A holistic approach ensures all relevant issues were covered 3. Conceptual models used to structure and integrate large and complex arrays of data and information 4. Flow requirements were assessed with regard to the whole flow regime
	Environmental Flows Assessment Program (EFAP) (Cottingham et al., 2005)	• Provide the science to inform an assessment of how water resource plans (WRPs) meet ecological outcomes • Measures performance in achieving stated ecological outcomes	1. WRPs are developed and amended 2. Monitoring of selected ecological assets through targeted monitoring projects 3. Reporting 4. Review (back to 1 for continual improvement)
Australian Capital Territory	Water Resource Environmental Flow Guidelines (ACT Government, 2013)	• To use the holistic approach for setting environmental flow guidelines • Recognizes the natural flow regimes as a guide to flow requirements of a system • To maintain healthy aquatic ecosystems	1. Identify essential features of the flow regime 2. Identification of flow components on the influence of ecosystem components 3. An adaptive management process is used to refine flow requirements when more information becomes available

flows are below a threshold discharge for a specified duration) and high flows pulses (periods when flows exceed a threshold discharge for a specified duration).

Table 2 provides examples of each type of flow recommendation and identifies the environmental objectives that they contribute to. There are four flow components recommended for this reach, as follows:

- A summer period (December–May) cease-to-flow event to occur on average every second year for a duration of between 13 and 32 days,
- A summer period (December–May) median flow of 10 ML/d every year (excluding the cease to flow period),
- A winter period (July–November) flow fresh where flows exceed 90 ML/d for at least ten days to occur with a frequency of one in two years,
- A winter period (July–November) minimum median flow of 90 ML/d.

Note that some recommended flow components contribute to multiple environmental objectives (listed in the fifth column of Table 2).

A limitation of the FLOWS method is the difficulty of assessing the marginal changes in risk for each of the environmental objectives when water allocation scenarios do not deliver on the full environmental flow recommendations. This is not possible with the standard FLOWS approach because the method provides a single recommended set of flow components rather than a model for assessing trade-off decisions. Furthermore, the different flow requirements to achieve the individual environmental objectives are aggregated to just a few recommended flow components, making it impossible to assess

Table 2 Examples of Environmental Flow Recommendations for a Reach on the Avoca River, Victoria

River	Avoca River			Reach	Amphitheatre—Coonooer
Compliance Point		Coonooer		Gauge No.	408200
Flow				**Rationale**	
Season	Magnitude	Frequency	Duration	Objective	Evaluation
Summer (Dec–May)	Cease to Flow 0 ML/day	1 in 2 years	13–32 days	10	0 ML/day flows recorded
	Minimum median flow 10 ML/day	Annually	Dec–May (excluding specified CTF)	1a, 2a, 4a, 5a, 5b, 6	Self-sustaining populations of small bodied fish
July–Nov	>90 ML/day	2 annually	10 days	1b, 2b, 4b	Self-sustaining populations of fish (Mountain Galaxias, Blackfish, Golden Perch)
July–Nov	Minimum median flow 90 ML/day	Annually	July–Nov	9	Self-sustaining populations of large bodied fish Off-stream habitats wetted

how each individual objective might be affected when the flow recommendations are not achieved. This is a serious limitation because it makes it difficult for decision makers to evaluate a trade-off between environmental and other water uses. Another limitation is that the environmental flow recommendations refer to average frequencies and single duration and discharge thresholds to specify recommendations for each flow component, but variability in these characteristics may be desirable.

The Flow Events Method (Stewardson & Gippel, 2003) overcomes these limitations using a more a detailed treatment of the eco-hydraulic linkages between flow objectives and environmental objectives and providing multiple risk-based flow recommendations. The Flow Events Method was used to extend the FLOWS procedure in a study of the Goulburn River (Cottingham et al., 2010).

MODELING ECOLOGICAL RESPONSES

In practice in Australia, it has been rare for environmental water provisions agreed through water planning to meet the full environmental flow requirements established using holistic methods. This is invariably the case where there is a history of water resource development for consumptive use. Water planners must consider the trade-off between environmental and human consumptive use of the resource in deciding on the water sharing arrangements. Understanding the ecological risks of marginal changes in environmental water provisions requires the translation of hydraulic model outputs, typically considered in holistic studies in Australia, to predictions of ecological responses.

As described previously, across the Murray-Darling Basin, each state has its own methods for environmental flows assessment, but all have common characteristics. Some of these commonalities lie around the process used to convert scenarios of flow delivery to predictions of ecological response. Generally speaking, all methods use hydraulic models to convert the outputs of water resource models into predictions of habitat available under different flow scenarios, and from here they make predictions of ecological response. Such an approach implicitly invokes the *field of dreams hypothesis*—the assumption that restoration of habitat is sufficient to cause an ecological response (Palmer et al., 1997). Though this assumption underpins the vast majority of ecological restoration projects, there is very little evidence for the assumption (Sudduth et al., 2011). An example where habitat alone may not be sufficient is an environmental flow scenario where winter flows are improved to provide deep-water pools as habitat for large-bodied native fish species. However, if there is insufficient primary and secondary production in the system, then there will be insufficient resources to support improved native fish populations and the habitat restoration will be ineffective.

This criticism of environmental flows assessment approaches is not confined to Australia. The majority of habitat-based (e.g., Thomas & Bovee, 1993) and even holistic (sensu Tharme, 2003) environmental flows assessment methods make an implicit or explicit assumption that the primary ecological benefits of environmental flows are realized through the provision of improved habitat.

RELIANCE ON EXPERT OPINION FOR PREDICTING ECOLOGICAL RESPONSES

The methods used to derive the relationship between hydraulic habitat and ecological response are generally quite vaguely specified. In earlier work (Stewardson & Webb, 2010), we contended that expert opinion was overwhelmingly being used to specify flow-response relationships.

An expert opinion may be very good, since generally the members of environmental flows assessment panels are specialists with long histories and profound expertise in the area which they are contributing to the assessment. Their predictions might be very sound. However, expert opinions suffer from two main weaknesses. First, they are rarely transparent. The expert opinion is the implicit synthesis of the years of experience, knowledge of the system being studied, and knowledge of how other systems have responded. It is difficult or impossible to pick from this the primary pieces of information used to inform the prediction. Second, because of the implicit and expert-centered synthesis of information and experience, these predictions are not repeatable. Faced with the same information, a different expert is likely to arrive at a different opinion. Thus, environmental flow recommendations depend to some extent on the makeup of the expert panel.

However, this situation is improving. For example, in the updated version of the FLOWS method in Victoria, the environmental flows assessment panel is required to record what information was used to make a prediction of ecological response to different flow events, and also to specify their level of confidence in the predictions (DEPI, 2013). This is an improvement over the previous version of the method, which had no such stipulation (DNRE, 2002). Such a requirement improves transparency of the environmental flows assessment, although it may not improve repeatability.

In general, however, it is still largely the case that methods used to make predictions of ecological responses to changing availability of hydraulic habitat are loosely specified and rely heavily on the particular specialist on the environmental flows assessment panel.

IMPROVING QUALITY OF ECOLOGICAL PREDICTIONS

The primary reason given for the reliance of environmental flows assessment panels on expert opinion is that there is a lack of sufficient data to derive empirical relationships (Stewardson & Webb, 2010). This will most likely be true for data directly collected in a river undergoing an environmental flows assessment, as few individual rivers have been studied in sufficient depth to draw out these kinds of relationships. However, if data are drawn from a wider range of sources, then it is often the case that the claim of insufficient data no longer stands up.

The Ecological Limits of Hydrological Alteration (ELOHA; Poff et al., 2010) framework is a holistic method for environmental flows assessment designed for implementation in areas where there is not a large amount of empirical information to inform flow recommendations. It has attracted adherents since its release and case studies are now starting to find their way into the grey (Arthington et al., 2012) and peer-reviewed (McManamay et al., 2013) literature.

ELOHA is mentioned here because it has a strong emphasis on developing flow-response relationships to underpin the derivation of flow recommendations. Although the framework does not actively specify how such relationships should be formed, emerging case studies are often locating sufficient relevant data to develop empirical relationships (Arthington et al., 2012; Kendy et al., 2012; McManamay et al., 2013; Wilding & Poff, 2009). The success of ELOHA case studies in developing empirical relationships between changes in flow regimes and ecological condition points towards a brighter future for data-driven environmental flows assessments.

At the same time as more data become available, there is an increasing array of statistical modeling tools that can be used to derive flow-response relationships to inform environmental flows assessments (Webb et al., 2017). These range in complexity from simple linear models that nearly any practitioner could undertake, through to machine-learning based approaches such as artificial neural networks that are, as yet, the domain of specialized research efforts. Webb et al. (2017) speculate that the availability of these more advanced tools might necessitate a new role on environmental flows assessment panels—an ecological modeling specialist—to take advantage of the latest developments in this space.

However, herein lies a potential problem. The collection and analysis of substantial data sets as part of an environmental flows assessment would result in an increased work load and hence expense. This increase would be magnified if an assessment employs one of the more specialized methods for flow-response modeling via the services of an ecological modeling specialist. Existing methods for environmental flows assessment across the Murray-Darling Basin enjoy broad support among stakeholders and it may be difficult to prosecute the argument for increased investment in this phase.

Previously it was noted that, while expert predictions are widely practiced and have proven to be an effective approach, they can lack transparency and are not repeatable. They certainly do not conform to scholarly definitions of *evidence-based practice*, where they are consistently ranked as the least reliable form of evidence for informing decision making (Burns et al., 2011). It is conceivable that an environmental flows program could be challenged in court (such action was mentioned during the early days of the *Guide to the Basin Plan*). Under those circumstances, the evidentiary weaknesses of expert-based assessments would be exposed. Such a critical incident would be a sad pathway towards improved use of empirical evidence and models for informing environmental flows assessments.

INTEGRATING ENVIRONMENTAL FLOWS IN WATER RESOURCE MODELING
WATER RESOURCE MODELS

Environmental flow demands can be incorporated into a water resource plan to assist in water resource planning and assessing compliance of environmental water delivery (Fig. 1). There has been a long evolution and development process for water resource models of the Murray-Darling Basin, with initial models developed in the 1960s (Close, 2010). These models have undergone many iterations to address water resource management or water sharing changes (such as the setting of the Cap), new system objectives (such as water quality and environmental flows) and new compliance and assessment requirements (such as water accounting).

The Murray-Darling Basin is covered by a number of separate water resource models, managed by different custodians. Broadly, there are individual tributary models that feed into a monthly simulation model (MSM) of the River Murray, and subsequently the daily routing model BIGMOD. Each model has been developed and maintained by the relevant jurisdiction that manages the resource. Queensland and New South Wales have favored IQQM as the modeling platform, Victoria has used REALM as the modeling platform and the Murray-Darling Basin Authority uses MSM-BIGMOD (Table 3).

There is increasing complexity in the interactions between river valleys with policies, such as water trade and environmental flow delivery, affecting a number of basins simultaneously. The discrete nature of the available water resource models, with varying modeling platforms and assumptions, complicates efforts to understand how the system behaves and responds basinwide. This is particularly relevant for environmental water delivery, where large volumes of water are held across different

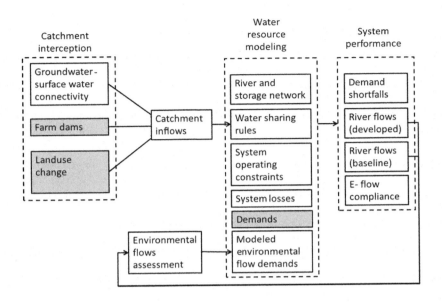

FIG. 1

The role of water resource modeling in environmental flows assessments, planning and review. The shaded cells represent anthropogenic impacts on the system. The water resource model can be run with specific scenarios to define developed and baseline river flows and environmental flows assessment sites. Once an environmental flows assessment study has been completed, the environmental water requirements can be translated into model inputs to assist in long-term water resource planning and the assessment of environmental water delivery compliance (ACT Government, 2013; Arthington et al., 1998; Brizga, 2007; Cottingham et al., 2005; DELWP, 2015; Favier et al., 2000; Shields & Good, 2002; WaterSelect, 2015).

catchments and release decisions are based on outcomes in multiple downstream catchments. The size and complexity of the water resource modeling to inform the Basin Plan demonstrates this (MDBA, 2012).

In recognition of this, an extensive research and implementation program has seen the development of a new modeling tool, SOURCE (eWater, 2013). SOURCE aims to provide a nationally consistent modeling base that is scalable and customizable, and most importantly, better able to represent the complex policy and management settings that now exist, and the range of water resource demands (including the environment). The development of SOURCE has involved 45 different partners across public and private industry and academia, development funding of $9 million (eWater, 2007) and adoption funding of $7.8 million (eWater, 2012). The eWater Cooperative Research Centre (CRC) was initially funded in 2006, and this has now been a 10-year process to transition to the new modeling platform. Importantly, the relevant jurisdictions and custodians have committed to implementing the SOURCE model across catchments of the Murray-Darling Basin. These models are now well underway, with many in the review stage. The SOURCE tools have deliberately been designed to be customizable, to ensure they remain relevant as new policy settings evolve into the future.

Table 3 River System Models in the Murray-Darling Basin

Model	Description	Rivers Modeled
IQQM	Integrated Quantity-Quality Model: hydrologic modeling tool developed by the NSW Government for use in planning and evaluating water resource management policies.	Paroo, Warrego, Condamine-Balonne (Upper, Mid, Lower), Nebine, Moonie, Border Rivers, Gwydir, Peel, Namoi, Castlereagh, Macquarie, Marthaguy, Bogan, Lachlan, Murrumbidgee, Barwon-Darling
REALM	Resource Allocation Model: water supply system simulation tool package for modeling water supply systems configured as a network of nodes and carriers representing reservoirs, demand centers, waterways, pipes, etc. Further details can be found on the DELWP website (DELWP, 2015)	Ovens (Upper, Lower), Goulburn, Wimmera, Avoca, ACT water supply.
MSM-BigMod	Murray Simulation Model and the daily forecasting model BigMod: purpose-built by the Murray-Darling Basin Commission to manage the Murray River system. MSM is a monthly model that includes the complex Murray accounting rules. The outputs from MSM form the inputs to BigMod, which is the daily routing engine that simulates the movement of water	Murray
WaterCress	Water Community Resource Evaluation and Simulation System: PC-based water management platform incorporating generic and specific hydrological models and functionalities for use in assessing water resources and designing and evaluating water management systems. Further details can be found on the WaterCress website (WaterSelect, 2015)	Eastern Mt Lofty Ranges (six separate catchments)
SMHS	Snowy Mountains Hydroelectric Scheme model: purpose-built by Snowy Hydro Ltd to guide the planning and operation of the SMHS.	Snowy Mountains Hydroelectric Scheme

Adapted from CSIRO. (2007). Overview of project methods. A report to the Australian Government from the CSIRO Murray-Darling Basin Sustainable Yields Project. Canberra, ACT, Australia, p. 7.

Many of the initial water resource models were developed with monthly time steps for the purposes of understanding system reliability. Environmental flows assessment requires an understanding of daily flow patterns to assess dry and wet spells, and to model the effects of short duration flow events. While some models either existed or have since been converted to a daily time step, other models require postprocessing to disaggregate outputs to daily flows. The methods for this vary for individual systems, but are broadly based on daily patterns from gauged flows (developed scenarios) or transposed flows (for reference conditions).

REPRESENTATION OF ENVIRONMENTAL WATER

Water resource models were for the most part developed prior to the widespread acceptance of the concept of environmental flows in water resources management. This has required retrofitting of existing models to accommodate the concept of environmental flows. There are two main ways that environmental water requirements can be incorporated within water resource models. The first is to preprocess an environmental demand time-series and link this to a particular location in the river network. The second is to create a series of rules that define the environmental demand.

The current water resource models (based in REALM, IQQM and MSM-BigMod) all rely on a preprocessed time series of environmental flow demand at a particular location. This is usually calculated based on the environmental flow recommendations for the frequency, timing and magnitude of particular flow events and the reference series occurrence of such events in the system. Where an environmental flow is rules based (i.e., it is provided as a flow target that a storage operator must comply with), it is entered into the model as a node or arc with a high penalty if not achieved. Where environmental water is provided through held environmental entitlements, a separate environmental water account is required within the model to meet a downstream environmental demand node. The challenge of this approach is that for every different climate scenario considered, a preprocessing step must occur. This complicates the process of assessing management outcomes across a range of scenarios.

The SOURCE model has approached the inclusion of environmental demands differently, responding to the additional flexibility required when managing environmental water entitlements. Environmental Demand Models (EDMs) can be used to represent environmental flow requirements for both instream and floodplains. The EDMs are effectively a series of rules that represent different elements of the environmental flow requirements (including different timing needs), and can be in the form of (eWater, 2013):

- *Flood or Fresh*—specifies a flood fresh, usually associated with a recruitment event such as to trigger fish movement or water floodplain vegetation,
- *Flow pattern*—specifies a pattern of flow used to define multipeak events,
- *Minimum flow*—specifies a minimum flow usually applied to maintain minimum habitat requirements,
- *Translucency*—specifies the flow requirements in terms of some other time series, usually the release from a dam based on the inflow to the dam.

Individual flow rules can be grouped or prioritized; for example, one flow component may only be required if another can also be met. This allows more complex representation of environmental flow requirements, and aligns more closely to the ecological information that informs the use of environmental entitlements. When deciding to release environmental water, SOURCE is programmed to only attempt an environmental release if a successful event is likely to be achieved (eWater, 2013). Again, this better mimics the way environmental water managers make release decisions.

While SOURCE goes a long way to addressing many of the current challenges in modeling environmental water demands, the reality is that it has not yet been widely used in planning processes. The real benefits (and any limitations) of SOURCE will become apparent as the various jurisdictions move in earnest to the new modeling platform. That said, SOURCE does provide the flexibility for users to build specific plug-ins to meet new management needs or challenges as they arise. This will allow the water resource models to more rapidly move with new environmental water management challenges as they occur.

CONCLUSION

Australia has a deserved international reputation of being at the forefront of the implementation of environmental flows. This includes the use of hydrological approaches for regional assessments of flow stress to identify catchments which are flow stressed, may require a Cap of sustainable diversion limit, and warrant a further, more detailed, environmental flows assessment. Holistic approaches have been used for these more detailed studies with several different methods developed for use in different jurisdictions, but all follow the same general principles. Methods for making the links between hydraulics and ecological response are probably the weakest part of the environmental flows assessment process, but there are signs of improvement (e.g., with the updated FLOWS method). There are now enough data, and enough modeling approaches, that we could be using more robust methods for deriving flow ecology relationships in environmental flows assessments in the Murray-Darling Basin. However, to do so will require additional resources for the process, and a change in attitude necessary to justify this extra investment.

REFERENCES

Acreman, M., Arthington, A. H., Colloff, M. J., Couch, C., Crossman, N. D., Dyer, F., et al. (2014). Environmental flows for natural, hybrid, and novel riverine ecosystems in a changing world. *Frontiers in Ecology and the Environment, 12*, 466–473.

ACT Government. (2013). *Water resources environmental flow guidelines*. Canberra, ACT, Australia: Minister for Environment and Sustainable Development.

Arthington, A. H., Brizga, S. O., & Kennard, M. J. (1998). *Comparative evaluation of environmental flow assessment techniques: Best practice framework*. Canberra, ACT, Australia: Land and Water Research and Development Corporation, Occasional Paper No 25/98.

Arthington, A. H., Bycroft, B., & Conrick, D. (1992). *Environmental study Barker-Barambah Creek*. Brisbane, QLD, Australia: Water Resources Commission, Department of Primary Industries, and Centre for Catchment and In-stream Research, Griffith University.

Arthington, A. H., Mackay, S. J., James, C. S., Rolls, R. J., Sternberg, D., Barnes, A., et al. (2012). *Ecological-limits-of-hydrologic-alteration: a test of the ELOHA framework in south-east Queensland*. Canberra, ACT, Australia: National Water Commission.

Blöschl, G., Sivapalan, M., Wegener, T., Viglione, A., & Savenije, H. (2013). *Runoff prediction in ungauged basins: Synthesis across processes, places and scales*. Cambridge, UK: Cambridge University Press.

Brizga, S. O. (2007). The first decade of water resource planning in Queensland: What have we learned about environmental flows? In: *Proceedings of the 5th Australian stream management conference. Australian rivers: Making a difference* (pp. 37–42). Thurgoona, NSW, Australia: Charles Sturt University.

Brown, A. E., Zhang, L., McMahon, T. A., Western, A. W., & Vertessy, R. A. (2005). A review of paired catchment studies for determining changes in water yield resulting from alterations in vegetation. *Journal of Hydrology, 310*, 28–61.

Burns, P. B., Rohrich, R. J., & Chung, K. C. (2011). The levels of evidence and their role in evidence-based medicine. *Plastic and Reconstructive Surgery, 128,* 305.

Close, A. (2010). *Presentation: River modelling for the Murray-Darling Basin.* Canberra, ACT, Australia: Murray Darling Basin Authority.

Cottingham, P., Quinn, G. P., King, A., Norris, R. C., Chessman, B. C., & Marshall, C. (2005). *Environmental flows monitoring and assessment framework.* Canberra, ACT.

Cottingham, P., Stewardson, M., Roberts, J., Oliver, R., Crook, D., Hillman, T., et al. (2010). Ecosystem response modeling in the Goulburn River: How much water is too much. In N. Saintilan & I. Overton (Eds.), *Ecosystem response modelling in the Murray-Darling Basin* (pp. 391–409). Canberra, Australia: CSIRO Publishing.

Davies, P., Stewardson, M., Hillman, T., Roberts, J. R., & Thoms, M. C. (2012). *Sustainable Rivers Audit 2: The ecological health of rivers in the Murray–Darling Basin at the end of the Millennium Drought (2008–2010). Vol. 1.* Canberra, ACT, Australia: Prepared by the Independent Sustainable Rivers Audit Group for the Murray Darling Basin (ISRAG).

DELWP. (2015). *REsource ALlocation Model (REALM) webpage.* Department of Environment, Land, Water and Planning. Available at: http://www.depi.vic.gov.au/water/water-resource-reporting/surface-water-modelling/resource-allocation-model-realm Accessed 02.11.16.

DEPI. (2013). *FLOWS—A method for determining environmental water requirements in Victoria: Edition 2.* Melbourne, Victoria, Australia: Department of Environment and Primary Industries.

DNRE. (2002). *The FLOWS method: A method for determining environmental water requirements in Victoria.* Consulting report by Sinclair Knights Mertz, and the CRC for Freshwater EcologyVictorian Department of Natural Resources and Environment.

eWater. (2007). *National water commission funding.* eWater. Available at http://ewater.org.au/news/media/?news=10 Accessed 02.11.16.

eWater. (2012). *Government funding for Source as national platform.* Available at http://ewater.org.au/news/media/?news=207 Accessed 02.11.16.

eWater. (2013). *Source user guide.* Canberra, ACT, Australia: eWater.

Favier, D., Rixon, S., & Scholz, G. (2000). *A river management plan for the Wakefield Catchment.* Adelaide, SA, Australia: Environmental Protection Agency.

Gippel, C. J., & Stewardson, M. J. (1995). Development of an environmental flow management strategy for the Thomson River, Victoria, Australia. *Regulated Rivers Research & Management, 10,* 121–135.

Hart, B. T. (2015a). The Australian Murray-Darling Basin Plan: Challenges in its implementation (part 1). *International Journal of Water Resources Development.* http://dx.doi.org/10.1080/07900627.2015.1083847.

Hart, B. T. (2015b). The Australian Murray-Darling Basin Plan: Challenges in its implementation (part 2). *International Journal of Water Resources Development.* http://dx.doi.org/10.1080/07900627.2015.1084494.

Hart, B. T. (2016). The Australian Murray-Darling Basin Plan: Factors leading to its successful development. *Ecohydrology and Hydrobiology.* http://dx.doi.org/10.1016/j.ecohyd.2016.09.002.

Horne, A., O'Donnell, E., & Tharme, R. (2017a). Mechanisms to allocate environmental water. In A. Horne, J. A. Webb, M. Stewardson, B. Richter, & M. Acreman (Eds.), *Water for the Environment: From policy and science to implementation and management.* Cambridge, MA: Elsevier.

Horne, A. C., Webb, J. A., Stewardson, M. J., Richter, B. D., & Acreman, M. (Eds.), (2017b). *Water for the environment: From policy and science to implementation and management.* Cambridge, MA: Elsevier.

Kendy, E., Apse, C., & Blann, K. (2012). *A practical guide to environmental flows for policy and planning.* Arlington, VA: The Nature Conservancy.

Lloyd, L. N., Anderson, B. G., Cooling, M., Gippel, C. J., Pope, A. J., & Sherwood, J. E. (2012). *Estuary environmental flows assessment methodology for Victoria: Report to the Department of Sustainability and Environment, Melbourne Water and Corangamite CMA.* Colac, Victoria, Australia: Lloyd Environmental Pty Ltd.

Lowe, L., & Nathan, R. J. (2006). Use of similarity criteria for transposing gauged streamflows to ungauged locations. *Australian Journal of Water Resources, 10*, 161–170.

McManamay, R. A., Orth, D. J., Dolloff, C. A., & Mathews, D. C. (2013). Application of the ELOHA framework to regulated rivers in the Upper Tennessee River Basin: A case study. *Environmental Management, 51*, 1210–1235.

MDBA. (2011). *The proposed 'environmentally sustainable level of take' for surface water of the Murray-Darling Basin: Methods and outcomes.* Canberra, ACT, Australia: Murray-Darling Basin Authority.

MDBA. (2012). *Hydrologic modelling to inform the proposed Basin Plan: Methods and results.* MDBA Publication no: 17/12 Canberra, ACT, Australia: Murray-Darling Basin Authority.

NWC. (2014). *Australian environmental water management: 2014 Review.* Canberra, ACT, Australia: National Water Commission.

Palmer, M. A., Hakenkamp, C. C., & Nelson-Baker, K. (1997). Ecological heterogeneity in streams: Why variance matters. *Journal of the North American Benthological Society, 16*, 189–202.

Poff, N. L., Richter, B. D., Arthington, A. H., Bunn, S. E., Naiman, R. J., Kendy, E., et al. (2010). The ecological limits of hydrologic alteration (ELOHA): A new framework for developing regional environmental flow standards. *Freshwater Biology, 55*, 147–170.

Poff, L. N., Tharme, R., & Arthington, A. (2017). Evolution of environmental flows assessment science, principles and methodologies. In A. Horne, J. A. Webb, M. J. Stewardson, B. Richter, & M. Acreman (Eds.), *Water for the Environment: From policy and science to implementation and management.* Cambridge, MA: Elsevier.

Richter, B., Baumgartner, J., Wigington, R., & Braun, D. (1997). How much water does a river need? *Freshwater Biology, 37*, 231–249.

Saintilan, N., & Overton, I. (Eds.), (2010). *Ecosystem response modelling in the Murray-Darling Basin.* Canberra, ACT, Australia: CSIRO Publishing.

Shields, J., & Good, R. (2002). Environmental water in a regulated river system: The Murrumbidgee River planning approach to the determination of environmental needs. *Water Science and Technology, 45*, 241–249.

SKM. (2005). *Development and application of a flow stressed ranking procedure.* Melbourne, Victoria, Australia: Sinclair Knight Merz.

Stewardson, M. J., Acreman, M., Costelloe, J., Fletcher, T., Fowler, K., Horne, A., et al. (2017). Understanding hydrological alteration. In A. Horne, J. A. Webb, M. J. Stewardson, B. Richter, & M. Acreman (Eds.), *Water for the Environment: From policy and science to implementation and management.* Cambridge, MA: Elsevier.

Stewardson, M. J., & Gippel, C. J. (2003). Incorporating flow variability into environmental flow regimes using the flow events method. *River Research and Applications, 19*, 459–472.

Stewardson, M. J., & Webb, J. A. (2010). Modelling ecological responses to flow alteration: Making the most of existing data and knowledge. In N. Saintilan & I. Overton (Eds.), *Ecosystem response modelling in the Murray-Darling Basin* (pp. 37–49). Melbourne, Victoria, Australia: CSIRO Publishing.

Sudduth, E. B., Hassett, B. A., Cada, P., & Bernhardt, E. S. (2011). Testing the field of dreams hypothesis: Functional responses to urbanization and restoration in stream ecosystems. *Ecological Applications, 21*, 1972–1988.

Swirepik, J., Burns, I., Dyer, F., Neave, I., O'Brien, M., Pryde, G., et al. (2016). Establishing environmental water requirements for the Murray–Darling Basin, Australia's largest developed river system. *River Research and Applications, 32*, 1153–1165.

Tharme, R. E. (2003). A global perspective on environmental flow assessment: Emerging trends in the development and application of environmental flow methodologies for rivers. *River Research and Applications, 19*, 397–441.

Thomas, J. A., & Bovee, K. D. (1993). Application and testing of a procedure to evaluate transferability of habitat suitability criteria. *Regulated Rivers Research & Management, 8*, 285–294.

WaterSelect. (2015). *The WaterCress Model website*. Water Select. Available at http://www.waterselect.com.au/watercress/watercress.html Accessed 02.11.16.

Webb, J. A., Arthington, A. H., & Olden, J. D. (2017). Models of ecological responses to flow regime change to inform environmental flows assessments. In A. C. Horne, J. A. Webb, M. J. Stewardson, B. D. Richter, & M. Acreman (Eds.), *Water for the Environment: From policy and science to implementation and management*. Cambridge, MA: Elsevier.

Wilding, T. K., & Poff, N. L. (2009). *Flow-ecology relationships for the watershed flow evaluation tool*. Denver, CO: Colorado Water Conservation Board.

FURTHER READING

CSIRO. (2007). *Overview of project methods. A report to the Australian Government from the CSIRO Murray-Darling Basin Sustainable Yields Project*. Canberra, ACT, Australia.

THE ROLE OF ECONOMIC ANALYSIS IN WATER RESOURCE MANAGEMENT—THE MURRAY-DARLING EXPERIENCE

D. James
Consultant, Whale Beach, NSW

CHAPTER OUTLINE

Decision Making in Water Resources Policy and Management. http://dx.doi.org/10.1016/B978-0-12-810523-8.00009-4

INTRODUCTION

This chapter discusses the economic analysis conducted in the triple bottom line (TBL) decision-making process to develop the Murray-Darling Basin Plan, MDBA (2011a, 2011b). The main aim of the Basin Plan is to rebalance the overallocated systems to achieve "a healthy, working Murray-Darling Basin with productive and resilient industries, confident communities and healthy and resilient rivers, wetlands and floodplains" (MDBA, 2015).

Background to the socioeconomic and ecological characteristics of the Basin, the institutional arrangements, and the recent major policy reforms—the Water Act 2007 (Aust Govt, 2014) and the Basin Plan (Aust Govt, 2012)—can be found in Chapters 1 and 13 of this book.

This chapter discusses applications of economic analysis in developing the Basin Plan and in monitoring and evaluating its progress. A vast amount of literature has been written on this subject, and only an overview can be presented here. Comprehensive reviews have been undertaken by the BDA Group (2010), Connell and Grafton (2011), KPMG (2011), and Wittwer (2012), among others. Two key documents prepared by the Murray-Darling Basin Authority summarize much of the economic research underpinning the MDBA (2011a, 2011b). This chapter draws extensively on those two reports.

ECONOMIC ANALYSIS FOR THE BASIN PLAN: TOOLS AND TECHNIQUES

The Basin Plan (Aust Govt, 2012) has several key mechanisms designed to give effect to the objects of the Water Act (see Chapter 13). Key roles for economic analysis in relation to the Basin Plan have been to

- Monitor the socioeconomic effects of the Basin Plan, distinguishing the effects of the Plan from other factors leading to socioeconomic changes in subregions and local communities in the Basin.
- Assess the behavior of markets for tradeable entitlements and water allocations in terms of the efficiency of water use, the allocation of water to the highest valued uses, reliability of supply and implications for water-dependent industries and communities.
- Estimate the economic costs and benefits and impacts attributable to the Plan, in particular for buybacks and investments in infrastructure projects aimed at water recovery and subsequent reallocation to the environment.

Challenges faced in choosing appropriate tools and conducting economic analysis include

- The scale (economic and spatial) at which analysis is required (farm, local community, subregion, region, state, or national).
- Compilation of datasets and modeling capability designed to establish economic profiles, analyze general trends in economic activity and identify the drivers producing change at predefined socioeconomic scales.
- Determining "baselines" against which past or proposed water resource management actions may be assessed.
- Identifying the impacts of water reform on different socioeconomic groups within the Basin.
- Incorporating nonmarket values (predominantly environmental) in the analysis.

- Predicting, monitoring and evaluating the effects of policies, plans and management actions, in retrospect or prospectively.
- Combining economic information with information from other disciplines for the purpose of assessing the efficiency of resource use, trade-offs and optimization of social, economic and environmental outcomes.
- Communicating the methods and results of economic analysis to the community.

In preparing the Basin Plan, the MDBA commissioned 22 social and economic reports. Other noncommissioned reports were also taken into consideration. The main economic tools, models, and techniques applied in formulating the Basin Plan are discussed in the following sections.

BENEFIT-COST ANALYSIS

In assessing the benefits of the Basin Plan, MDBA considered it important to express them as far as possible in monetary terms to facilitate a comparison of benefits with the economic costs. The main tool of analysis incorporating such values is benefit-cost analysis (BCA).

BCA originated in the United States *Flood Control Act 1936* as a means of evaluating flood control projects, which required that "the benefits, to whomsoever they may accrue, are in excess of the costs." It is based on the principles and ethical judgments of neoclassical welfare economics and focuses on the efficiency with which resources are used to support the wellbeing of the community. There are now numerous texts containing guidelines for BCA (ADB, 1997; Boardman et al., 2011; Commonwealth of Australia, 2006; James & Francisco, 2015; OECD, 1994; Pearce & Nash, 1981).

In a BCA the preferences and values of individuals are expressed in terms of their willingness to pay (WTP). WTP values exceeding prices actually paid are known as "consumer surplus" and are included as benefits. Net benefits accruing to producers, defined as "producer surplus," are also included in BCA evaluations.

Benefits and costs in a BCA are typically identified and valued in terms of the incremental change between alternative scenarios: a base-case or business-as-usual scenario, and the relevant policy or management scenario. Scenarios are compared in terms of the net present value, benefit cost ratio, and internal rate of return. Secondary impacts, such as economy-wide flow-on effects on output, employment or income, are usually not included in a BCA.

Two important value judgments are required in a BCA. The first is the assumption that all values can be assessed within the context of the prevailing distribution of income. The second relates to the weights that are applied to gains and losses at different points in future time through the technique of discounting. There is ongoing debate about how to choose appropriate discount rates, especially with regard to the welfare of future generations. The latest official BCA guidelines issued by the UK Treasury prescribe the use of declining discount rates to address this issue, depending on how far into the future the analysis extends (HM Treasury, 2003).

Stochastic processes in the environment, such as variations in rainfall over time and their effects on benefits and costs, can be handled using techniques such as Monte Carlo simulation, use of payoff matrices, risk preferences, and expected value analysis (ADB, 1997; Dixon et al., 1989; Pearce & Nash, 1981).

A report by the Centre for International Economics for MDBA provided assessments of the economic benefits and costs of different water-recovery targets of 3000, 3500, and 4000 GLy^{-1} incorporating values associated with irrigated production and values of the environment (CIE, 2011). A discount rate of 7% was adopted. The study produced estimates of losses in producer surplus of $1–4.5 billion with a water-recovery scenario of 3000 GLy^{-1} and a range of $1.3–7.3 billion with a scenario of 4000 GLy^{-1}. Nonuse environmental values (see later) ranged from $3 to $5 billion, increasing to more than $8.5 billion when including environmental improvement benefits for the Coorong. Recreation values were estimated at $500–600 million. The report commented that nonuse environmental values in the analysis were largely attributable to respondents in capital cities in the relevant states.

A review of the CIE study by Frontier Economics (2011) noted that the study was based on average rather than incremental values, simplistic ecological response functions and did not consider local and regional costs and benefits.

MDBA concluded, on the basis of extensive modeling results, that the benefits of the Basin Plan can be expected to exceed the costs. However, the "costs" underpinning this statement were interpreted as the estimated loss in gross regional product (GRP) associated with the Basin Plan, comprising only 1% of GRP for the Basin as whole, rather than a cost in terms of resource commitment.

While the conceptual basis of BCA offers sound principles on which to evaluate the Basin Plan and any of its variants, the tool can be considered most useful for evaluations at the project or program scale, such as investments in infrastructure, productivity improvements in irrigated agriculture, and management of particular environmental assets or sites.

In the case of the Basin Plan, as with any policy initiative, there are gainers and losers. From a social equity perspective, who gains and who loses is a critical policy determinant. Considerations of "fairness" may well override principles of resource use efficiency.

VALUING THE ENVIRONMENT

It has become increasingly apparent that goods and services yielded by the environment contribute significantly to community wellbeing. There is explicit recognition of this in the MDBA framework for evaluating progress of the Basin Plan (MDBA, 2014).

Assessing the benefits provided by natural ecosystems is now the focus of numerous studies and guidelines. The Millennium Ecosystem Services Assessment represented a global effort to assess the value of ecosystems and their contribution to human wellbeing (MEA, 2005). The MEA framework itself, focusing on the relationships between ecosystems and community wellbeing, can be adapted to incorporate scientific judgments and community preferences (Maynard et al., 2015).

Worldwide interest in ecosystem services was stirred with a pioneering study aimed at estimating economic values of ecosystem services globally (Costanza et al., 1997). Extensive work on valuing ecosystem services has since been conducted under The Ecosystems and Biodiversity (TEEB) project (Sukhdev et al., 2010; TEEB, 2010). Attention has recently been paid to ecosystem services specifically associated with water-resource systems (Martin-Ortega et al., 2015; Vermaat et al., 2016).

As part of the planning process for the Basin Plan, a report was prepared for MDBA by CSIRO (Crossman et al., 2011) based on the MEA and the TEEB project. The CSIRO study highlighted declining environmental quality in the Basin and the need to develop and apply an ecosystem services framework related to the Basin's water-resource system.

TOTAL ECONOMIC VALUE

A widely accepted categorization by economists of the values associated with the environment is total economic value (TEV; Pearce & Turner, 1989). TEV itself can be subdivided into separate categories: use and nonuse values. *Use values* refer to direct uses of the environment and may be consumptive, such as diversions of in-stream flows for irrigation or town water supplies, or nonconsumptive, for example, nature-based tourism or water-dependent recreation. The environment may also perform functions that indirectly benefit the community, such as the assimilation of wastes, the provision of shade and shelter, or scenic amenity.

Nonuse values occur mainly outside the market system, have no explicit prices and are determined mostly in implicit or "shadow price" form. They are usually categorized as "existence" values (the importance of knowing that certain species or ecosystems continue to exist); bequest values (the value of passing on environmental assets to future generations); vicarious use values (sharing in absentia the environmental experiences of others); option values (having the opportunity to interact with the environment at some future time of choosing); and quasioption values (the value of delaying exploitation or use of the environment in order to learn more about its qualities or behavior for the purpose of adaptive management).

TECHNIQUES FOR VALUING THE ENVIRONMENT

Techniques applied by economists to derive environmental values date back to the 1970s (Freeman, 1979; Sinden & Worrell, 1979) and are now exhaustively documented in the academic literature (ADB, 1996; Bateman et al., 2003; Bennett, 2011; Freeman, 1993; Haab & McConnell, 2002; Hanley & Barbier, 2009; Smith, 1996). They include market-based methods, suitable for valuing the effects of environmental change in terms of consumer and producer surplus; surrogate market techniques such as travel cost modeling to value consumer surplus associated with free visits to natural areas, or property value differentials reflecting the quality of the environment; and stated preference models involving surveys that ask respondents hypothetically what they would be willing to pay for particular environmental attributes, functions or scenarios. Stated preference modeling is advocated as the only technique capable of deriving nonuse values.

STATED PREFERENCE MODELS

There are basically two kinds of stated preference models. The first comprises *contingent valuation*, where only one environmental scenario is postulated (Bjornstad & Kahn, 1996; Mitchell & Carson, 1989). A NOAA panel of Nobel laureates in economics (Arrow et al., 1993), among others, has recommended guidelines for sound practice in contingent valuation.

The second stated preference method is *choice modeling*, in which respondents are asked about their preferences for different combinations of environmental attributes together with a postulated cost or price (Bennett & Blamey, 2001; Louviere et al., 2000).

BENEFITS TRANSFER TECHNIQUE

Where it is not feasible due to limitations of data, time, skills, or research resources to work with primary data and valuation techniques, a fall-back is to apply the "benefits transfer" technique, in which values are adopted that have already been derived for similar sites or circumstances.

DIRECT USES OF THE BASIN ENVIRONMENT

Direct uses of the water-resource system in the Murray-Darling, apart from irrigated agriculture, include commercial and recreational fishing, and tourism.

Assessments of tourism have been made by Tourism Research Australia of the effects of drought in the Murray region extending from 1999 to 2008 (DRET, 2010). The research indicates that in 2008, as a consequence of drought, direct tourism expenditure in the Murray region was $69.9 million lower (in 2008 dollars) compared with a no-drought situation. Over the 14-year period covered by the study, the total reduction in spending was estimated at $351.4 million. Flow-on impacts to local communities were estimated to result in a reduction in GRP of $461 million and a loss of 596 full-time jobs.

Work was conducted for MDBA by Deloitte Access Economics to estimate fisheries benefits of a $2750 \, \text{GLy}^{-1}$ water-recovery scenario (DAE, 2012). The research indicated that by 2020, provided the anticipated outcomes of the Environmental Watering Plan are realized, there would be an increase in consumer surplus for recreational fishing of $9.1 million annually and an increase in producer surplus for commercial fisheries of $254,000 annually. It was assumed, however, that there would be effective conservation of native fish and aquatic ecosystems for these benefits to be realized.

ENVIRONMENTAL VALUES IN THE BASIN PLAN

In formulating the Basin Plan, the main source of information on economic environmental values was a study by Morrison and Hatton McDonald (2010). This study produced indicative estimates of expected improvements in the environment resulting from the Basin Plan relative to a base case, encompassing the benefits of water-related recreation, native vegetation, native fish, waterbird breeding and waterbirds and other species. The approach involved reviewing the values derived in existing studies, applying the benefits transfer technique, and assigning them to 19 subregions in the Basin. Travel cost modeling was the technique mostly used in deriving recreation benefits, whereas stated preference modeling, namely contingent valuation and choice modeling, produced nonuse values incorporated in the study.

Nonuse values were presented for each region for a 1% increase in native vegetation, 1% increase in native fish populations, a 1-year increase in the frequency of waterbird breeding and a unit increase in the number of waterbirds and other species. Environmental values were estimated as present values. For many of the source studies a discount rate of 28% was assumed. Aggregate estimates for the Basin as a whole were not produced, but these were calculated subsequently by MDBA, using different assumptions regarding the weights attached to different States (MDBA, 2011b). The calculations suggest that aggregate environmental benefits range from $132 to $187 million for a 1% increase in native vegetation, $95 to $117 million for a 1% increase in native fish populations, $563 to $770 million for a 1-year increase in the frequency of waterbird breeding and $44 to $77 million for a unit increase in the number of waterbirds and other species present.

LIMITATIONS OF MONETARY ENVIRONMENTAL VALUATION

Accounting for monetized environmental values in the evaluation of policies, programs, and projects is challenging. To some extent the reasons are philosophical or ethical. Different groups in society have different perceptions of the environment and place different weights on its importance relative to other things affecting their wellbeing. It is often difficult for governments to achieve an appropriate balance of outcomes where environmental values and ordinary commercial values are in apparent conflict with each other.

There are serious limitations, in particular to stated preference valuation methods. Care must be taken in designing an effective questionnaire, as various potential biases may arise (Mitchell & Carson, 1989). One challenge is specifying the relevant scenario or attributes to be valued, avoiding the "embedding effect" (Kahneman & Knetsch, 1992a, 1992b). Biases may also arise with choice of a payment vehicle: for example, a tax, environmental charge, contribution to an environmental fund or an increased indirect cost such as higher electricity bills. People's WTP, including protest responses, can be profoundly influenced by the payment vehicle that is presented.

More generally, there may be strong community objections to any endeavors to place monetary values on the environment. One reason is that the WTP for environmental attributes or processes is dependent also on the ability to pay, and although one of the axioms of BCA is that individual preferences count, they are likely to count more for people on high incomes compared with those on low incomes.

It is worth noting that MDBA itself expressed strong reservations regarding monetary valuation of the environment in its overview of the economic analysis, stating that

> While "non-use" values are likely to form a significant component of the benefit of policy interventions relating to improved natural resource management, there is significant scepticism about the robustness of their valuation. This is an important point because the CIE report highlighted that most of the benefit of the Basin Plan would be comprised of these "non-use" values. There is also significant scepticism as to whether "non-use" values which accrue to the population outside the Basin should carry equivalent weighting with "non-use" values—or, indeed costs—which accrue within the Basin.
>
> **(MDBA, 2011a, p. 116)**

COST-EFFECTIVENESS AND COST TRADE-OFF ANALYSIS

If deriving monetary values for the environment is problematic, what are the alternatives? One approach is to apply *cost-effectiveness analysis* (CEA) and *cost trade-off analysis*, making use of environmental indicators based on scientific evidence and expert judgment rather than monetary representations of environmental benefits.

COST-EFFECTIVENESS ANALYSIS

CEA is a widely accepted and applied technique. In CEA, the usual aim is to achieve maximum economic efficiency: namely, selecting the option that achieves the same objective at least cost or, alternatively, selecting the option that achieves the maximum indicator or index benefits for any given level of cost. CEA works most easily when there is only one target indicator, but composite indicators (indices) may also be constructed and used (OECD, 2008).

COST TRADE-OFF ANALYSIS

Trade-off analysis based on CEA involves examining different levels of environmental achievement and comparing them with the corresponding levels of cost. The shape of the cost curve often suggests logical thresholds or cut-off points as increasing amounts of funding are allocated to environmental improvement.

It is important when conducting a CEA to explore win-win outcomes and incorporate cost offsets in the estimation of costs. For example, environmental watering may produce cobenefits such as increased net returns to floodplain graziers resulting from overbank flows and more regular in-stream water availability. Water-quality improvements can enhance the productivity of irrigated crops and orchards, as well as lowering treatment costs for town water supplies.

TRADE-OFF ANALYSIS IN THE BASIN PLAN

In formulating the Basin Plan, it is evident that MDBA has implicitly adopted a cost trade-off approach, comparing the potential environmental benefits of different water-resource management regimes with the associated costs. Extensive ecological research has been conducted for the Environmental Watering Plan and development of the Ecological Elements Method to identify environmental benefits in scientific terms of different levels of water recovery and flow management regimes (Overton et al., 2014). Specific environmental targets or outcomes have been formulated by scientists as key objectives of the Basin Plan, based on vegetation, native fish and waterbird populations. Site-specific flow indicators have also been adopted as proxies for environmental outcomes in different parts of the Basin.

The main costs relating to the achievement of environmental targets in the Basin Plan considered by MDBA have been the costs of buybacks of entitlements, costs of investments in infrastructure, and the costs of removing constraints to improved environmental flows. Other costs, incurred by the private sector and water-dependent local communities, comprise potential losses in net returns in irrigated agriculture resulting from reduced water availability, as well as indirect economic effects in dependent economic sectors and communities (Adamson et al., 2011; Mallawaarachchi et al., 2010). MDBA has also interpreted reductions in GRP, income, and employment as costs of the Basin Plan.

ECONOMIC IMPACT ANALYSIS

Economic impact modeling focuses on the effects of exogenous changes, relative to a base case, on key economic indicators such as output, income, and employment. General-equilibrium models designed to assess economic impacts can take different forms, but the main aim is to assess impacts at an economy-wide scale. Such models were used extensively in formulating the Basin Plan.

INPUT-OUTPUT MODELS

The simplest kind of general-equilibrium model is an *input-output* (I-O) model in which all economic interactions are represented by fixed coefficients (Leontief, 1986). The main advantage of an I-O model is its simplicity. Such models lend themselves to detailed analyses at a regional or subregional scale.

Several studies conducted for the Basin Plan relied on I-O modeling or the assumption of fixed coefficients. Arche Consulting (2012) used I-O analysis, based on employment data for the 2006 census, to assess the effects of a reduction in water extraction by local communities, down to a local government area scale. Drivers in the model were changes in SDLs and investments proposed in the Water for the Future program. Responses were calculated in terms of direct effects on agricultural production as well as flow-on regional impacts using gross regional output, value-added, income and employment as the relevant indicators.

Simple linear relationships were assumed in a noncommissioned study by Judith Stubbs and Associates, JSA (2010) assessing the effects on employment and income of changes in water availability at the enterprise, local, Basin, and national levels. A reduction of 10% in water availability was calculated to result in a reduction in total employment of 6000 persons, together with a reduction in productivity of $0.6 billion per year. A 50% reduction in water availability would reduce total employment by 28,000 and a reduction in productivity of $2.76 billion per year. However, as linear functions were used in the calculations, the results tend to overstate the impacts that could be expected or have been observed.

Neither of these studies allowed explicitly for possible adaptations by farmers, such as adopting water-saving technology, altering crop mixes, switching to dryland grazing or cropping, or changing farm-management practices. Other sectors could also be expected to adapt to changes in economic circumstances resulting from flow-on effects associated with agricultural production.

LIMITATIONS OF I-O MODELS

I-O models have distinct limitations. The assumption of linear relationships between inputs and outputs means that changes between variables are equiproportional. A second limitation is that, subsequent to some exogenous change, producers may not continue using the same production technology. A wide range of options is usually available for adaptive management of water-dependent production systems. For example, if water availability decreases, instead of reducing output by the same percentage, crop producers may introduce water-saving technology such as a switch from overhead to drip irrigation or the introduction of drought-tolerant crops. A third limitation is that they do not track the dynamics of adjustments to exogenous shocks. The typical I-O model is a static equilibrium model, suitable only for simulating short-term impacts. Dynamic forms of an I-O model can be constructed, making use of a dynamic inverse matrix representing capital formation (Leontief, 1970) and other models may treat household income and consumption expenditure as endogenous (Jensen et al., 1979), but again such models usually rely on the assumption of fixed coefficients.

The consensus regarding the use of I-O models is that they tend to overstate impacts and focus mainly on short-run outcomes. At best, they indicate only the upper or lower bounds of likely impacts. In particular, multipliers derived from such models may give a misleading picture of what might be expected to occur in reality as a consequence of exogenous shocks to the economy.

COMPUTABLE GENERAL-EQUILIBRIUM MODELS

Most economy-wide economic impact assessments are now conducted using *computable general-equilibrium (CGE) models*. With the development of detailed economic databases, software packages and advances in computer technology, it has become possible to solve large systems of simultaneous equations, both linear and nonlinear, involving hundreds of variables, functional relationships and identities.

Extensive development of CGE models has occurred since the 1980s, such as conversion of static-equilibrium or comparative statics models to a dynamic form, disaggregation of the economy into multiple regions, subdivision of labor into different occupational categories, and specification of interconnections between the economy and natural resources or the environment. Comprehensive interconnected subregional or spatial models may be constructed adopting either a top-down or bottom-up approach. Some modelers have relied on microsimulation or optimization modeling of water-resource use and have combined it with broader frameworks for the economy as a whole.

CGE MODELING FOR THE BASIN PLAN

Several studies of the economic impacts associated with different water-recovery scenarios were conducted for MDBA using CGE modeling. A two-stage approach was adopted by the Australian Bureau of Agricultural Economics (ABARE) and Bureau of Rural Sciences (BRS) (ABARE-BRS, 2010a, 2010b). The first stage involved a static partial-equilibrium model (the Water Trade Model) with 11 land-use activities and 24 regions that simulates the effects of changes in water availability, including the operation of markets, on irrigated agriculture. Outputs from the model were then entered into ABARE's AusRegion dynamic CGE model, which simulates impacts at the regional, Basin, and national scale. AusRegion has 7 Basin regions, 31 commodities, and 4 primary factors of production: land, labor, capital, and natural resources. Data to populate the model were for years 2000–01 and 2005–06. Volumes assumed in the water-recovery scenarios were 3000, 3500, and 4000 GLy^{-1}. The reductions in water use corresponding to these levels ranged from 2666 to 3507 GLy^{-1}, resulting in a decrease in the gross value of irrigated agricultural production (GVIAP) ranging from $805 to $1075 million y^{-1}, and a decrease in profits ranging from $123 to $183 million y^{-1}.

Later work by ABARES (2011a) assumed government purchase of entitlements amounting to $3.1 billion and investments in infrastructure improvements of $4.4 billion. Various improvements were introduced to the AusRegion model, including modifying the dynamic properties of the model to allow for "stickiness" in the responses of labor and capital to changes in agricultural production. Combining all these modifications, the decrease in GVIAP ranged from $524 million y^{-1} for the 3000 GLy^{-1} scenario to $995 million y^{-1} for the 4000 GLy^{-1} scenario, and a decrease in the gross value of agricultural production (GVAP) of $496 to $921 million y^{-1}. MDBA, however, considered that insufficient allowance was made for a likely transition from irrigated to dryland agriculture in response to reductions in water availability.

A further modeling exercise was conducted by ABARES (2011b) for a water-recovery scenario of 2800 GLy^{-1}. It is worth noting that this scenario is the closest to the one ultimately adopted in the final Basin Plan (2750 GLy^{-1}). Details of the modeling conducted by ABARE-BRS and ABARES are summarized in MDBA (2011b).

MDBA additionally commissioned the Centre for Policy Studies (COPS) at Monash University to conduct impact analysis using their TERM-H2O model (Wittwer, 2010, 2011). TERM is a multiregional dynamic CGE model incorporating water accounts. It has 35 industry sectors, including 17 farm and 10 irrigation sectors, 28 commodities, and 22 regions. As well as the usual characteristics of a dynamic CGE model, it integrates dryland agricultural production and water-dependent agricultural production within the model. As a consequence, the impacts obtained with TERM-H2O are less severe than those obtained by ABARES. Water-recovery scenarios modeled by COPS were 2400, 2800, and 3200 GLy^{-1}. Assumptions were adopted relating to the expenditure of funds received by sellers of entitlements, as well as government investments in water-related infrastructure. GRP was predicted to fall by 0.05% by 2020. A shift from irrigated to dryland agriculture would involve an increase of 0.14% in GRP as an offset to the decrease that otherwise would have occurred. Employment impacts of the Basin Plan were also found to be small, with only three regions (the Border, Moonie, and Condamine-Balonne regions) facing minor impacts.

In summary, MDBA considered that the CGE modeling of the impacts of the Basin Plan was the best available. However, strong caveats were placed on interpreting the results, as all the modeling made use of ABS Agricultural Census data for 2005–06, one of the worst years of the Millennium drought. A general coverage of the Australian experience in CGE modeling of water resources appears in Wittwer (2012).

IMPACT ANALYSIS AT LOCAL COMMUNITY SCALE

Detailed community-scale economic impact modeling of water recovery was conducted for the Basin Plan. It has also been extensively applied in the Northern Basin Review (see the following).

SOURCES AND PATTERNS OF LOCAL IMPACT

The two main sources of impact in water-dependent Basin communities have been the buyback of water entitlements and investments in infrastructure modernization and on-farm technology designed to improve water-use efficiency.

The ways in which entitlements are purchased or investments in infrastructure are made can have significant economic impacts at the local community scale. The direct impacts largely depend on the relative importance of irrigated agriculture in the local economy, but flow-on effects to other sectors of the local economy resulting from changes in irrigated agriculture have the potential to exacerbate the direct economic effects and social consequences of changes in agricultural production.

Impacts associated with buy-backs of entitlements depend on the volume of entitlements purchased, the type of entitlement, the timing of purchases, and the number, diversity, and size of enterprises that choose to sell. Concerns expressed by water-dependent communities relate to the possibility of steep increases in the price of entitlements or allocations, volatility, or speculation in water markets, and the "Swiss-cheese" effect whereby the loss of permits within particular catchments may require remaining water holders to pay higher fees to water-supply corporations to cover the costs of infrastructure and management.

The use of funds received by sellers of entitlements has important implications for community economic impacts. Potentially adverse impacts may be ameliorated if the funds are expended in the local area, but if they flow elsewhere, local communities may be significantly affected. Even if sellers feel sufficiently compensated by selling their entitlements, any resulting reduction in irrigated agricultural activity may indirectly affect local agricultural service and product suppliers, as well as other economic activities such as wholesale and retail trade, professional services, and facilities engaged in processing agricultural commodities. Similar considerations apply to the impacts associated with investments in infrastructure, especially the ways in which infrastructure investment funds flow throughout local communities.

COMMUNITY IMPACT MODELING FOR THE BASIN PLAN

Research into community-scale impacts associated with the Basin Plan was conducted by Marsden, Jacobs Associates and others (MJA, 2010). The conceptual framework underpinning the assessments applied by MJA involved several components: determination of community exposure to an adverse process or event; the sensitivity of community responses to that exposure; the adaptive capacity of the community to respond; and the vulnerability of the community, comprising the remaining effects after coping and adaptive responses have taken place.

Impacts were predicted by MJA for different forms of agricultural production, economic sectors, and subregions within the Basin. Key subregions were classified in terms of the GVIAP and the percentage of the population regarded as disadvantaged according to their SEIFA (Socio-Economic Indexes for Areas) scores.

A follow-up study was conducted in 2011 (EBC, 2011). A vulnerability matrix was constructed, establishing four categories of towns in the Basin according to their agricultural dependence and their population. The main indicator used to assess the dependence on water was the percent of total employees in agriculture, processing, transport, and storage in 2006. Highly vulnerable towns were those with high levels of dependence on water and small populations, whereas towns with low risk of adverse impact were those with low dependence on water and large population. Large towns typically are characterized by more diverse economies and a greater capacity to respond to external shocks.

OPTIMIZING ECONOMIC, SOCIAL, AND ENVIRONMENTAL OUTCOMES

The Water Act 2007 requires that the MDBA must "optimize economic, social, and environmental outcomes" when establishing sustainable diversion limits—but what does this mean? The discussion presented in this chapter makes it clear that, even with the best scientific evidence and economic modeling results, *subjective judgments* are unavoidable in interpreting and applying any concept of optimization. Economic analysis and results can be only one aspect of integrated decision making for such a complex resource system as the Murray-Darling Basin.

The TBL (triple bottom-line) is a convenient way of categorizing the three broad classes of effect in the Water Act. A distinction can, however, be drawn between TBL decision making and TBL accounting. TBL decision making refers to the process of evaluating options and suggesting desirable action, while TBL accounting refers to the interpretation and reporting of economic, social, and environmental information to support the decision-making process.

Formal frameworks for evaluating and ranking options involving TBL outcomes include multi-criteria analysis models and decision support systems (Janssen, 1992). In such models, whatever indicators are selected to represent economic effects, it is essential that they are consistently applied across all options to facilitate a sound comparison of policy options. It is also important not to double-count representations of outcomes. If physical ecological indicators are chosen to represent environmental effects, they should not be valued and double-counted as economic values.

THE NORTHERN BASIN REVIEW

A Northern Basin Review is being carried out at the time of writing to revisit the recommended water-recovery target of 390 GLy^{-1} established in the Basin Plan and recommend to the Ministerial Council whether a change in the target is deemed desirable. Several optional recovery targets (scenarios) are under consideration, ranging from 278 to 415 GLy^{-1}.

A TBL approach is being applied, aimed at optimizing social, economic, and environmental outcomes, in which different socioeconomic and environmental indicators are represented in an effects matrix of modeled outcomes.

The economic modeling component focuses on the economic impacts of the different water-recovery scenarios in 21 local communities in the Northern Basin. The dependence on irrigation water by these local communities varies. Interconnections between water recovery and impacts are analyzed using a hydrological and land-use model, connected to detailed econometric models of local community impacts. The main indicators of impact comprise changes in areas of irrigated agriculture, direct effects on employment in irrigated agriculture and flow-on employment effects in other economic sectors within each local community. More comprehensive socioeconomic information relating to each of the communities, including the potential vulnerability to changes in irrigated agriculture, has been compiled in a related set of community profiles.

Additional modeling undertaken for the Northern Basin Review consists of a detailed rules-based farm-scale simulation model, constructed collaboratively by MDBA and farmers in the Lower Balonne catchment that assesses the benefits of overbank flows to the grazing industry resulting from different return flows to the environment under each of the water-recovery scenarios.

GENERAL CONCLUSIONS

A wide range of economic modeling techniques has been applied in formulating the Basin Plan, recognizing that economic appraisals of options for managing water resources in the Basin involve several different kinds of economic indicators. In recommending a particular level of water recovery, MDBA has taken into account detailed economic information relating to the efficiency of resource use, the magnitude and pattern of macroeconomic impacts, and the incidence of benefits and costs for different industries and communities within the Basin. It is not possible to say that any single approach to the economic analysis is the best, as there is clearly no all-embracing method that can offer an unequivocal guide to final decisions. Rather, a diverse range of models and results must be relied upon, based on the best environmental science and economic modeling techniques.

Assessments of trade-offs may be undertaken making use of formal decision support systems in which outcomes and importance weights are explicitly displayed, or through more informal methods based on predetermined targets, bounds of acceptability and notions of fairness about the magnitudes and relative importance of different kinds of effects. What is not clear from the large volume of economic results obtained is how MDBA combined them with other indicators of community wellbeing and environmental restoration to determine the optimum balance of social, economic, and environmental outcomes.

Perhaps the most important conclusion reached by MDBA is that the Basin Plan must be viewed as an evolving blueprint based on the concepts and practice of adaptive management and that close collaboration between government and stakeholders and local communities is essential in achieving success.

REFERENCES

ABARE-BRS. (2010a). *Environmentally sustainable diversion limits in the Murray-Darling Basin: Socio-economic analysis.* Report by the Australian Bureau of Agricultural Economics and Bureau of Rural Sciences for the Murray-Darling Basin Authority, Canberra (132 p.).

ABARE-BRS. (2010b). *Assessing the regional impact of the Murray-Darling Basin plan and the Australian government's water for the future program in the Murray-Darling Basin.* Report by the Australian Bureau of Agricultural Economics and Bureau of Rural Sciences for the Department of Sustainability, Environment, Water, Population and Communities, Canberra (49 p.).

ABARES. (2011a). *Economic analysis of alternative sustainable diversion limits options.* Report by the Australian Bureau of Agricultural Economics and Bureau and Sciences for Marsen Jacobs and Associates, Canberra (82 p.).

ABARES. (2011b). *Modelling the economic effects of the Murray-Darling Basin Plan.* Report by the Australian Bureau of Agricultural Economics and Bureau and Sciences for the Murray-Darling Basin Authority, Canberra (109 p.).

Adamson, D., Quiggin, J., & Quiggin, D. (2011). *Water supply variability and sustainable diversion limits: Issues to consider in developing the Murray-Darling Basin.* Report for ABARES by the Risk and Sustainable Management Group. Brisbane: School of Economics, University of Queensland. 56 p.

ADB. (1996). *Economic evaluation of environmental impacts: A workbook.* Manila: Asian Development Bank.

ADB. (1997). *Guidelines for economic analysis of projects.* Manila: Asian Development Bank.

Arche. (2012). *Basin case studies: The socio-economic impacts of sustainable diversion limits and water for the future investments: An assessment at a local scale.* Final report by Arche Consulting in Association with Gillespie Economics for the Department of Sustainability, Environment, Water, Population and Communities, Canberra (92 p.).

Arrow, K., Solow, R., Portney, P., Leamer, E., Radner, R., & Schuman, H. (1993). Report of NOAA panel on contingent valuation. *Federal Register*, 58, 4601–4614.

Aust Govt. (2012). *Water Act 2007—Basin Plan 2012, extract for the Federal Register of Legislative instruments.* Canberra: Office of Parliamentary Counsel, Australian Government. 245 p. 28 November.

Aust Govt. (2014). *Water Act 2007 (with amendments).* Canberra: Office of Parliamentary Counsel, Australian Government 571 p.

Bateman, I. J., Lovett, A. A., & Brainard, J. S. (2003). *Applied environmental economics: A GIS approach to cost-benefit analysis.* Cambridge: Cambridge University Press.

BDA Group. (2010). *Review of social and economic studies in the Murray-Darling Basin.* Report by BDA Group for the Murray-Darling Basin Authority, Canberra (130 p.).

Bennett, J. (2011). *The international handbook on non-market environmental valuation.* Cheltenham: Edward Elgar.

Bennett, J., & Blamey, R. (2001). *The choice modelling approach to environmental valuation.* Cheltenham: Edward Elgar.

Bjornstad, D. J., & Kahn, J. R. (Eds.), (1996). *The contingent valuation of environmental resources: Methodological issues and research needs.* Cheltenham: Edward Elgar.

Boardman, A. E., Greenberg, D. H., Vining, A. R., & Weimer, D. L. (2011). *Cost-benefit analysis: Concepts and practice* (4th ed.). Upper Saddle River, NJ: Prentice Hall.

CIE. (2011). *Economic benefits and costs of the proposed Basin Plan: Discussion and issues*. Report by the Centre for International Economics for the Murray-Darling Basin Authority, Canberra (166 p.).

Commonwealth of Australia. (2006). *Handbook of cost benefit analysis*. Canberra: Department of Finance and Administration. 180 p.

Connell, D., & Grafton, P. (Eds.), (2011). *Basin futures: Water reform in the Murray-Darling Basin*. Canberra: ANU ePress.

Costanza, R., d'Arge, R., de Groot, R., Farber, S., Grasso, M., Hannon, B., et al. (1997). The value of the world's ecosystem services and natural capital. *Nature, 387*(6630), 253–260.

Crossman, N. D., Rustomji, P., Brown, A., Pollino, C., Colloff, M., Lester, R., et al. (2011). *Status of the aquatic ecosystems of the Murray–Darling Basin and a framework for assessing the ecosystem services they provide*. Interim Report to the Murray–Darling Basin Authority from the CSIRO Multiple Benefits of the Basin Plan Project, CSIRO Water for a Healthy Country National Research Flagship, Canberra (120 p.).

DAE. (2012). *Benefits of the Basin Plan for the fishing industries in the Murray-Darling Basin*. Report by Deloitte Access Economics for the Murray-Darling Basin Authority, Canberra (72 p.).

Dixon, J. A., James, D. E., & Sherman, P. B. (1989). *The economics of dryland management*. London: Earthscan Publications.

DRE.T (2010). *Destination visitor survey strategic regional research report, NSW, Victoria and SA, impact of drought in the Murray River region*. Canberra: Tourism Research Australia, Department of Resources, Energy and Tourism. 12 p.

EBC. (2011). *Community impacts of the guide to the proposed Murray-Darling Basin Plan*. Report by for the Murray-Darling Basin Authority, Canberra (133 p.).

Freeman, A. M. (1979). *The benefits of environmental improvement: Theory and practice*. Baltimore, MD: Resources for the Future, Johns Hopkins University Press.

Freeman, A. M. (1993). *The measurement of environmental and resource values: Theory and methods*. Washington, DC: Resources for the Future.

Frontier Economics. (2011). *The potential use of benefit-cost analysis in developing the Basin Plan*. Report by Frontier Economics, Commonwealth of Australia, Canberra (20 p.).

Haab, T. C., & McConnell, K. E. (2002). *Valuing environmental and natural resources: The econometrics of non-market valuation*. Cheltenham: Edward Elgar.

Hanley, N., & Barbier, E. B. (2009). *Pricing nature: Cost-benefit analysis and environmental policy*. Cheltenham: Edward Elgar.

HM Treasury. (2003). *The green book: Appraisal and evaluation in central government*. London: The Stationery Office.

James, D. E., & Francisco, H. A. (Eds.), (2015). *Cost-benefit studies of natural resource management in Southeast Asia*. Singapore: Springer.

Janssen, R. (1992). *Multi-objective decision support for environmental management*. Dordrecht: Kluwer Academic Publishers.

Jensen, R. C., Mandeville, T. D., & Kurunaratne, N. D. (1979). *Regional economic planning*. London: Croom Helm.

JSA. (2010). *Report 4: Exploring the relationship between community resilience and irrigated agriculture in the Murray-Darling Basin: Social and economic impacts of reduced irrigation water*. Report by Judith Stubbs and Associates, Cotton Catchment Communities Cooperative Research Centre, Narrabri.

Kahneman, D., & Knetsch, J. L. (1992a). Valuing public goods: The purchase of moral satisfaction. *Journal of Environmental Economics and Management, 22*, 57–70.

Kahneman, D., & Knetsch, J. L. (1992b). Contingent valuation and the value of public goods: Reply. *Journal of Environmental Economics and Management, 22*, 90–94.

KPMG. (2011). *Review of the MDBA's socio-economic impact modelling*. Report by KPMG for the Murray-Darling Basin Authority, Canberra (54 p.).

Leontief, W. (1970). The dynamic inverse. In A. P. Carter & A. Brody (Eds.), *Contributions of input-output analysis* (pp. 17–46). Amsterdam: North-Holland Publishing.

Leontief, W. (1986). *Input-output economics* (2nd ed.). New York, NY: Oxford University Press.

Louviere, J. J., Hensher, D. A., & Swait, J. D. (2000). *Stated choice methods: Analysis and application.* Cambridge: Cambridge University Press.

Mallawaarachchi, T., Adamson, D., Chambers, S., & Schrobback, P. (2010). *Economic analysis of diversion options for the Murray-Darling Basin Plan: returns to irrigation under reduced water availability.* Report by the Risk and Sustainable Management Group, School of Economics, University of Queensland, for the Murray-Darling Basin Authority, Brisbane (55 p.).

Martin-Ortega, J., Ferrier, R. C., Gordon, I. J., & Khan, S. (Eds.), (2015). *Water ecosystem services: A global perspective.* Cambridge: UNESCO and Cambridge University Press.

Maynard, S., James, D., & Davidson, A. (2015). Determining the value of multiple ecosystem services in terms of community wellbeing: Who should be the valuing agent? *Ecological Economics, 115,* 22–28.

MDBA. (2011a). *Socio-economic analysis and the Draft Basin Plan: Part A, overview and analysis.* Canberra: Murray-Darling Basin Authority. 162 p. MDBA Publication no. 52/12.

MDBA. (2011b). *Socio-economic analysis and the Draft Basin Plan, Part B, commissioned and non-commissioned reports which informed the MDBA's socio-economic analysis.* Canberra: Murray-Darling Basin Authority. 219 p. MDBA Publication no. 52/12.

MDBA. (2014). *Murray-Darling Basin reforms: Framework for evaluating progress.* Canberra: Murray-Darling Basin Authority. MDBA Publication no. 9/14.

MDBA. (2015). *Towards a healthy, working Murray-Darling Basin: Basin Plan Annual Report 2014–15.* Canberra: Murray Darling Basin Authority. 49 p.

MEA. (2005). *Millennium ecosystem assessment: Ecosystems and human wellbeing: A framework for assessment.* Washington, DC: World Resources Institute and Island Press.

Mitchell, R. C., & Carson, R. T. (1989). *Using surveys to value public goods: The contingent valuation method.* Washington, DC: Resources for the Future.

MJA. (2010). *Economic and social profiles and impact assessments for the Murray-Darling Basin Plan: Synthesis report,* Report by Marsden Jacobs and Associates, RMCG, EBC Consultants, DBM Consultants, Australian National University, McCleod, G., and Cummins, T. for the Murray-Darling Basin Authority, Canberra (244 p.).

Morrison, M., & Hatton McDonald, D. (2010). *Economic valuation of environmental benefits in the Murray-Darling Basin.* Report by the Institute for Land, Water and Society, Charles Sturt University and CSIRO Ecosystems Sciences for the Murray-Darling Basin Authority, Canberra (46 p.).

OECD. (1994). *Project and policy appraisal: Integrating economics and environment.* Paris: Organisation for Economic Co-operation and Development.

OECD. (2008). *Handbook on constructing composite indicators: Methodology and user guide.* Paris: Organisation for Economic Co-operation and Development.

Overton, I. C., Pollino, C. A., Roberts, J., Reid, J. R. W., Bond, N. R., McGinness, H. M., et al. (2014). *Development of the Murray-Darling Basin Plan SDL adjustment ecological elements method.* Report by CSIRO for the Murray-Darling Basin Authority, Canberra (190 p.).

Pearce, D. W., & Nash, C. A. (1981). *The social appraisal of projects: A text in cost-benefit analysis.* London: Macmillan.

Pearce, D. W., & Turner, R. K. (1989). *The economics of natural resources and the environment.* New York, NY: Harvester Wheatsheaf.

Sinden, J. A., & Worrell, A. C. (1979). *Unpriced values: Decisions without market prices.* New York, NY: Wiley.

Smith, V. K. (1996). *Estimating economic values for nature: Methods for non-market valuation.* Cheltenham: Edward Elgar.

Sukhdev, P., Wittmer, H., Schröter-Schlaack, C., Nesshöver, C., Bishop, J., ten Brink, P., et al. (2010). *Mainstreaming the economics of nature: A synthesis of the approach, conclusions and recommendations of TEEB*. Nairobi: United Nations Environment Programme.

TEEB. (2010). In P. Kumar (Ed.), *The economics of ecosystems and biodiversity: Ecological and economic foundations*. Washington, DC: Earthscan.

Vermaat, J. E., et al. (2016). Assessing the societal benefits of river restoration using the ecosystem services approach. *Hydrobiologica, 769*, 121–135.

Wittwer, G. (2010). *The regional economic impacts of sustainable diversion limits*. Report by the Centre for Policy Studies, Monash University for the Murray-Darling Basin Authority, Canberra (59 p.).

Wittwer, G. (2011). *Basin Plan CGE modelling using TERM-H20*. Report by the Centre for Policy Studies, Monash University for ABARES on behalf of the Murray-Darling Basin Authority, Canberra (42 p.).

Wittwer, G. (Ed.), (2012). *Economic modelling of water: The Australian CGE experience*. Dordrecht: Springer.

CHAPTER

SOCIAL-ECOLOGICAL ANALYSES FOR BETTER WATER RESOURCES DECISIONS

S.A. Bekessy, M.J. Selinske
RMIT University, Melbourne, VIC, Australia

CHAPTER OUTLINE

INTRODUCTION

Environmental management typically occurs in complex sociopolitical settings and water resource management is no exception. Decisions often involve multiple, potentially competing stakeholders, made in highly politically charged settings and often under extreme uncertainty. Water resource management decisions can have profound impact on freshwater, marine and terrestrial ecosystems over large temporal and spatial scales and can underpin critical outcomes for communities, including

Decision Making in Water Resources Policy and Management. http://dx.doi.org/10.1016/B978-0-12-810523-8.00010-0

threatening lives and livelihoods. There is increasing awareness that it's the dynamic connections between the social and ecological systems that are critical to understand for effective policy design (Liu et al., 2007).

The Murray-Darling Basin, one of Australia's largest lowland river systems, is an example of a water resources management setting where the interplay between social and ecological systems has largely driven outcomes. Whalley et al. (2011) describe key transformations in the basin wetlands since European colonization in the 19th century, relating to shifts in the dynamics between the ecological, economic and social drivers. Examples include grazing of former wetlands following damming of the river for irrigation, and the availability of machinery capable of cultivating the heavy textured soils leading to the emergence of dryland cropping as a major enterprise in the basin.

The interplay between social, economic, and ecological drivers is typically complex and can result in unintended feedbacks (Larrosa et al., 2016). The Murray-Darling River Basin again provides a good example, where responses to the Millennium drought were complex and sometimes counterintuitive. While irrigators were overtly negative toward proposals to cap water extraction (including public gatherings to burn the Guide to the Basin Plan) practices shifted regardless, leading to substantial efficiency gains (Pittock, 2016). Some of the mechanisms that were nominally rejected by landholders, including trading of water rights, became widespread as the drought became more severe. This example highlights that responses to policy interventions are not always straightforward.

Understanding the links between social and ecological systems and predicting the impact of interventions designed to improve environmental outcomes requires a highly interdisciplinary approach. Decision support and analysis tools utilized in social-ecological systems (SES) (Berkes et al., 2000) can help reduce the uncertainty and increase the transparency of complicated water management decisions. Many tools and approaches exist for helping to understand the interdependence between ecological and social dynamics and the impact of nonlinear behaviors, to predict the future impacts of different decisions, unpack the uncertainties in water resource systems (both social and ecological) and find ways to address them. Systematic approaches can assist with eliciting desired futures, representing alternative stakeholder perspectives, analyzing scenarios and exploring the impact of social networks.

In this chapter we describe key frameworks for understanding SES and models for analyzing the consequences of different decisions. We provide examples of the application of these frameworks and modeling tools in a variety of water resource management contexts. We conclude by critically analyzing the frameworks and tools available in terms of their capacity to support water resource management decisions and describe pathways to mainstream the use of these approaches.

FRAMEWORKS FOR UNDERSTANDING SOCIAL-ECOLOGICAL SYSTEM (SES)

SES are complex, adaptive systems defined by interactions between the social and ecological spheres and at various scales are both spatial, temporal, and social structures (Berkes et al., 2000). Numerous frameworks have been proposed to represent these systems with the aim of understanding critical elements and their dynamic interactions (Binder et al., 2013). In this section we describe three key frameworks that have usefully informed water resource management dilemmas: Ostrom's Social-Ecological Systems framework (SESF), the Driver-Pressure-State-Impact-Response Framework (DPSIR) and Structured Decision Making.

OSTROM'S SES FRAMEWORK

Ostrom's Nobel prize–winning Social-Ecological Systems Framework (SESF—Ostrom, 2007) is one of the most highly utilized interdisciplinary frameworks for representing and understanding natural resource management problems, providing a multileveled structure for organizing subsystems and variables of SES. The approach originates from political sciences and is based on theories of collective choice and common-pool resources.

The SES framework was developed to allow for comparison between different contexts and identify the principal variables within a system that act as leverage points. Depending on context, different variables take precedence in relevance, allowing the framework to be fitted to the natural resource problem. The SES framework has been widely used to understand fisheries, forestry, grazing lands, and water commons resources. Fig. 1 presents the structure and elements of the framework that are typically used to understand SES.

Pahl-Wostl et al. (2012) provides an example of the use of Ostrom's (2007) framework in the context of water management. In this study, the framework was used to describe the water governance regimes in 29 river basins and to analyze their social and environmental performance. The analysis demonstrated the value of polycentric governance regimes, which are characterized by a distribution of power between independent actors, but with effective coordination under a general system of rules. This type of governance generally delivered better social and environmental outcomes than other regimes, in particular in the capacity to respond to challenges from climate change and dealing with uncertainty (Pahl-Wostl et al., 2012).

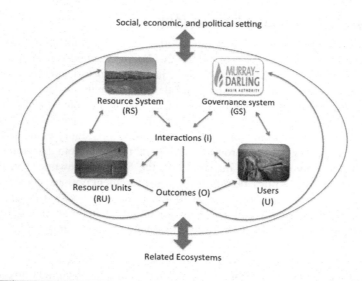

FIG. 1

Graphical representation of Ostrom's Social Ecological Systems Framework.

Adapted from Ostrom, E. (2007). A diagnostic approach for going beyond panaceas. Proceedings of the National Academy of Sciences of the United States of America, 104, *15181–15187.*

DRIVER-PRESSURE-STATE-IMPACT-RESPONSE FRAMEWORK

The DPSIR originates from integrated environmental assessment and systems science and is a policy-oriented, causal framework for describing the interactions between society and the environment. The DPSIR framework has been adopted by several international organizations such as the OECD, the US Environmental Protection Agency and the European Union to understand and report on environmental matters (Patrício et al., 2016).

The DPSIR framework describes a chain of causal links starting with "driving forces" (human activities) through "pressures" (e.g., emissions or waste) to "states" (physical, chemical, and biological) and "impacts" on ecological and social systems, eventually leading to political "responses" (management actions, prioritization, or target setting). As a first step in applying the DPSIR framework, data and information on all the different elements in the chain are collected and then possible connections between these elements are proposed. The aim of the exercise is to gauge the effectiveness of the management responses put in place.

Patrício et al. (2016) undertook a systematic review of the use of DPSIR in marine and coastal zones and found the framework to be a useful tool for risk assessment of a wide variety of environmental problems and capable of linking disciplines, academic policy makers and practitioners (although stakeholder engagement is often poor). However, the focus of DPSIR on one-to-one relationships can disregard the complex interactions between multiple pressures, activities, the environment, and society. Hence, the DPSIR framework has been criticized for being overly simplistic, and may be inadequate in addressing state change. Patrício et al. (2016) recommend moving from generic/conceptual application of DPSIR toward models that are specific and quantified (Fig. 2).

STRUCTURED DECISION-MAKING FRAMEWORK

Structured decision making (SDM) is a formal method for identifying and assessing creative management options for complex problems (Gregory et al., 2012). The method is designed to help with the analysis of decisions involving diverse stakeholders, large amounts of uncertainty and challenging trade-offs. SDM guides clear objective setting and delivers insights to decision makers regarding the performance of different management options in terms of their capacity to deliver on objectives. The aim is to improve communication between stakeholders, clarify trade-offs between management options and identify win-win solutions for multiple stakeholders (Gregory et al., 2012).

Structured decision making is based around theories and methods from decision theory (Bellman, 1957). In the context of environmental management, the SDM framework typically includes the following elements: objective setting, identification of management options, construction of a model of the system behavior (linked back to responses to management actions), a program of monitoring to track progress against objectives and a system for choosing optimal management actions that satisfy objectives (McCarthy & Possingham, 2007) (see Fig. 3 for a graphical description of this process).

A good example of SDM in the context of a water resources dilemma is an investigation of how to build conflicting values into decisions about nonnative fish control in the Grand Canyon, undertaken by the USGS (Runge et al., 2011). In this case, SDM was used to elicit values and objectives and develop management scenarios that were palatable for traditional tribes, natural resource managers and industries invested in the region. While the traditional tribes disliked the mechanical removal of exotic trout

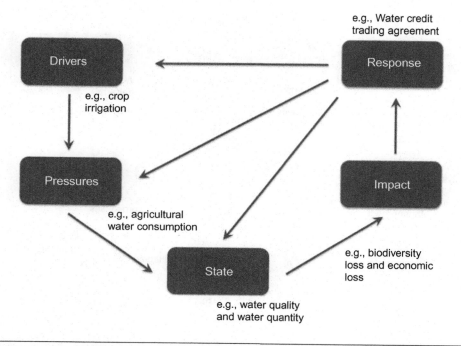

FIG. 2

A representation of the DPSIR framework as applied to a water resources dilemma.

Adapted from European Environment Agency (2016).

in sacred areas, the presence of this species impacts on other endangered fish species. Multicriteria analysis using objective, cultural and hypothesis weightings helped identify an adaptive strategy with potential solutions. The SDM process helped clarify the diversity of objectives for the area and investigated key uncertainties and the value of monitoring information to resolve uncertainty.

CHARACTERISTICS OF USEFUL SOCIOECOLOGICAL FRAMEWORKS FOR WATER RESOURCES MANAGEMENT

While the frameworks described previously take very different approaches and are derived from highly diverse disciplines, all assist in simplifying SES, identifying key drivers of change and leverage points. From the perspective of water resource managers, the usefulness of social-ecological frameworks should also extend to assisting in decision making. Which governance approach is most likely to succeed? How should the decision problem be specified? Where are the key points in the system that policy interventions should target? Depending on the decision context, frameworks that truly integrate social and ecological information (Binder et al., 2013), involve stakeholders (Patrício et al., 2016), help resolve competing objectives, can deal with uncertainty and identify tangible pathways forward (action-oriented—Binder et al., 2013) will likely be most useful.

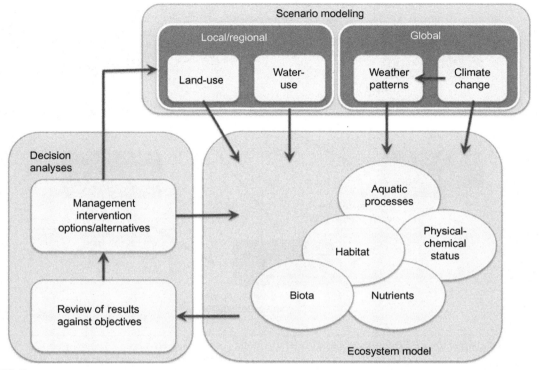

FIG. 3

Conceptual outline of the structured decision making framework as applied to an environmental decision problem in a riverine system.

Adapted from Anthony et al. (2014).

DECISION TOOLS FOR MODELING AND ANALYZING SES

The outcomes of interactions between social and ecological systems are typically complex and multifaceted. Therefore, conceptual and quantitative models are important for understanding and characterizing SES, and for making predictions about the outcomes of management. The transdisciplinary nature of SES, which makes them complex to understand, also brings in tools from a wealth of fields including economics, political science, mathematical modeling and behavioral economics. The types of tools that are likely to be useful can deal with nonlinear dynamics, cross-scale interactions and uncertainty and be capable of dealing with a variety of data sources, both qualitative and quantitative. While the range of tools that we describe in the following sections is by no means comprehensive, we have tried to focus on the most commonly used tools for analyzing SES, with potential application to water resources management.

AGENT-BASED MODELING

Agent-based modeling (ABM) is a computational approach that represents an individual's decision making over time and space (Grimm et al., 2005). The ABM approach involves simulations of a group of semiautonomous entities (the "agents") interacting with each other and their environment. The modeling approach is general with agents potentially representing people, organizations or animals. The microscale actions of agents are aggregated in the model revealing macroscale phenomena and patterns.

ABM is able to reflect the heterogeneity of individual decision makers and is one of the few modeling tools that allows for adaptive responses to emergent endogenous or exogenous phenomena. This makes it useful in SES where resource-users' decision making is varied among individuals and dynamic. The internal complexity of the individual agents can also vary, ranging from simple reactive behaviors through to proactive agents that evolve, learning from past experiences (Miller & Page, 2009). ABMs are also used in ecology, typically known as individual-based models, allowing for easy integration of ecological and social models.

Agent-based modeling is well established in land-use dynamics and is also used commonly in water resource (e.g., Schlüter & Pahl-Wostl, 2007) and fisheries decision problems (e.g., Little et al., 2009). Schlüter and Pahl-Wostl (2007) used an ABM to model the resilience of the Amudarya River Delta in Central Asia under centralized and decentralized governance regimes. Policy scenarios were created to optimize multiple ecosystem services including those that support fisheries and irrigated agriculture within the river basin. The ABM represented agents within each of the governance systems responding to water availability. The model showed that livelihood diversification under a decentralized authority increased the livelihood and resource resilience in the river basin.

The ABM approach is highly flexible, with the richness of models expressing agents' internal structures and their interactions limited only by the programming ability of the researchers and the data available to parameterize models. At the same time, often multiple ad hoc assumptions are required to build the model and it is easy to construct overly complex ABMs that overfit any data used to parameterize them. Furthermore, ABMs are more difficult to analyze, understand and communicate than many other analytical models (Grimm et al., 2005).

BACKCASTING AND FORECASTING

Backcasting is a collaborative planning tool that starts by asking stakeholders to identify desirable future conditions and then works backwards to identify policies and programs that will connect the future to the present (Gleick, 1998). Backcasting provides a structured method for exploring the trade-offs between different policy interventions and their likelihood of achieving desirable futures in the face of uncertainty (Gordon, 2015). Outputs of the exercise include short-term targets and multiple pathways as steps to reach desired states from the present.

The Texas Water Development Board used backcasting to inform sustainable groundwater management (Gleeson et al., 2012). In this case, the desired future conditions identified by stakeholders were defined by the level of groundwater, storage volume or spring flow. Groundwater models were then used to identify maximum pumping rates that maintain the desired groundwater conditions into the future.

While backcasting works to identify steps to achieve a desirable future, forecasting predicts the future based on current trend analysis. Backcasting can work together with forecasting to help identify effective water resources management strategies. Mitchell and White (2003) describe the synergistic use of these tools in the context of urban water management to identify short-term actions through forecasting to "pick the low hanging fruit" (p. 25), while challenging existing assumptions and identifying longer-term strategies using backcasting. Mitchell and White (2003) demonstrate the use of these tools in evaluating water delivery infrastructure for a greenfield development in Sydney, Australia.

BAYESIAN BELIEF NETWORKS

Bayesian belief networks (BBNs) represent another promising tool for integrating social and ecological information for water resources decision making. Outputs from a wide variety of model types can be combined and communicated using the common language of probability theory (Pearl, 1988). In a BBN, each node represents a key process and links show causal relationships between nodes. Thus the networks provide a graphical demonstration of cause and effect. Prototype BBNs can be established using both existing data sets and expert knowledge, and then more exact probability distributions can be determined using discipline-specific submodels for each node. As the BBN approach relies on estimates of the probability distributions for various cause and effect relationships, it is not bound to a specific discipline. Hence, this method can work as an interdisciplinary bridge, allowing quantitative estimates for how various social and ecological processes interact and impact policy outcomes. Another advantage is that uncertainty is built into the approach right from the start, as nodes take probability distributions to represent causal links to other nodes. A drawback is that most BBNs are static networks where causal influences are acyclic, and thus do not allow feedback between nodes, though approaches to deal with this limitation are being developed (Lähdesmäki & Shmulevich, 2008).

Chan et al. (2012) use BBN models to inform decisions regarding water extraction and environmental flows in the Daly River, Northern Territory, Australia. Using a combination of quantitative ecological data and expert knowledge, a BBN was parameterized to represent changes in hydrology and related impacts on two native fish species under a range of water extraction scenarios. While this particular case study focused on ecological input and outcomes, BBNs are capable of including social and economic data and can incorporate submodels from a range of different disciplines (see e.g., Torabi et al., 2016).

BIOECONOMIC MODELING

Bioeconomic modeling represents an extension of economic modeling to the problem of natural resource management, including both economic and biophysical components (Kragt, 2012). Bioeconomic models demonstrate optimal levels of resource use determined by social and biological resource constraints (Ruben et al., 1998). They can be both static and dynamic, with feedbacks represented, and commonly used to represent agricultural systems and fisheries (Schlüter et al., 2012).

In the context of water resources management, bioeconomic modeling has been used to assess the impact of agriculture in watersheds on water availability and nutrient loads (Graveline, 2016). Belhouchette et al. (2011) use bioeconomic modeling to represent the impact of a nitrate reduction policy in the mid-Pyrenees region of France. To model the agricultural system they used the social variables of farm scale and income in conjunction with nitrate loads, erosion and irrigation under a nitrate tax policy scenario. The analysis demonstrated that stricter nitrate policy reduced erosion but increased water use and did not impact nitrogen leaching. This was a result of a shift in crop production away from spring crops, which require more nitrogen but less water, to maize, which is a summer crop. Despite the switch to low nitrogen fertilizer, maize intensifies nitrogen leaching as a result of increased water needs. In this context bioeconomic modeling was helpful in anticipating unintended consequences resulting from a well-intentioned water pollution policy.

Ideally, bioeconomic models reflect the trade-offs between natural and socioeconomic systems, to help evaluate how management actions affect different policy objectives. However, the outputs from bioeconomic models are only as good as the data and assumptions upon which the models are based. Bioeconomic models have been criticized for their tendency to utilize simplistic ecological components (Kragt, 2012) and to poorly integrate with socioeconomic components. Furthermore, bioeconomic models tend to focus on productive (marketable) environmental goods and services, but typically don't incorporate intangible ecosystem goods and services, such as recreation opportunities or esthetic amenity (Kragt, 2012).

GAME-THEORETIC MODELING

Game theory is a mathematical representation of rational decision makers in conflict and cooperation (von Neumann & Morgenstern, 1944). Multiple types of "games" (such as the prisoner's dilemma; Axelrod, 2006) have been developed to understand the development of cooperation among system actors in economic and resource dilemmas. Game theoretical modeling is used to model natural resource utilization in a variety of common property policy and management contexts (Anderies et al., 2004). Despite the shortcoming of modeling intelligent, rational agents, game theory is able to provide general rules of logic and guide the development of cooperative agreements by interpreting the self-interest behaviors of actors in a system rather than behaviors that reflect the system's objectives (Madani, 2010).

In water resources decision making, game theory has been applied to irrigated agriculture, hydro-electric power, water pollution, groundwater management, international water conflicts, and water-quality management (Parrachino et al., 2006). Strategic game theory was used to model the potential outcomes of ecosystem restoration in the Upper Sacramento River Valley (United States) (Buckley & Haddad, 2006). Cooperative and noncooperative models of resource user behavior are used because it was found that competing resource users, specifically farmers, can have opposite goals to the conservation plans and ecosystem service goals for NGOs. The model illustrates that optimal ecological restoration strategies in the Sacramento River Valley, and likely other contexts, should account for potential negative social feedbacks.

SOCIAL NETWORK ANALYSIS

Increasingly, social network analysis (SNA) is used as a tool in conservation and social-ecological decision making to identify structures and actors that manage a natural resource or across a landscape (Bodin et al., 2014). SNAs provide a snapshot of interconnecting nodes that represent the actors and their links between other actors. This is used to analyze different scales of a system and allow decision makers to observe linkages within the systems (Guerrero et al., 2013). Network analysis can be used to combine both the social and ecological aspects of a conservation problem by linking social actors with ecological structures (Janssen & Ostrom, 2006). SNA is used to identify key actors (both individuals and organizations) of a system—those that exert influence within the network or bridge disparate parts of the network. A challenge to the usefulness of SNAs is that the analysis lacks temporal dynamism (Snijders, 2001) and may not capture the influences of shadow networks that exert influence over the decision-making process (Olsson et al., 2006).

Given that water resource systems are highly complex, linking multiple ecosystems and governance regimes across vast geographic spaces, SNA is a useful tool to avoid scale mismatches and aid in identifying system leverage points. SNA was used by Stein et al. (2011) to identify governance gaps among 70 organizations and user groups in the Mkindo catchment of Tanzania. Their research found that poor coordination of water resources management within the catchment and governing institutions had failed to build upon existing, informal structures. Rather than utilizing emergent governance structures, other models were introduced that organized the stakeholders (universities, NGOs, and local governments) into hierarchal top-down structures that were ultimately less effective.

SCENARIO ANALYSIS AND PLANNING

Scenario analysis and planning is a useful tool for exploring plausible futures of SES (Bengston et al., 2012). The role of scenario analysis and planning is to look at various future states of a system operating under uncertainty and generate strategies to meet potential management challenges (Peterson et al., 2003). Often the development of scenarios is used as an engagement process and its success depends on collaboration between diverse stakeholders. In this way, it does not predict but anticipates multiple futures. Similar to some of the other tools outlined here, scenario analysis allows for the incorporation of both quantitative and qualitative data. Once scenarios and strategies are produced they can be combined with other modeling tools such as Bayesian belief networks or agent-based models to explore future implications of policy outcomes given a variety of plausible scenarios. A comprehensive futures study would then incorporate the output of the models into the scenario analysis and planning in an iterative process.

Carpenter et al. (2015) developed scenarios based on ecosystem service modeling and stakeholder interviews for the Yahara Watershed (United States). Socioecological and biophysical trends were used to illustrate the current context and near future projections for the Yahara Watershed. The social input of over 80 stakeholders were integrated with local and global trends to generate a series of potential drivers of change. These were then used to develop four potential future scenarios. These scenarios informed the development of ecosystem service models to understand future management challenges and opportunities.

DISCUSSION AND CONCLUSIONS

As for many environmental issues, water management problems are typically situated in highly complex SES with profound uncertainties and multiple interactions across spatial and temporal scales (Schlüter et al., 2012). Complexities such as nonlinear behaviors and unintended feedbacks between social and ecological systems are commonplace. Making good decisions under these circumstances necessarily depends on decision support tools; relying on intuition is unlikely to be a robust strategy and will lack the transparency that those affected by such decisions will demand.

Here we have presented a highly diverse range of frameworks and methodologies that support decision analysis of SES and have argued that they can help with the identification, development, and evaluation of management interventions. Frameworks such as structured decision making can effectively engage stakeholders in the process of defining objectives and assessing management options; tools such as backcasting and scenario analysis can assist policy makers in analyzing the consequences of different management interventions; and modeling approaches such as ABMs and game theory can help explore the behavior of individuals and their interactions.

Despite their potential, the application of social-ecological research to real management problems is limited (Leenhardt et al., 2015). While impressive advances have been made in social-ecological research, significant challenges remain in the translation to applied domains, and in particular in providing decision-relevant information to policy makers (Leenhardt et al., 2015).

A key constraint is that integration of social and ecological aspects in SES research is often limited (Rissman & Gillon, 2016). The most common approach is still to focus on single hypotheses, based in one or two disciplines (Leenhardt et al., 2015). Models are often insufficiently integrated as a result of greater focus on either the biophysical or the socioeconomic components (Kragt, 2012). Rissman and Gillon (2016) found that in the majority of SES research, ecological variables (and particularly biodiversity) are included in a trivial way or not at all. These are important considerations because research that meaningfully integrates ecological and social components generally provides more specific and useful policy recommendations (Rissman & Gillon, 2016).

A further barrier to the mainstream adoption of social-ecological analyses relates to challenges incorporating the range of behaviors and responses from social systems and its feedback into ecological systems (Schulter et al., 2016). Capturing the dynamic nature of social responses and selecting among the myriad of potentially relevant variables present major technical and theoretical challenges (Bodin & Tengö, 2012). Social simulation is arguably a transforming trend in 21st century social science, with great potential for modeling socioecological systems (Cioffi-Revilla, 2010), but this type of analysis is yet to be utilized extensively in environmental decision making.

Researchers with the multidisciplinary skills required to undertake integrated social-ecological analyses are rare. Hence, teams of researchers representing different disciplines may be required (Leenhardt et al., 2015), but this comes with its own challenges. Philosophical differences across the disciplines may act as a barrier to collaboration: ecologists tend to use models as inferential or predictive tools while social scientists more often use them as vehicles for communication (Drechsler et al., 2007). A fundamental obstacle to bridging these disciplines lies in translating the complexity of qualitative data typical of the social sciences into realistic, meaningful quantitative measures for use in formal, analytical models (Cooke et al., 2009).

While modeling SES is inherently difficult, we otherwise must rely on empirical understandings of current conditions, which can give a misleading representation of how the system will operate into the future. Without such models, *a priori* evaluations of policy approaches will remain simplistic and will continue to overlook key social or ecological processes that determine their success or failure (Bekessy & Cooke, 2011). Hence, there is substantial benefit to overcoming the substantial disciplinary, theoretical and methodological impediments to widespread adoption of socioecological methods to support water resources management.

ACKNOWLEDGEMENTS

Sarah Bekessy and Matthew Selinske were supported by an Australian Research Council Future Fellowship FT130101225 and the Australian Research Council Centre of Excellence for Environmental Decisions.

REFERENCES

Anderies, J. M., Janssen, M. A., & Ostrom, E. (2004). A framework to analyze the robustness of social-ecological systems from an institutional perspective. *Ecology and Society*, *9*, 18.

Axelrod, R. M. (2006). *The evolution of cooperation*. New York, NY: Basic Books.

Anthony, K., Wolff, N., Mumby, P., McDonald-Madden, E. & Devlin, M. (2014). Informing Management Decisions To Reduce GBR Coral Vulnerability Under Environmental Change. Retrieved from http://www.nerptropical.edu.au/publication/informing-management-decisions-reduce-gbr-coral-vulnerability-under-environmental-change.

Bekessy, S., & Cooke, B. (2011). The social and cultural drivers behind the success of PES. *Payments for ecosystem services and biodiversity* (pp. 141–169). Rome: FAO.

Belhouchette, H., Louhichi, K., Therond, O., Mouratiadou, I., Wery, J., Van Ittersum, M., et al. (2011). Assessing the impact of the Nitrate Directive on farming systems using a bio-economic modeling chain. *Agricultural Systems*, *104*, 135–145.

Bellman, R. (1957). *Dynamic programming*. Princeton, NJ: Princeton University Press.

Bengston, D. N., Kubik, G. H., & Bishop, P. C. (2012). Strengthening environmental foresight: Potential contributions of futures research. *Ecology and Society*, *17*, 10.

Berkes, F., Folke, C., & Colding, J. (2000). *Linking social and ecological systems: Management practices and social mechanisms for building resilience*. Cambridge, UK: Cambridge University Press.

Binder, C. R., Hinkel, J., Bots, P. W., & Pahl-Wostl, C. (2013). Comparison of frameworks for analyzing social-ecological systems. *Ecology and Society*, *18*, 26.

Bodin, Ö., & Tengö, M. (2012). Disentangling intangible social–ecological systems. *Global Environmental Change*, *22*(2), 430–439.

Bodin, Ö., Crona, B., Thyresson, M., Golz, A. L., & Tengö, M. (2014). Conservation success as a function of good alignment of social and ecological structures and processes. *Conservation Biology*, *28*, 1371–1379.

Buckley, M., & Haddad, B. M. (2006). Socially strategic ecological restoration: A game-theoretic analysis shortened. *Environmental Management*, *38*, 48–61.

Carpenter, S. R., Booth, E. G., Gillon, S., Kucharik, C. J., Loheide, S., Mase, A. S., et al. (2015). Plausible futures of a social-ecological system: Yahara watershed, Wisconsin, USA. *Ecology and Society*, *20*, 10.

Chan, T. U., Hart, B. T., Kennard, M. J., Pusey, B. J., Shenton, W., Douglas, M. M., et al. (2012). Bayesian network models for environmental flow decision making in the Daly River, Northern Territory, Australia. *River Research and Applications*, *28*, 283–301.

Cioffi-Revilla, C. (2010). Computational social science. *Computational Statistics*, *2*, 259–271.

Drechsler, M., Grimm, V., Mysiak, J., & Wätzold, F. (2007). Differences and similarities between ecological and economic models for biodiversity conservation. *Ecological Economics*, *62*(2), 232–241.

European Environment Agency (2016). The DPSIR framework. Retrieved from http://www.eea.europa.eu/publications/92-9167-059-6-sum/page002.html.

Gleeson, T., Alley, W. M., Allen, D. M., Sophocleous, M. A., Zhou, Y., Taniguchi, M., et al. (2012). Towards sustainable groundwater use: Setting long-term goals, backcasting and managing adaptively. *Ground Water*, *50*, 19–26.

Gleick, P. H. (1998). Water in crisis: Paths to sustainable water use. *Ecological Applications*, *8*, 571–579.

Gordon, A. (2015). Implementing backcasting for conservation: Determining multiple policy pathways for retaining future targets of endangered woodlands in Sydney, Australia. *Biological Conservation*, *181*, 182–189.

Graveline, N. (2016). Economic calibrated models for water allocation in agricultural production: A review. *Environmental Modelling & Software*, *81*, 12–25.

Gregory, R., Failing, L., Harstone, M., Long, G., McDaniels, T., & Ohlson, D. (2012). *Structured decision making: A practical guide to environmental management choices*. West Sussex, UK: Wiley Blackwell.

Grimm, V., Revilla, E., Berger, U., Jeltsch, F., Mooij, W. M., Railsback, S. F., et al. (2005). Pattern-oriented modeling of agent-based complex systems: Lessons from ecology. *Science*, *310*, 987–991.

Guerrero, A. M., McAllister, R. Y. A. N., Corcoran, J., & Wilson, K. A. (2013). Scale mismatches, conservation planning, and the value of social-network analyses. *Conservation Biology*, *27*, 35–44.

Janssen, M. A., & Ostrom, E. (2006). Governing social-ecological systems. *Handbook of Computational Economics*, *2*, 1465–1509.

Kragt, M. (2012). *Bioeconomic modeling: Integrating economic and environmental systems? In: R. Seppelt, A. A. Voinov, S. Lange, & D. Bankamp (Eds.), Proceedings of international congress on environmental modelling and software managing resources of a limited planet*. Leipzig, Germany: International Environmental Modelling and Software Society (iEMSs). http://www.iemss.org/society/index.php/iemss-2012-proceedings.

Lähdesmäki, H., & Shmulevich, I. (2008). Learning the structure of dynamic Bayesian networks from time series and steady state measurements. *Machine Learning*, *71*, 185–217.

Larrosa, C., Carrasco, L. R., & Milner-Gulland, E. J. (2016). Unintended feedbacks: Challenges and opportunities for improving conservation effectiveness. *Conservation Letters*, *9*, 316–326.

Leenhardt, P., Teneva, L., Kininmonth, S., Darling, E., Cooley, S., & Claudet, J. (2015). Challenges, insights and perspectives associated with using social-ecological science for marine conservation. *Ocean and Coastal Management*, *115*, 49–60.

Levin, S., Xepapadeas, T., Crépin, A. S., Norberg, J., De Zeeuw, A., Folke, C., et al. (2013). Social-ecological systems as complex adaptive systems: Modeling and policy implications. *Environment and Development Economics*, *18*, 111–132.

Little, L. R., Punt, A. E., Mapstone, B. D., Begg, G. A., Goldman, B., & Williams, A. J. (2009). An agent-based model for simulating trading of multi-species fisheries quota. *Ecological Modelling*, *22*, 3404–3412.

Liu, J., Dietz, T., Carpenter, S. R., Alberti, M., Folke, C., Moran, E., et al. (2007). Complexity of coupled human and natural systems. *Science*, *317*, 1513–1516.

Madani, K. (2010). Game theory and water resources. *Journal of Hydrology*, *381*, 225–238.

McCarthy, M. A., & Possingham, H. P. (2007). Active adaptive management for conservation. *Conservation Biology*, *21*, 956–963.

Miller, J. H., & Page, S. E. (2009). *Complex adaptive systems: An introduction to computational models of social life*. Princeton, NJ: Princeton University Press.

Mitchell, C. A., & White, S. (2003). Forecasting and backcasting for sustainable urban water futures. *Water*, *30*, 25–30.

Olsson, P., Gunderson, L. H., Carpenter, S. R., Ryan, P., Lebel, L., Folke, C., et al. (2006). Shooting the rapids: Navigating transitions to adaptive governance of social-ecological systems. *Ecology and Society*, *11*, 18.

Ostrom, E. (2007). A diagnostic approach for going beyond panaceas. *Proceedings of the National Academy of Sciences of the United States of America*, *104*, 15181–15187.

Pahl-Wostl, C., Lebel, L., Knieper, C., & Nikitina, E. (2012). From applying panaceas to mastering complexity: Toward adaptive water governance in river basins. *Environmental Science and Policy*, *23*, 24–34.

Parrachino, I., Dinar, A., & Patrone, F. (2006). *Cooperative game theory and its application to natural, environmental, and water resource issues: 3. Application to water resources: World Bank Policy Research Working Paper, Washington*.

Patrício, J., Elliot, M., Mazik, K., Papadopoulou, K., & Smith, C. J. (2016). DPSIR—Two decades of trying to develop a unifying framework for marine environmental management? *Frontiers in Marine Science*, *14*, 177.

Pearl, J. (1988). *Probabilistic reasoning in intelligent systems*. San Mateo, CA: Morgan Kaufmann.

Peterson, G. D., Cumming, G. S., & Carpenter, S. R. (2003). Scenario planning: A tool for conservation in an uncertain world. *Conservation Biology, 17*, 358–366.

Pittock, J. (2016). The Murray–Darling Basin: Climate change, infrastructure and water. In C. Tortajada (Ed.), *Increasing resilience to climate variability and change* (pp. 41–60). Singapore: Springer.

Rissman, A. R., & Gillon, S. (2016). Where are ecology and biodiversity in social-ecological systems research? A review of research methods and applied recommendations. *Conservation Letters*. http://dx.doi.org/10.1111/conl.12250.

Ruben, R., Moll, H., & Kuyvenhoven, A. (1998). Integrating agricultural research and policy analysis: Analytical framework and policy applications for bio-economic modeling. *Agricultural Systems, 58*, 331–349.

Runge, M. C., Bean, E., Smith, D. R., & Kokos, S. (2011). *Non-native fish control below Glen Canyon Dam—Report from a structured decision-making project: US Geological Survey open-file report number 2011–1012*. http://pubs.usgs.gov/of/2011/1012/ [74 pp.].

Schlüter, M., Baeza, A., Dressler, G., Frank, K., Groeneveld, J., Jager, W., et al. (2017). A framework for mapping and comparing behavioural theories in models of social-ecological systems. *Ecological Economics, 131*, 21–35.

Schlüter, M., Mcallister, R. R. J., Arlinghaus, R., Bunnefeld, N., Eisenack, K., Hölker, F., et al. (2012). New horizons for managing the environment: A review of coupled social-ecological systems modeling. *Natural Resource Modeling, 25*, 219–272.

Schlüter, M., & Pahl-Wostl, C. (2007). Mechanisms of resilience in common-pool resource management systems: An agent-based model of water use in a river basin. *Ecology and Society, 12*, 4.

Snijders, T. A. B. (2001). The statistical evaluation of social network dynamics. *Sociological Methodology, 31*, 361–395.

Stein, C., Ernstson, H., & Barron, J. (2011). A social network approach to analyzing water governance: The case of the Mkindo catchment, Tanzania. *Physics and Chemistry of the Earth, Parts A/B/C, 36*, 1085–1092.

Torabi, N., Mata, L., Gordon, A., Garrard, G., Wecott, W., Dettmann, P., et al. (2016). The money or the trees: What drives landholders' participation in biodiverse carbon plantings? *Global Ecology and Conservation, 7*, 1–11.

von Neumann, J., & Morgenstern, O. (1944). *Theory of games and economic behavior*. Princeton, NJ: Princeton University Press.

Whalley, R. D. B., Price, J. N., Macdonald, M. J., & Berney, P. J. (2011). Drivers of change in the social-ecological systems of the Gwydir Wetlands and Macquarie Marshes in northern New South Wales, Australia. *The Rangeland Journal, 33*, 109–119.

ASSESSING AND MANAGING THE SOCIAL EFFECTS OF WATER REFORM IN AGRICULTURAL AREAS

10

J. Schirmer
University of Canberra, Canberra, ACT, Australia

CHAPTER OUTLINE

INTRODUCTION

WHY WATER REFORM NEEDS TO BE SOCIALLY ACCEPTABLE

Water-reform processes involve changing how people access, use and interact with water resources, whether the reforms are implemented in response to a drying climate, increasing climatic variability, or historical overallocation of water. The centrality of human action to water policy and management means that, in order to succeed, water-reform processes need to be socially acceptable. Attempts to implement water-reform policies or programs that have low social acceptability commonly result in social conflict (Gross, 2008). Social acceptability thus needs to be included alongside other core considerations when designing and implementing water-reform processes:

> The choice and development of institutional structures supporting water and irrigation policy is ultimately a compromise between the physical nature of the resource, human reactions to policies and competing social objectives ... Policy implementation is a complex socio-political process of balancing the requirements of political and social acceptability, economic and technical feasibility and administrative reality.
>
> **Shi and Meyer (2009), p. 767**

Water reform will be socially acceptable only if policies and programs use processes and achieve outcomes that are just and fair (Gross, 2008; Trawick, 2002), successfully minimize or mitigate negative impacts on humans, and provide opportunities for positive socioeconomic outcomes (Vanclay et al., 2015). However, consideration of these social effects of water reform often occurs relatively late in the process of developing water reforms, after addressing other aspects such as technical feasibility and ecological outcomes, and in many cases social assessments are not utilized as part of decision-making processes for water reform (Baldwin et al., 2009). This (often unintentional) treatment of social effects as a secondary consideration can reduce the likelihood of water reform achieving any of its objectives, whether social, environmental, or economic.

There is growing recognition of the importance of early and ongoing consideration of social effects when designing policies and programs for any kind of natural resource management. In part due to recognition of this need, the idea of integrated water resource management (IWRM) has gained currency internationally, based on the principle that successful water management requires integrated assessment of both environmental health and societal needs. IWRM has become a basis for water governance and management in many catchments across multiple countries since it first emerged in the 1990s, and typically involves collaborative water-resource planning by multiple stakeholders, ensuring different interests are included in decision-making processes.

However, despite its increasing popularity, successfully implementing IWRM remains a challenge. In particular, many IWRM approaches fail to conduct adequate social impact assessment (SIA) or to bring findings from this assessment to bear as part of their decision-making processes (Hooper, 2011). Addressing this gap requires integrating best-practice approaches to conducting SIAs, also often called social assessments, into water-resource management processes.

ASSESSMENT OF SOCIAL EFFECTS (SIA) IS CRITICAL TO ACHIEVING SOCIAL ACCEPTABILITY

The assessment of social effects is essential to achieving social acceptability of water reform. However, to achieve this, the assessment needs to be a central part of the process of designing and implementing reforms. The field of SIA emerged in the 1960s to complement environmental impact assessment. In its early days, SIA typically involved a single predictive assessment of social impacts prior to implementing a policy or program, and often had relatively little influence on design or implementation. As a consequence, it often failed to ensure social acceptability of the policy or program (Esteves et al., 2012). SIA has since evolved into an ongoing process of assessing social effects that occurs prior to and throughout implementation of a program or policy, enabling better consideration of social effects at all points from initial design through to implementation. Simultaneously, the goal of SIA has evolved from a focus on mitigating harm, to proactively designing policies, programs or projects to increase their social benefits as well as minimize social costs:

> [There is an] increasing focus in SIA on enhancing the benefits of projects to impacted communities. Although the need to ensure that the negative impacts are identified and effectively mitigated remains, also of value is revising projects and ancillary activities to ensure greater benefits to communities. This is necessary for the project to earn its 'social licence to operate'; and also because attempting to minimise harm (the traditional approach in SIA) does not ensure that the project will be considered acceptable by local stakeholders … Important to earning a social licence to operate is to treat communities with respect. Meaningful, transparent and ongoing community engagement practices from the earliest stages of any intervention are essential to build trust and respect.
>
> **Vanclay et al. (2015), p. iv**

This shift to supporting positive outcomes is consistent with growing international recognition of the importance of building human capacity to adapt successfully to changes such as water reform. For example, climate-change adaptation approaches emphasize the importance of enabling individuals and communities "…*to moderate or avoid harm or exploit beneficial opportunities*" (IPCC, 2014, p. 118) via building their adaptive capacity.

Successful use of social assessment in water-resources management requires both identifying robust methods of social assessment, and ensuring the findings of social assessment are integrated into decision-making processes about water-resource policies and management. This may require challenging existing assumptions about what should be assessed, what is a robust and defensible method of assessing it, and the values and principles that should underpin decision making about the management of water resources. For example, if decision makers are used to making decisions based on assessing the overall opportunity cost or cost-benefit of a policy, rather than assessing distribution of impacts for different people and stakeholder values, attempting to use results of social assessment may require substantial change in thinking and perspectives. On this point, Syme et al. (2008) noted that the use of self-report measures, a common tool in social assessment, is often rejected by policymakers, and that this acts as a barrier to using results of social assessments in water policy and management.

Australian experiences of water reform provide many lessons that can inform achieving successful integration of social assessments in water-reform processes. This chapter focuses on lessons learned from the development and early years of implementation of the Murray-Darling Basin Plan (Basin Plan), to identify three key principles for successfully assessing and managing social effects in water reform processes. These are: (a) the importance of good public participation, (b) ensuring robust methods are used to assess the social outcomes of water reform throughout design and implementation, and (c) ensuring social effects are considered when selecting and designing the policy instruments used to enact water reform.

THE MURRAY-DARLING BASIN AND THE BASIN PLAN

The Murray and Darling rivers, located in southeastern Australia, form Australia's longest river system. Their catchment—the Murray-Darling Basin—covers 14% of Australia's land mass (Balcombe et al., 2011), and its water flows have the highest variability of those on any continent on earth (Quiggin et al., 2010). The Basin crosses multiple jurisdictions, containing parts of four Australian states and one territory. It is environmentally significant, containing more than 30,000 wetlands, 16 listed as internationally important under the Ramsar Convention on Wetlands (ABS, 2008). Also, 39% of Australia's agricultural produce (ABS, 2008) comes from the Basin, much of it from the approximately 2% of irrigated land in the Basin (Wei et al., 2011). The expansion of agriculture in the Basin was in part driven by government-sponsored irrigation schemes, involving large programs of dam building and diversion of water flows to create the Basin's many irrigation areas (Quiggin et al., 2010).

From the 1980s there was growing concern about overallocation of water for agricultural purposes in the Basin, and the consequences of overallocation for both the health of key wetland and water-dependent ecosystems, and for ensuring water supply to irrigators during the droughts that periodically occur in the Basin (Quiggin et al., 2010). A series of policy reforms were enacted to help address these problems. These included the separation of water rights from land rights, creation of water markets to enable water to go to the most valuable and efficient uses, imposing an upper limit on the volume of surface water extracted for agriculture, improving the efficiency of water use by irrigators, and increasing water flows to some key environmental sites in the Basin (Grafton & Horne, 2014). However, of these reforms, "none … provided an enduring solution" (Quiggin, 2012).

The need for water reform was given further impetus by the Millennium drought, one of the severest and longest droughts recorded since European colonization of Australia. Lasting from around 1997 to 2009 in southeastern Australia[1] (Gale et al., 2014), Basin water inflows reached historical lows in 2007. The then federal government implemented a Commonwealth Water Act that established the Murray-Darling Basin Authority (MDBA). The MDBA had responsibility for developing "a Basin Plan that would set sustainable diversion limits (SDLs) for surface and groundwater, basin-wide environmental objectives, and measures to improve the security of water entitlements," among other objectives (Grafton & Horne, 2014).

For the next 3 years, the Basin Plan was developed, a process that was often accompanied by high levels of public controversy and debate about the likely socioeconomic effects of the Plan, described in further detail subsequently (see also Chapters 6, 11 and 13). In Nov. 2012, the Murray-Darling Basin Plan was passed by the Federal Parliament. From 2019, SDLs will apply that reduce water available for consumptive use in the Basin by 2750 gigaliters (GL)/y, representing a reduction of around 20% compared to 2009 water volumes (Horne, 2013). The period of 7 years between passage of the Plan and the implementation of the SDLs was put in place to enable gradual achievement of the 2750 GL/y reduction through a mix of instruments included in the Plan (Horne, 2013).

Three principal instruments were put in place to "recover" water that could then be used for "environmental watering" to support environmental health in the Basin[2]:

- Water purchase by the government: irrigators could voluntarily sell part or all of their water entitlements to the government
- On-farm infrastructure improvement: government grants supported irrigators to improve the water-use efficiency of their on-farm irrigation infrastructure, with resulting water savings shared between irrigators and the government

---

[1]Exact dates are contested, and varied in different regions.
[2]In addition to these three measures, water trading reforms have sought to increase transparency and ease of water trading, providing a market mechanism intended to support water being used where it is able to produce best value.

- Off-farm infrastructure improvement: government grants were provided to irrigation infrastructure operators to enable modernization of off-farm irrigation infrastructure, meaning the irrigation channels and other infrastructure used to deliver water to farms; this reduced water use through factors such as reduced water evaporation.

By Jun. 2016, these instruments had resulted in 72%, or 1981 GL, being "recovered" of the target 2750 GL (MDBA, 2016). Of this, 1168 GL was recovered through direct sale of water entitlements to the government by irrigators, 764 GL through infrastructure modernization (both on-farm and off-farm), and 49 GL through other means.

UNDERSTANDING SOCIAL EFFECTS OF THE PLAN

The Basin Plan involves multiple actions, and for this reason a given person or community may experience very different social effects as a result of the Plan, depending on how many of the actions related to the Plan directly or indirectly affect their lives. The Plan can have effects both through the *processes* used to develop and implement the Plan, and through the specific *mechanisms* put in place, including the measures to reduce consumptive water use described previously, and the increases in delivery of water to environmental sites (environmental watering) that also form a key part of the Plan. Together, these processes and mechanisms have a range of effects, that contribute to overall experiences of the Plan and whether it is considered a socially acceptable form of water reform.

To fully understand the complex social effects of developing and implementing the Basin Plan, it is important to first consider the nature of "social impacts."[3] A social effect, or social impact, is sometimes defined broadly to simply mean "everything that affects people" (Vanclay et al., 2015). More specifically, it is often defined as "the lived experience of social change" (Schirmer, 2011).

Importantly, social change is not the same as social impact. Each of the actions implemented as part of the Basin Plan can lead to a range of changes: for example, changes in agricultural production, in health of local environmental sites, in efficiency of irrigation on farms, in labor time required on farms, to name a few. These changes can all be experienced in different ways: a reduction in job hours, for example, may be experienced as a benefit by a person who is financially well off and seeking to reduce their workload, and as a cost by a person who needs the income they were previously receiving from higher work hours. This means that each of these changes will affect ("impact") people and communities in differing ways depending on the specific circumstances of those people and communities.

We need to both understand the changes resulting from the Basin Plan, *and* to assess how people experience these changes (the social "impacts" of the changes) (Van Schooten et al., 2003). These changes can be analyzed at the scale of a household or an entire community. For example, it is possible to identify the proportion of households experiencing an increase in work hours, or to model how much employment hours have increased across an entire community. Social impacts—the ways people experience these social changes—were defined by Vanclay (2002) as including the following:

- people's *way of life*—how they live, work, play, interact with one another;
- their *culture*—shared beliefs, customs, values, and language or dialect;
- their *community*—cohesion, stability, character, services, and facilities;

---

[3]In this document, the terms *social change*, *social impact* and *social effects* are used inclusively to refer to changes that are positive, negative or neutral for the people they affect.

- their *political systems*—are people able to participate in decisions that affect their lives;
- their *environment*—the quality of the air and water people use, and access to and control over resources;
- their *health and well-being*—physical, mental, social, and spiritual well-being;
- their *personal and property rights*—how people are economically affected, and how their rights are affected;
- their *fears and aspirations*—perceptions about safety, their future, their community's future.

Experiences to date of assessing social effects of the Basin Plan suggest that three key lessons emerge, each of which is examined in the following subsections. The first is that good public consultation processes can reduce the experience of negative social impacts and that, as a corollary, the process of assessing social effects can itself create effects. The second is that good social assessment goes well beyond public consultation, using robust methods to assess the social outcomes of water reform. The third is that, to be most effective, social assessment needs to inform the selection and design of policy instruments.

ROBUST PUBLIC CONSULTATION IS ESSENTIAL TO BOTH SOCIAL ACCEPTABILITY AND EFFECTIVE SOCIAL ASSESSMENT

To be effective, social assessments must examine the issues that matter to the people potentially affected by water reform, and must do so using methods trusted by these people. Robust public consultation processes are critical to achieving both these things, and a lack of public consultation can itself cause negative social impacts. This is highlighted by a key period of public controversy that emerged during the development of the Basin Plan (see also Chapter 11). As part of development of the Plan, a Guide to the proposed Plan was released publicly in 2010 (MDBA, 2010). The Guide was intended as a consultation document, and proposed multiple scenarios for change, including some that proposed a much larger reduction in consumptive water use than the 2750 GL in the eventual Basin Plan. The Guide was met with strong criticism and opposition from many irrigation-dependent communities in the Basin, including what became an infamous demonstration in which irrigators burned copies of the Guide (Gale et al., 2014).

Criticisms of the Guide focused on two aspects: a view that social and economic issues had not been adequately considered as part of developing the scenarios, and concern about lack of community consultation, illustrated by the following quotes:

> … the Murray–Darling Basin Authority's Guide to the Proposed Basin Plan was … a plan proposing water cuts that would have, if not outright killed irrigation areas and their towns, at the very least caused them to atrophy, because social and economic factors were not considered. A plan that was assembled without consultation with farmers and scientists from the irrigation areas. … This is not a storm in a teacup: over 2 million people live in the basin, which produces 40% of our agriculture.
>
> **Jennings (2011)**

> The Guide to the proposed Basin Plan has, for the most part, been designed by scientists, engineers and other technical experts, with little input from planners and social scientists.
>
> **Boully and Maywald (2011)**

Multiple analyses have concluded that the widely reported public backlash to the Guide was in large part due to these two factors, and have suggested that the solution to preventing similar future public outrage involves stronger consultation prior to releasing an "options" style document, and better social and economic assessment. However, criticisms of the Guide involved more than simply a call for more consultation and more assessment. Many of those who criticized the Guide specifically criticized the approach to social and economic assessment that was included in the Guide.

The Guide did in fact include some assessment of economic outcomes of the proposed Basin Plan scenarios. These initially concluded that, in the long run, 800 jobs would be lost across the Basin as a result of the Plan, while acknowledging that "short run impacts" might involve greater numbers of job losses. The Guide also pointed to the potential for more significant negative impacts in several irrigation-dependent regions within the Basin. As Mulligan (2011) pointed out, however, the assessments that were included were generalized across the whole Basin, looked only at the long-term outcomes, and while mentioning the potential for some communities to experience more negative impacts, did not examine in detail what might be experienced by specific individuals and communities—or how to mitigate these effects. Gross (2011) further highlighted that lack of good process surrounding the social and economic assessments that had been undertaken led to a strong sense of injustice among those likely to be affected by the Plan. A subsequent examination of community impacts of the Plan, commissioned by the MDBA, found that the socioeconomic assessment included as part of the Guide itself had negative social impacts, reducing confidence of irrigation communities in their future: "Communities felt that they had been abandoned. The figure of only 800 jobs being lost was considered an insult to their intelligence" (EBC et al., 2011).

The experience of the release of the Guide highlights what is well known in SIA: the processes used to develop a policy and associated programs of action themselves have social effects. Work on "anticipatory impacts" suggests that negative social impacts, particularly fear and anxiety, are experienced most intensely during the period in which a policy or action is being planned. During this pre-implementation phase, people and communities know that there will be change, but do not yet know the nature of the changes, as they have not yet been finalized. This anticipatory period is characterized by high uncertainty about the impacts of the development on a person or community's future, and is typically accompanied by high levels of concern about the potential risks (Loxton et al., 2013; Vanclay, 2012; Walker et al., 2000).

Anticipatory impacts are important: experiencing uncertainty can be associated with many negative effects. For example, a person who is uncertain about the effects of the Basin Plan on their community may hold off investing in expanding their business, buying a house, or may elect to take a job offer in another community. Thus uncertainty about the future can itself lead to social and economic change, and social impacts.

The release of the Guide thus increased the experience of anticipatory impacts, and in itself created negative social impacts in terms of reduced confidence of many living in irrigation-dependent communities about their future. This potential for anticipatory impacts did not end when the Basin Plan was signed into law. The 7-year period of transition between signing the Plan and full implementation of the 2750 GL reduction also represents an anticipatory period. This means there is ongoing anticipatory impact, in which the quality of the process of decision making can have an important effect on how people experience the Basin Plan.

Reducing anticipatory impacts requires increasing the level of certainty people feel about their future. This means building the confidence of people living in Basin communities dependent on irrigated agriculture that they know how the Basin Plan will affect them (including positive and negative effects), that they can cope with negative effects and take advantage of any opportunities offered by the Plan, and that they can communicate concerns and engage in decision-making processes about the Plan if they wish to.

To better understand the experience of anticipatory impacts, questions were asked about this as part of the University of Canberra's *Regional Wellbeing Survey*, an annual survey started in large part to enable assessment of the social impacts of the multiple changes typically being experienced by rural and regional Australians at any given point in time—one of which is water reform (Schirmer & Berry, 2014). This annual survey aims to make data available that can support robust SIA, recognizing that a key gap in assessment of social effects of water reform and other changes affecting rural communities is a lack of available data (Baldwin et al., 2009).

By 2014, 2 years into implementation of the Plan, data from the Regional Wellbeing Survey identified that most Basin residents who had an interest in the Basin Plan felt they could access information about the Plan (Fig. 1). However, only just over one in three felt they knew how the Plan would affect them, or that they could communicate their views to decision makers. Further, one in five or fewer felt that local knowledge had informed the Plan, that the Plan increased their confidence in the future, or that decision makers would listen to their views. Lack of confidence in being listened to is not unusual: for example, data from the same survey used to produce Fig. 1 found that most rural and regional Australians did not feel their views on a wide range of issues would be listened to (Schirmer & Berry, 2014). These results suggest there is a high potential for ongoing negative anticipatory impacts from the Basin Plan. It suggests in particular a potential trade-off between having an extended period of implementation with flexibility in how water reductions will occur, and the uncertainty felt by residents who find it complex to understand what the consequences will be for their community. These findings emphasize the challenge of implementing adaptive management while also providing adequate certainty for Basin residents to plan for their futures.

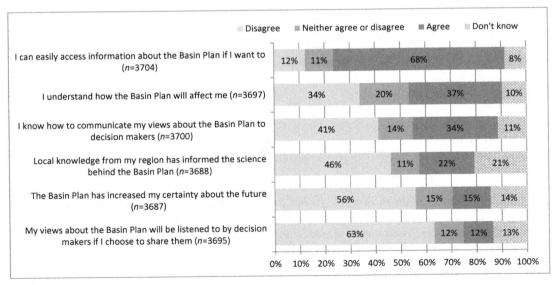

FIG. 1

Views of Basin Plan processes: confidence in accessing information, being heard, and certainty about the future.

Modified from Schirmer, J., & Berry, H. (2014). People and place in Australia: The 2013 Regional Wellbeing Survey. *Canberra: University of Canberra.*

Good social assessment should involve assessing the quality of community and public participation processes, and helping decision makers improve the quality of these processes. Good assessment should also itself use appropriate consultation processes: social assessments should include consultation of people and communities affected by the Plan, as well as decision makers charged with implementing the Plan. This consultation is necessary to identify the issues that need to be assessed, and the types of information needed by people, communities and decision makers. Only if this is done will an assessment focus on the right issues and be able to make a genuine contribution to achieving more positive social outcomes.

ROBUST SOCIAL ASSESSMENTS MUST GO BEYOND PUBLIC CONSULTATION

Social assessment and public consultation are often conflated: some assume that SIA is simply a process of consultation. As stated previously, public consultation is key to good social assessment. If the concerns of those potentially impacted are not understood, the assessment will fail to examine what really matters to communities, and will not be able to identify the right measures for mitigating negative impacts and supporting positive outcomes. However, robust social assessment involves much more than public consultation. Ideally, it involves producing the information needed to inform ongoing consultation and decision-making processes. For example, public consultation may identify concerns that the Basin Plan could result in job losses, population decline, or increased stress for farmers. During the design of the Plan, good social assessments will use robust methods to predict the likelihood of each of these outcomes, and most importantly, whether the Plan can be designed and implemented in ways that minimize the risk of these outcomes. During implementation of the Plan, ongoing social assessments should examine whether these outcomes are occurring, and whether any change is needed to the mechanisms put in place to reduce likelihood of these outcomes.

Doing this requires using methods that go beyond consulting with stakeholders, to collecting evidence that enables assessment of outcomes, and making this evidence available to inform subsequent discussions. This type of assessment, however, presents a wide range of challenges. The first and most common is that the data needed for good social assessments are typically not readily available: new data typically need to be collected in order to properly assess social effects (Baldwin et al., 2009). Even once appropriate data are collected, there are key challenges in analyzing these data: two of the principal considerations are assessing cause-and-effect relationships, and assessing the distribution of effects and cumulative effects.

Analyzing cause and effect in the "real world"

At any given point in time communities will typically be experiencing a number of different types of change. For example, an irrigation-dependent community in the Basin might be experiencing changes caused by shifting agricultural markets, climatic variability (e.g., drought, floods), changes in local governance, services or infrastructure, or the Basin Plan, to name just a few. The complexity of the changes occurring in a community increases the possibility that residents will not be aware of all the factors contributing to particular social changes they are observing, and will "misattribute" these social changes as being largely or solely caused by whichever factor is receiving the greatest public attention (Williams & Schirmer, 2012). Good social assessment can produce information that helps residents examine the relative effects of different changes occurring in their community, and better identify what is likely to be causing both negative change and positive opportunities. This in turn can help ensure that the right actions are taken to address negative social impacts, through ensuring their causes are correctly identified.

Studies of social impact undertaken before implementation of a policy or program typically use models to predict impacts: these models assume away much of the complexity of multiple changes occurring in the real world. They are highly useful, and form an important part of impact assessment. However, they may produce results that are not then easily observed when the project is actually implemented. For example, a model may predict that the Basin Plan will result in 100 job losses in a given community. Three years into implementation of the Plan, data may show that the community has in fact lost 500 jobs, triggering debate about whether the additional job losses are due to the Plan, or due to other changes that are simultaneously happening in the community, such as a downturn in agricultural markets. Trust in social assessments based purely on modeling is likely to be low in this situation, as members of the community question whether the initial modeling adequately accounted for all the effects of the Plan. Good social assessment therefore needs to go beyond predictive or oversimplified models, and track "real-world" outcomes, including being able to answer the question of whether the Plan or other factors caused the 500 job losses. This requires tackling the challenge of untangling likely cause-and-effect relationships using real-world data.

Using real-world data to identify the likely links between water reform ("cause") and social impacts ("effect") is challenging. The term "likely" is used here as, from a scientific point of view, it is not possible to definitively prove the exact amount of social change or impact resulting from the Basin Plan. Strict causal proof, in statistical terms, can only be established when there has been a clear experimental design in which the only factor that varies is the Basin Plan. For example, this would require comparing communities in which every factor influencing social change was identical, with the exception of the Basin Plan. Such a study design is not possible in the real world, and for this reason multiple alternative approaches have been developed (Schirmer, 2011).

Studies that need to examine real-world cause-and-effect relationships, such as those conducted in the field of epidemiology when tracking outbreaks of particular diseases, often draw on multiple criteria to establish the presence of a likely cause-effect relationship. In the absence of the ability to perform true experimental study design, these should ideally include as many as possible of the following "Bradford Hill" criteria (Lucas & McMichael, 2005): strength of associations between hypothesized causes and effects; consistency of observed associations; specificity of effects—for example, a particular social change only occurring among irrigators given on-farm infrastructure grants as part of the Plan; temporality (cause preceding effect); gradient (whether social effects follow a "dose-response" relationship with greater effects experienced by those who were more exposed to an activity occurring under the Plan); plausibility of cause-effect relationships; and coherence (consistency with known factors).

This can appear highly complex, but in reality can be assessed by tracing a coherent logic in terms of a chain of cause-and-effect events. This can be illustrated by examining the effects of government investment in on-farm infrastructure, one of the programs put in place to help achieve reductions in consumptive water use that form part of the Basin Plan.

To assess the social effects of infrastructure investment on irrigators, initial consultation was undertaken with a number of irrigators and farmer representative organizations. This consultation identified the types of cause-and-effect outcomes irrigators, water policy makers, and other stakeholders with an interest in water reform believed might be resulting from on-farm infrastructure investment (Schirmer, 2017). The following possible outcomes were identified for irrigators (the direct recipients of grants). Some of these contradict each other, as not all those consulted had the same views about likely outcomes:

- *Positive outcomes*—Increased flexibility of water delivery and use on farm, enabling more cost efficient production, higher profitability and hence higher household well-being due to higher income; reduced on-farm labor time; better ability to cope with periods of low water availability; reduced infrastructure maintenance costs,

- *Negative outcomes*—Increased power/electricity costs, increased farm debt (farmers had to contribute to costs of the upgrade).

A subsequent survey of irrigators then asked both (a) whether those who had received grants felt any of the hypothesized positive and negative outcomes had occurred, and (b) whether those who received on-farm grants were more likely to report experiencing any of the hypothesized positive and negative outcomes when compared to those who had not received grants.

Irrigators assessed the outcomes as being largely positive. While some negative impacts on electricity costs and debt were reported, the overwhelming majority of farmers felt the on-farm infrastructure had positive effects for their farm as a whole (Fig. 2). However, on its own this does not establish a clear cause-and-effect relationship beyond the perception of the farmer.

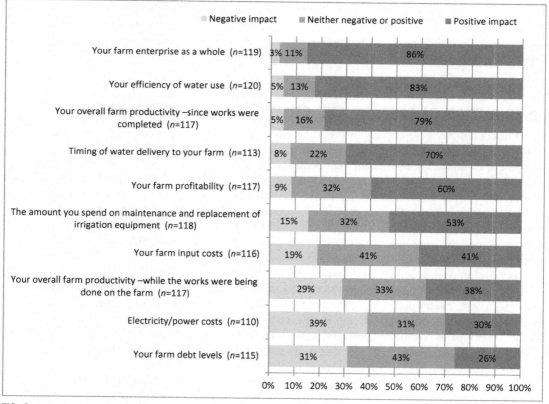

FIG. 2

Outcomes of upgrading/expanding on-farm infrastructure for irrigators who received grants as part of the Basin Plan.

Source: *2015 Regional Wellbeing Survey.*

To better establish whether on-farm infrastructure modernization had outcomes that were different from those that would have occurred in the absence of the works, irrigators who had and had not modernized their on-farm water infrastructure were compared (Schirmer, 2017). This analysis identified that Basin irrigators who had upgraded on-farm infrastructure were slightly more likely to report making a financial surplus on their farm in 2014–15 (54%) compared to those who had not upgraded since 2008 (48%). They were also more likely to be increasing the area they irrigated (26% compared to 12% of those who had not upgraded), and less likely to be decreasing either the area of land irrigated (24% compared to 33%) or reducing their overall farm production (9% compared to 15%). These relationships were typically stronger for irrigators who had upgraded on-farm infrastructure in recent years, and weaker for those who had upgraded several years previously, suggesting that they were likely to be a direct outcome of upgrading on-farm infrastructure (effects of infrastructure modernization are most likely to be observed in the initial years after infrastructure works are completed).

By comparing those who had and had not modernized infrastructure, and by comparing outcomes over time, there is reasonably high confidence that on-farm infrastructure has led to positive outcomes in terms of farm production and the farm enterprise. This can reasonably be assumed to have had a positive impact on the farmer's life overall. However, when the overall well-being of farmers who had and had not modernized infrastructure was compared, there were no significant differences, suggesting modernization is positive for most farmers, but does not cause changes substantial enough to be observable as changes in their overall well-being (Schirmer, 2017).

DISTRIBUTION OF EFFECTS AND CUMULATIVE EFFECTS MATTER

Social effects of any given policy or program will differ for different people and communities, and robust social assessment should examine the distribution of effects, not simply the "net" effect, of water reform. For example, an assessment that identifies the Basin Plan will result in net job losses of 800 across the Basin says little about the true social impacts of this change. A net job loss of 800 might involve 800 people choosing to retire early with money received from selling water entitlements, and few negative impacts. Or it might be the outcome of 3000 job losses in irrigated agriculture, concentrated in a small number of communities, and 2200 new jobs being created in different communities to those where jobs were lost. In this latter case, the net job loss figure of 800 is associated with significant negative social impacts for those who lost employment. Assessment of the distribution of effects is therefore just as important as assessment of the net outcome (Fig. 3).

Additionally, the social effects of water reform will differ depending on how the water reform interacts with other events occurring at the same time. For example, the effects of water reform might be different if water reform occurs during a period of extended drought, compared to a period of good rainfall. This type of "cumulative-effects" assessment is essential to understanding the real-world social effects of implementation of any change in natural resource management (Franks et al., 2010a, 2010b; Loxton et al., 2013), including water reform (Baldwin et al., 2009).

This is illustrated by assessment of the outcomes associated with on-farm water infrastructure modernization undertaken as part of the Basin Plan. As described earlier, many irrigators have received grants to facilitate modernization of their on-farm irrigation infrastructure, and this was associated with slightly higher on-farm profitability when all irrigators who had modernized infrastructure were compared to all those who had not.

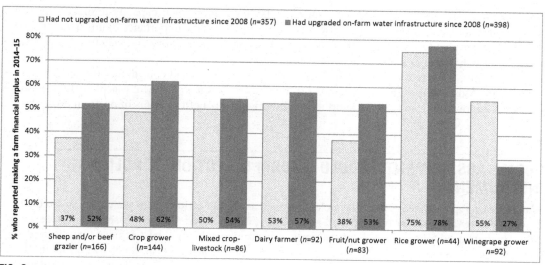

FIG. 3

Proportion of Basin irrigators who reported making a farm financial surplus in 2014–15: comparison of those who had upgraded on-farm infrastructure and those who had not, by farm type.

Source: *2015 Regional Wellbeing Survey.*

However, when the profitability of different *types* of irrigated agricultural producers was examined, not all were experiencing increased profitability as a result of on-farm infrastructure investment. While most types of producers were more likely to report making a profit if they had modernized on-farm water infrastructure than if they had not, the opposite was true for wine grape growers. Those who had modernized on-farm water infrastructure were substantially less likely to report making a profit on their farm in 2014–15 (Schirmer, 2017).

This finding was explored with representatives of wine-grape growing industries, who identified that in the period examined, many wine grape growers had been unable to take advantage of the water efficiencies of their upgraded infrastructure, as a severe downturn in the market for wine grapes in recent years had meant many wine grape growers were reducing production, and hence not fully utilizing their on-farm water infrastructure. Additionally, the poor market conditions meant many growers were selling their water allocation on the water trade market instead of using it to grow grapes. Those who had upgraded on-farm infrastructure with a government grant had lower volumes to sell, as they had transferred some of their water entitlements back to the government as part of the conditions of the grant, and hence had less opportunity to profit from water trading during the market downturn affecting the grape-growing sector.

In this example, the effects of water reform have differed for different irrigators. A policy instrument that achieves water savings—on-farm water infrastructure improvement—has had positive effects for profitability of irrigators who have experienced normal market conditions, as it has facilitated more efficient production requiring less water (and hence lower water costs). However, in the case of wine grape growers, the same instrument, through reducing the volume of water entitlements an irrigator has to sell on water markets, has reduced their capacity to make a profit by selling water during a period of market downturn. This highlights the importance of understanding how any policy instrument may interact with factors such as market conditions and water trade options, to result in very different outcomes for different irrigators.

Importantly, the effect on wine grape growers was short term, and considered likely to change in future years. Most of these growers felt their on-farm infrastructure investment was likely to be positive for their farm overall despite in the short term not bringing a benefit in terms of on-farm profitability.

SOCIAL ASSESSMENT SHOULD INFORM SELECTION OF POLICY INSTRUMENTS

Water reform typically seeks to achieve changes such as greater efficiency of water use, or a reduction in water use by a particular type of user, such as irrigation farmers. In most cases, the key decisions policy makers need to make are about the types of instruments that will best achieve their desired outcome. For example, a reduction in irrigated water use can be achieved in many ways, ranging from compulsory resumption of irrigator's water rights by the government, to voluntary sale of entitlements to government by irrigators, or water rights being transferred to government in return for irrigators receiving grants to increase on-farm water-use efficiency. Each of these instruments is likely to have very different social effects, and to have different levels of social acceptability. They are also likely to vary substantially in their overall cost to the government.

Water reform is not likely to be viewed as socially acceptable unless those affected see that social considerations have been included in the process of selecting policy instruments. Engaging in good public consultation and robust assessment is not enough on its own. Multiple authors have highlighted that consultations and assessments fail if they are not then used to inform decision making about how water reform will work in practice (Alston & Mason, 2008). For social assessment to inform the selection of policy instruments, however, the consideration of social effects must be a key part of the process of selecting these instruments, in addition to other critical considerations of cost-effectiveness and effectiveness in achieving change in water use.

The Murray-Darling Basin Plan provides useful illustrations of the challenge of doing this. An ongoing debate about implementation of the Basin Plan has focused on whether the water savings desired under the Plan should be achieved by investing in improving the water-use efficiency of irrigation infrastructure, with a goal of achieving water savings while enabling maintenance of agricultural production, or by simply purchasing water entitlements from irrigators who wish to sell them, with those irrigators then choosing how they wish to spend the money received from the sale.

Economic assessments have identified that if the objective is to return water to the environment in a cost-effective manner, then the market-based approach, direct purchase of water entitlements from irrigators, is the optimal policy instrument, costing less per gigaliter of water returned to the environment (Grafton, 2010):

> ... if the $3.1 billion allocated to buying water entitlements and the $5.8 billion targeted for water infrastructure subsidies under Water for the Future were combined to purchase water entitlements from willing sellers, the Commonwealth Government would be much more likely to achieve healthy working rivers within the Basin, and for no extra cost.
>
> **Grafton (2010)**

Grafton (2010) also argues that market-based instruments afford irrigators greater flexibility in using funds to put to the uses that can most benefit them, whereas infrastructure investment assumes benefit is achieved through maintaining existing agricultural production with lower water volumes.

In reality, both these instruments have been used to achieve the objectives of reduced irrigated water use in the Basin. However, debate about the relative effects of each for both the irrigators who choose to sell water or upgrade infrastructure, and for the communities in which the instruments are applied, has continued through implementation of the Basin Plan. Schirmer (2017) highlights that irrigators report differing direct effects from each. As shown in Fig. 2, 86% of those who modernized on-farm infrastructure felt it was positive for their farm enterprise overall and 60% felt it improved their farm profitability, with very few reporting negative impacts. However, when those who had sold entitlements to the government were asked the same questions about effects, sale of entitlements was considered positive for the farm by only 47% of those who had sold some of their water entitlements and remained in irrigated agriculture, while 33% reported that it had an overall negative impact on their farm enterprise; 41% felt farm profitability improved while 35% felt their profitability was negatively impacted by their choice to sell entitlements.

Even these relatively simple statistics highlight that the different instruments can have quite different on-ground outcomes for farmers. The flow-on effects are equally important: irrigators who had modernized on-farm infrastructure were typically maintaining or increasing their farm production, something which is likely to minimize potential for negative impacts on irrigation communities, where many jobs depend on the flow-on activities generated by agricultural production. Sale of water entitlements, meanwhile, was more often associated with a decline in on-farm employment and plans to downsize the farm, actions more likely to have negative flow-on effects into communities (Schirmer, 2017).

These findings from ongoing monitoring of social effects of the instruments used to achieve the Plan highlight the importance of understanding the different social effects that may result from use of different policy instruments. They suggest that investment in water-efficient infrastructure has reduced the negative social impacts of the Plan, despite having a higher economic cost per unit of recovered water. Actively considering social effects in water-reform decision-making processes ensures policy makers better understand the trade-offs they may be making when they choose one instrument over another, including the potential for costly social impacts that may not be well reflected in traditional economic assessments.

CONCLUSIONS

While the importance of assessing the social effects of water reform is well recognized, implementation of robust social assessment processes lags well behind this recognition. Australian experience, particularly in the development and implementation of the Basin Plan, highlights that for social assessment to be effective in increasing the social acceptability of water reform, it must be invested in as a meaningful and major component of the policy development and implementation process. This requires integration of social assessment into the main decision-making processes of water reform, rather than social assessment being undertaken in relative isolation from other decision-making processes.

Further, it requires good public consultation, robust methodologies for understanding cause-and-effect relationships, distribution and accumulation of social impacts, and for assessing the relative social effects of different policy instruments and approaches. The process of social assessment is as important as its outcomes. Poor processes can themselves have negative social impacts, and reduce the social acceptability of water reform. Good processes can ensure that decision makers consider social effects and take deliberate actions to minimize negative impacts and promote positive social outcomes in the design and implementation of water reform, and are seen to have done so.

REFERENCES

ABS. (2008). *Water and the Murray Darling Basin: A Statistical Profile, Australia 2000-01 to 2005-06*. Canberra, Australia: Australian Bureau of Statistics (Catalogue No. 4610.0.55.007).

Alston, M., & Mason, R. (2008). Who turns the taps off? Introducing social flow to the Australian water debate. *Rural Society, 18*, 131–139.

Balcombe, S. R., Sheldon, F., Capon, S. J., Bond, N. R., Hadwen, W. L., Marsh, N., et al. (2011). Climate-change threats to native fish in degraded rivers and floodplains of the Murray-Darling Basin, Australia. *Marine and Freshwater Research, 62*, 1099–1114.

Baldwin, C., O'Keefe, V., & Hamstead, M. (2009). Reclaiming the balance: Social and economic assessment—Lessons learned after ten years of water reforms in Australia. *Australasian Journal of Environmental Management, 16*, 70–83.

Boully, L., & Maywald, K. (2011). Basin Bookends, the community perspective. In D. Connell & R. Q. Grafton (Eds.), *Basin futures: Water reform in the Murray-Darling Basin* (pp. 101–114). Canberra: ANU E Press.

EBC, RMCG, Marsden Jacob Associates, EconSearch, McLeod, G., Cummins, T., et al. (May 2011). *Community impacts of the Guide to the proposed Murray-Darling Basin Plan. Volume 3: Community impacts. Report to the Murray-Darling Basin Authority, Canberra.* .

Esteves, A. M., Franks, D., & Vanclay, F. (2012). Social impact assessment: The state of the art. *Impact Assessment and Project Appraisal, 30*, 34–42.

Franks, D. M., Brereton, D., & Moran, C. J. (2010a). Managing the cumulative impacts of coal mining on regional communities and environments in Australia. *Impact Assessment and Project Appraisal, 28*, 299–312.

Franks, D. M., Brereton, D., Sarker, T., & Cohen, T. (2010b). *Cumulative impacts: A good practice guide for the Australian coal mining industry.* Brisbane: Centre for Social Responsibility in Mining & Centre for Water in the Minerals Industry, Sustainable Minerals Institute, The University of Queensland & Australian Coal Association Research Program.

Gale, M., Edwards, M., Wilson, L., & Greig, A. (2014). The boomerang effect: A case study of the Murray-Darling Basin Plan. *Australian Journal of Public Administration, 73*(2), 153–163.

Grafton, R. Q. (2010). How to increase the cost-effectiveness of water reform and environmental flows in the Murray-Darling Basin. *Agenda: A Journal of Policy Analysis and Reform, 17*, 17–40.

Grafton, R. Q., & Horne, J. (2014). Water markets in the Murray-Darling Basin. *Agricultural Water Management, 145*, 61–71.

Gross, C. (2008). A measure of fairness: An investigative framework to explore perceptions of fairness and justice in real-life social conflict. *Human Ecology Review, 15*, 130.

Gross, C. (2011). Why justice is important. In D. Connell & R. Q. Grafton (Eds.), *Basin futures: Water reform in the Murray-Darling Basin* (pp. 149–162). Canberra: ANU E Press.

Hooper, B. P. (2011). Integrated water resources management and river basin governance. *Journal of Contemporary Water Research and Education, 126*, 3.

Horne, J. (2013). Economic approaches to water management in Australia. *International Journal of Water Resources Development, 29*, 526–543.

IPCC. (2014). Annex II: Glossary. In: K. J. Mach, S. Planton, C. von Stechow, R. K. Pachauri, & L. A. Meyer (Eds.), (2014). *Climate change 2014: Synthesis report. Contribution of working groups I, II and III to the fifth assessment report of the intergovernmental panel on climate change references 18* (pp. 117–130). Geneva, Switzerland: IPCC.

Jennings, K. (2011). Water under the bridge: A guide to the Murray-Darling Basin. *The Monthly*, 34–38.

Loxton, E. A., Schirmer, J., & Kanowski, P. (2013). Exploring the social dimensions and complexity of cumulative impacts: A case study of forest policy changes in Western Australia. *Impact Assessment and Project Appraisal*, *31*, 52–63.

Lucas, R. M., & McMichael, A. J. (2005). Association or causation: Evaluating links between environment and disease. *Bulletin of the World Health Organization*, *83*, 792–795.

MDBA. (2010). *Guide to the proposed Basin Plan: Overview*. Canberra: Murray-Darling Basin Authority.

MDBA. (2016). *Progress on water recovery*. Canberra: Murray-Darling Basin Authority. http://www.mdba.gov.au/managing-water/environmental-water/progress-water-recovery (Accessed 9 September 2016).

Mulligan, M. (2011). Rethinking community in the face of natural resources management challenges. In D. Connell & R. Q. Grafton (Eds.), *Basin futures: Water reform in the Murray-Darling Basin* (pp. 135–148). Canberra: ANU E Press.

Quiggin, J. (2012). Why the guide to the proposed Basin Plan failed, and what can be done to fix it. In J. Quiggin, T. Mallawaarachchi, & S. Chambers (Eds.), *Water policy reform lessons in sustainability from the Murray-Darling Basin* (pp. 49–60). Cheltenham: Edward Elgar.

Quiggin, J., Adamson, D., Chambers, S., & Schrobback, P. (2010). Climate change, uncertainty, and adaptation: The case of irrigated agriculture in the Murray-Darling Basin in Australia. *Canadian Journal of Agricultural Economics/Revue canadienne d' agroeconomie*, *58*, 531–554.

Schirmer, J. (2011). Scaling up: Assessing social impacts at the macro-scale. *Environmental Impact Assessment Review*, *31*, 382–391.

Schirmer, J. (2017). *Social effects of water reform in the Murray-Darling Basin*. Canberra: University of Canberra.

Schirmer, J., & Berry, H. (2014). *People and place in Australia: The 2013 Regional Wellbeing Survey*. Canberra: University of Canberra.

Shi, T., & Meyer, W. (2009). Moving towards a policy proactive irrigation sector: Some Australian experiences. *Water Policy*, *11*, 763–783.

Syme, G. J., Porter, N. B., Goeft, U., & Kington, E. A. (2008). Integrating social well being into assessments of water policy: Meeting the challenge for decision makers. *Water Policy*, *10*, 323–343.

Trawick, P. (2002). The moral economy of water: General principles for successfully managing the commons. *GAIA-Ecological Perspectives for Science and Society*, *11*, 191–194.

Van Schooten, M., Vanclay, F., & Slootweg, R. (2003). Conceptualizing social change processes and social impacts. In H. A. Becker & F. Vanclay (Eds.), *The international handbook of social impact assessment: Conceptual and methodological advances* (pp. 74–91). Cheltenham: Edward Elgar.

Vanclay, F. (2002). Conceptualising social impacts. *Environmental Impact Assessment Review*, *22*, 183–211.

Vanclay, F. (2012). The potential application of social impact assessment in integrated coastal zone management. *Ocean & Coastal Management*, *68*, 149–156.

Vanclay, F., Esteves, A. M., Aucamp, I., & Franks, D. M. (2015). *Social impact assessment: Guidance for assessing and managing the social impacts of projects*. Fargo, North Dakota: International Association for Impact Assessment.

Walker, J. L., Mitchell, B., & Wismer, S. (2000). Impacts during project anticipation in Molas, Indonesia: Implications for social impact assessment. *Environmental Impact Assessment Review*, *20*, 513–535.

Wei, Y., Langford, J., Willett, I. R., Barlow, S., & Lyle, C. (2011). Is irrigated agriculture in the Murray Darling Basin well prepared to deal with reductions in water availability? *Global Environmental Change*, *21*, 906–916.

Williams, K. J., & Schirmer, J. (2012). Understanding the relationship between social change and its impacts: The experience of rural land use change in south-eastern Australia. *Journal of Rural Studies*, *28*, 538–548.

FINDING DIAMONDS IN THE DUST: COMMUNITY ENGAGEMENT IN MURRAY-DARLING BASIN PLANNING

P.-L. Tan*, K. Auty[†,‡]

Griffith Law School, Brisbane, QLD, Australia* Commission for Sustainability and the Environment, Canberra, ACT, Australia[†] *Honorary Professor with the University of Melbourne*[‡]

CHAPTER OUTLINE

INTRODUCTION

Effective community engagement is an essential component of any water resources reform process. This chapter reviews the community engagement efforts by the Murray-Darling Basin Authority (MDBA) during the 4-year period (2009–2012) spanning the controversial *Guide to the proposed*

Decision Making in Water Resources Policy and Management. http://dx.doi.org/10.1016/B978-0-12-810523-8.00012-4

Basin Plan (MDBA, 2010) up to the enacting of the Basin Plan by the Australian Parliament in Nov. 2012. This review has been used to identify the key elements that underpin an effective community engagement process.

The Murray-Darling Basin Plan (Aust Govt, 2012) is a formal document made under the Commonwealth Water Act 2007 (Aust Govt, 2014a, 2014b). In a first, the Basin Plan acknowledges the cultural, social, environmental, spiritual, and economic connections of Traditional Owners of the Murray-Darling Basin to their lands and waters, and endorses the words of Tom Trevorrow, explicitly supporting the role and philosophy of Traditional Owners.

In the Council of Australian Government (CoAG) water reforms during the decade 1994 and 2004, community engagement was regarded as vital for building public confidence in these reforms. This was articulated in the provisions of the National Water Initiative (NWI; NWC, 2004), where the terms "consultation" and "community input" are used (c 93, NWI). In this, the NWI is consistent with international practice, particularly the European Framework Directive (Mostert et al., 2007) and in natural resource management (Sayce et al., 2013).

For the Murray-Darling Basin, the water reforms initially aimed to recover an average of about 3000–4000 gigaliters (GL) each year to support the Basin's freshwater and estuarine ecosystems. As this represents over 20% of the consumptive water pool, the recovery efforts through basin planning were a crucial test of how stakeholders and the wider public would perceive the need to reduce water extractions.

For this chapter, we have separated the Basin planning engagement activities into two 2-year phases.

(a) *Phase 1* (2009–10)—relates to the period leading up to the publication and the immediate aftermath of the Guide in Oct. 2010. This series of documents focused on technical background, hydrologic indicator sites, profiles of irrigation districts and how regions would be affected. Initial efforts to engage stakeholders received trenchant criticism mainly for process and communication issues—both key components to effective stakeholder engagement (Boully & Maywald, 2011; Evans & Pratchett, 2013). A Parliamentary Enquiry affirmed the need for better local engagement (House of Representatives, 2011).

(b) *Phase 2* (2011–12) relates to the resetting of the engagement process by the MDBA following the issues with the Guide. This resulted in three revisions of a document called the Proposed Basin Plan, the first of which was published in Nov. 2011, and then, following a new engagement process, a revised third draft was published in May 2012. After further comments by the Ministerial Council and key stakeholder groups, a final Proposed Basin Plan was published in Aug 2012. The Commonwealth Minister accepted the document and tabled the Basin Plan in the Australian Parliament, where it was passed on 22 Nov. with bipartisan support.

In this chapter we review the community engagement in the southern part of the Murray-Darling Basin, as there were different issues at play in the northern Basin. We address four questions: (a) whether it was possible to begin the consultation process with an appreciation of community expectations; (b) what has the MDBA done in terms of engagement; (c) what has been learned, and (d) are the lessons portable, across time and space. We find that the period 2009–12 marks a watershed in the practice of community engagement, not only for the MDBA but more broadly where it provides insights into governance theory and practice.

IAP2'S public participation spectrum

The IAP2 Federation has developed the Spectrun to help groups define the public's role in any public participation process. The IAP2 Spectrum is quicly becoming an international standard.

Increasing impact on the decision

	Inform	Consult	Involve	Collaborate	Empower
Public participation goal	To provide the public with balanced and objective information to assist them in understanding the problem, alternatives, opportunities and/or solutions.	To provide the public feedback on analysis, alternatives and/or decisions.	To work directly with the public throughout the process to ensure that public concerns and aspirations are consistently understood and considered.	To partner with the public in each aspect of the decision including the development of alternatives and the identification of the preferred solution.	To place final decision making in the hands of the public.
Promises to the public	We will keep you informed.	We will keep you informed, listen to and acknowledge concerns and aspirations, and provide feedback on how public input influenced the decision. We will seek your feedback on drafts and proposals.	We will work with you to ensure that your concerns and aspirations are directly reflected in the alternatives developed and provide feedback on how public input influenced the decision.	We will work together with you to formulate solutions and incorporate your advice and recommendations into the decisions to the maximum extent possible.	We will implement what you decide.

FIG. 1

Public participation spectrum.

From IAPP (2015). Core values awards, International Association of Public Participation. *http://c.ymcdn.com/sites/www.iap2.org/ resource/resmgr/Core_Values/WEB_1510_IAP2_Core_Value_Awa.pdf.*

ENGAGEMENT AS A SPECTRUM AND PROCESS

The International Association of Public Participation has defined the public participation (engagement) process as a spectrum of activities starting with informing and then moving through consultation, involvement, collaboration, and finally empowerment, as illustrated in Fig. 1. Unless well managed, this engagement process can result in an expectation in some stakeholders that the decision making may be devolved to the public. However, along the engagement spectrum, reaching the "involve" or "collaborate" level of participation requires extensive effort and empowerment, particularly when applied to complex issues common in the context of river basin planning and management in Australia.

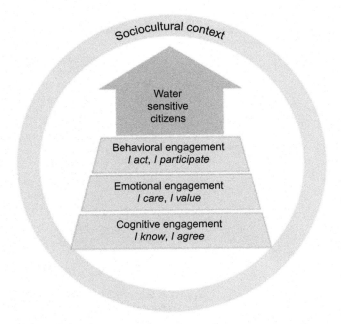

FIG. 2

Framework for assessing water-related engagement.

From Dean, A. J., Lindsay, J., Fielding, K. S., & Smith, L. D. G. (2016). Fostering water sensitive citizenship: Community profiles of engagement in water-related issues. Environmental Science & Policy, 55, *238–247.*

These issues are well known, but include geographical, environmental, cultural, social, and economic dimensions across a massive scale and in circumstances of prior environmental neglect.

Dean et al. (2016) identify a framework (Fig. 2) for identifying levels of citizen or community engagement in water-related planning. They recognize that a "community" rarely consists of a homogeneous group, and that within a community there are diverse and even competing attitudes and interests that need to be addressed. For these reasons, Dean et al. (2016) recommend that different engagement profiles and associated factors can be instrumental in determining "footholds" for intervention. The framework provides a helpful tool to track individual/community buy-in and involvement in any water-related community engagement process, acknowledging how demographic, household, and psychological factors play into aspects of engagement.

ENGAGEMENT LINKED TO GOVERNANCE AND OTHER THEORIES

Three main theories linking engagement and governance are relevant to this discussion. The term *community engagement* is used often interchangeably with *public participation* (IAPP, 2015). Current public discourse favors the former and considers it a broader concept, referring to longer-term arrangements, particularly when it relates to policy making. This derives from *governance theories* acknowledging that governments no longer can act alone to resolve complex problems (Biswas & Tortajada, 2010; Head, 2010). Ross et al. (2016) point to a further subtle distinction between the two terms, saying that engagement is seen as an ongoing, two-way (or multiway) process in which the focus is on relationship building rather than on the decisions.

The term *localism*, also drawn from governance theory, has more recently entered the sociopolitical discourse. Hildreth (2011) distinguishes three main types of localism, although Evans et al. (2013) argue it is the mix and balance of all three that is important rather than the types. They are:

- *Representative localism*—characterized by a hierarchical provision of powers and responsibility by national to lower levels of government,
- *Managerial localism*—differentiated by devolution of powers on condition that agreed objectives are met,
- *Community localism*—participatory in nature, where devolved powers are substantial and supported by rights and resources.

Engagement is also core to the theoretical basis of a *social license to operate*. This relates mainly to resource-based commercial enterprises, for example mining or forestry, and refers to society's expectations of these enterprises (Dare et al., 2014). An accepted social license confers multiple advantages to an organization, including a sound corporate image, reduced regulation, and better market access for goods. Again, a social license is built on robust relationships and the *perceived* acceptability of the enterprise. Dare et al. (2014) notes that a social license is not a single license, but a "continuum of multiple licenses achieved across various groups." Such a license relates to issues of scale (local, regional, national) and has normative elements of legitimacy, credibility and trust (Thomson & Boutilier, 2011). Obviously, there are overlaps between this particular construct and the governance theories examined earlier.

Because issues around sustainability can rarely be managed or resolved at an individual or organizational scale, *social learning* comes into relevance. Critical and profound learning, indeed transformational learning, is achieved by "triple-loop" learning, a concept borrowed from management theory (Mitchell, 2013; Pahl-Wostl, 2009). As opposed to "single-loop" learning, which merely questions *actions*, and "double-loop" learning, which questions governing *assumptions* and is therefore said to involve organizational learning, triple-loop learning occurs when the *context of governance* is examined, leading to lessons that permeate or are socialized into the community (Reed et al., 2010 as cited in Mitchell, 2013). Social learning results from processes that are inclusive, interactive, and flexible (Lacroix & Megdal, 2016; Tan et al., 2012).

Imagery for engagement ranges from ladders of participation (Arnstein, 1969), to toolkits (Reed, 2008), to wheels of engagement (Lacroix & Megdal, 2016; Tan et al., 2012) which emphasize the iterative nature of engagement. Lacroix and Megdal (2016) observe that ethical considerations are sometimes incorporated in stakeholder engagement practice, although management perspectives predominate. Each of these theoretical positions plays out in various ways in this discussion of the MDBA response to community interest in the planning process.

GENERAL COMMUNITY ENGAGEMENT
THE CONTEXT OF ENGAGEMENT EFFORTS

Australians have had nearly three decades of ambitious water reform (Tan et al., 2012; see also Chapters 1 and 13) made necessary because of earlier decisions that allowed consumptive water use largely without limits and without protecting freshwater ecosystems. Jurisdictional and institutional fragmentation sheltered this poor decision making.

Major planks in the reform policy documents of 1994 and 2004 were:

- Giving the environment a share of the water, and equal protection with consumptive use;
- Introducing tradable water entitlements, and establishing a national water market;
- Pricing of water to reflect cost;
- Statutory water planning that would provide an orderly, transparent, accountable process and one that consults the community;
- Institutional reform through the setting up of the National Water Commission, an independent statutory body to oversee reform.

These reform measures were implemented by the states with ensuing conflict, and were coupled initially with annual audits and financial productivity incentives.

Reform measures in the Murray-Darling Basin included:

- The Living Murray Initiative 2004, which promised to deliver 500 GL y^{-1} of water as a first step towards restoring the health of the River Murray, specifically to six icon sites in the Basin, and with a complementary package of infrastructure measures to recover the water (MDBA, 2011a);
- Institutional reform through the Water Act 2007 (C'th), which set up the Commonwealth Environmental Water Holder (CEWH) and the MDBA, both of which had statutory obligations (Aust Govt, 2014a). The MDBA has responsibility for many aspects of integrated planning for sustainable use of water in the Basin, while the CEWH is responsible for acquiring and managing environmental water to give effect to Australia's international obligations, and to give effect to the Basin Plan;
- Water buybacks in the Basin with the first purchases made in 2007 (Crase et al., 2009; NWC, 2008).

These initiatives met with criticism and anger from consumptive users. Community consultation around *The Living Murray* initiative was highly criticized for information failure, insensitivity to community costs of participation, and a perception by the public that decisions had been made before consultation commenced (Crase et al., 2005). A lack of trust by stakeholders in governance processes over fairness and equity, and how potential social outcomes would be mitigated, emerged as key issues and drove suspicion and anger (Alston et al., 2016).

ENGAGEMENT OVER THE GUIDE

In the development phase of the Murray-Darling Basin Plan, the Basin's irrigation community, particularly in NSW, had numerous negative perceptions. As noted by one participant:

> Many of the Catchment Management Authority [CMA] Chairs had extensive experience with the NSW water reforms during the 2000s and were fully aware of … engagement fatigue, despair and mistrust in community engagement delivering on their expectations. The Namoi groundwater plan was a case in point and their "compensation" proved to be nowhere near what the arrangements for buy-back was to be during the past 5–6 years. Their voluntary retirement, history of use arrangements and discounting meant that many groundwater users were not treated the same as the current process allows. There was a prescient sense of this from the users.
>
> **Personal Communication, Jim McDonald, Ex Chair Namoi CMA, Aug. 2016.**

Before the Guide was released, the NSW CMA chairs, sensing a distinct lack of engagement at the catchment level, approached the MDBA and offered to assist by partnering, whether formally or informally, with the MDBA. They were rebuffed (Personal Communication, Jim McDonald). In addition, although there was a communication and engagement strategy for the release of the Guide prepared by MDBA officers, the leadership in the Authority was not supportive of this and made a choice not to talk to journalists (Personal Communication, K. Maguire, Ex MDBA Executive, Aug. 2016; also see Wahlquist, 2011).

It was therefore unsurprising that, like the buyback scheme that started in the Millennium drought (Wittwer, 2011), initial efforts by the MDBA to engage with stakeholders in 2010 met with considerable hostility. Stressed Basin communities demanded a say in the planning process, but the engagement efforts for the 2010 Guide were typified by one-way communication by Basin officers (Evans & Pratchett, 2013).

Discussion on the Guide commenced with a large meeting of stakeholders in Canberra (attended by one of the authors) and involved state government officers, NGOs, consumptive users, and academics. Small groups were formed to discuss issues, and although the responses were measured, the feeling around this meeting was that there were no adequate answers to questions raised. Other meetings were also later held in major cities. At a Brisbane meeting (attended by one of the authors) it was clear that the meeting was about information giving and not consultation. As a technical document the Guide was seen to be "impressive" but the impression it conveyed to the community at large was "this is it" rather than "this is where we start a discussion" (R. Treweeke, Basin Community Committee, Sep. 2016)

The first regional meeting was held in Shepparton (Victoria) with the second in Deniliquin (NSW). At the third regional meeting in Griffith (NSW) there was heightened anxiety and heat. Winery workers—in high-visibility vests—were bussed to the Griffith meeting (Personal Communication, K. Maguire, Ex MDBA Executive, Aug. 2016). Copies of the Guide were burnt, making national news headlines.

It is clear from the discussions we have had with those who were involved that there was very poor engagement at the time; the Authority was struggling with a great deal of push back from interested parties; and the leadership of the Authority was not well prepared to deal with the level of tension. There is no definitive timeline given by the Guide, but it appears that the Basin Plan would have been finalized by 2011 and that States would start implementation by 2012. This factor would also weigh heavily on the approach of the Authority, both staff and Board members (R. Treweeke Sep. 2016). Writing of these early communication efforts, Evans and Pratchett (2013, p. 552) comment that the real purpose appeared "to have been an attempt to convince the participants of the 'science' rather than consulting the public about the plan." This "public" was a complex bundling of interests and oppositionists in towns and regions with vastly different aspirations.

ENGAGEMENT OVER PROPOSED PLAN 2012

In the fall-out from the unsuccessful launch of the Guide, the Authority had an opportunity for organizational renewal. Three members of the MDBA, including its Chair and CEO, left their positions in the 6-month period after the Guide was released (MDBA, 2011b). This presented the Authority with an opportunity to recast the process, consider its communication methodologies, change the leadership style and allow Authority staff to regroup around change.

During the Phase 2 resetting of the MDB planning process, the proposed sustainable diversion limit (SDL) was increased over an earlier value, meaning the recovery volume was reduced to 2750 GL y^{-1}; this resulted from a refined method of assessing environmentally sustainable levels of take (ESLT)

(Young et al., 2011, p. 26). The use of scientific information in the refined method was endorsed by a CSIRO-led review, although defensibility of hydrological and environmental modeling and analyses behind the SDL and ESLT were difficult to assess (Young et al., 2011). Fig. 3 shows the methodology for

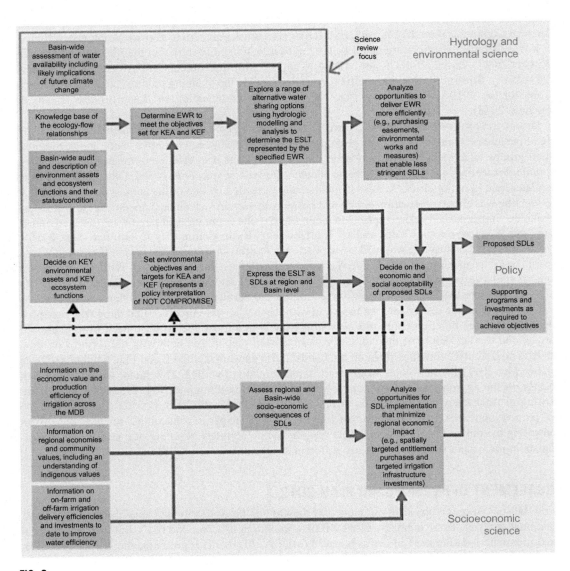

FIG. 3

Conceptual framework for surface-water sustainable diversion limit (SDL) determination.

From Young, W. J., Bond, N., Brookes, J., Gawne, B., & Jones, G. I. (2011). Science Review of the estimation of an environmentally sustainable level of take for the Murray-Darling Basin, final report to the Murray-Darling Basin Authority. *CSIRO Water for a Health Country Flagship Report to Canberra: Murray-Darling Basin Authority.*

determining SDLs including socioeconomic objectives which were "policy judgments … reflecting multiple trade-off decisions to meet requirements of the Water Act 2007" (MDBA, 2011c; Young et al., 2011, p. 14). The water recovery is occurring through a combination of investment in infrastructure efficiency and water buybacks (Aust Govt, 2014b).

Based on this conceptual framework, expertise from outside the Authority was commissioned to assess the impacts of the Guide/Proposed Plan with a view to externally and authoritatively auditing the work that had already been undertaken and to expand the analysis of the "needs" of the Basin and its communities to include social and economic considerations. Although there was considerable argument about whether the Basin Plan process was actually permitted to include these considerations, the Water Act 2007 is quite clear in its requirement that in seeking to achieve environmentally sustainable levels of extraction, the MDBA needs to do this "in a way that optimizes economic, social and environmental outcomes" (Aust Govt, 2014a, Part 1 3a).

The socioeconomic impact of the Guide/Proposed Plan on irrigated agriculture was assessed by Frontier Economics (2010) and reviewed by BDA Group (2010). In this period, four different assessments were done: for recreational boating industries (Marsden Jacob Associates, 2012), fishing industries (Deloitte Access Economics, 2012), primary producers (GHD, 2012), and a basin-wide assessment (CSIRO, 2012).

Environmental groups and irrigators continued to oppose the proposal. During this hiatus period, all interested parties continued to lobby for the outcomes they desired. Tensions remained, but the heat seemed to evaporate.

While Phase 1 largely focused on delivering information at selected rural towns by high-level officials at highly organized, large town-hall style meetings, those driving Phase 2 realized there was a much greater need for place-based, more localized and better-prepared community engagement across the whole Basin. Recalibration of community engagement under the new leadership unfolded as a matter of urgency across the whole Basin, but also with regard to hot spots, in the following manner:

- The new MDBA Chair (Hon. Craig Knowles) started working with regional powerbrokers to organize top level meetings in more than 20 towns to lay the foundations for future community engagement on the draft Basin Plan (MDBA, 2011a–2011d);
- The Basin Community Committee (BCC) was given very clear but respectful guidance about its roles and responsibilities, and the engagement team sought the involvement of BCC members in community meetings, especially in their local areas (BCC meeting summary 1 Dec., 2012);
- The regional meeting schedule involved MDBA staff members, who were able to answer operational questions while actively supported by the Chair;
- "Taking-the-time" was recognized as an important element of the new program and smaller roundtable discussions, which were hosted by highly interested local groups resourced to facilitate discussion of in-depth technical information in local contexts and with the input of local knowledge;
- MDBA established new advisory groups in 2011 to provide preliminary feedback on policy development, namely a seven-member Proposed Basin Plan Working Group, and a nine-member Proposed Basin Plan Testing Committee (MDBA, 2011a–2011d).

Notwithstanding these initiatives conflicted views continued to play out in the debate. In *South Australia* (*SA*), groups of environmentalists, fishers and an irrigator protested the Proposed Basin Plan,[1] and the

---

[1] http://www.abc.net.au/pm/content/2011/s3378328.htm.

Premier (Hon. Jay Weatherill) threated to take the matter to the High Court.[2] But, during the consultation period, negotiations succeeded in changing the tone of the discussion. Ironically, conflict led to better communication as the Local Government Association and their state representatives were engaged with member Councils and state and Commonwealth governments throughout the consultation period on the Proposed Basin Plan. In SA this culminated in a "highly successful" Local Government forum being held in Adelaide on 2 Mar. 2012 (LGASA, 2012). This forum, attended by planning and other staff, councilors and interested parties (not just departmental staff), provided the opportunity for Councils to discuss issues directly with the MDBA, as well as hearing the SA Government's position on the Proposed Basin Plan from the Premier. The emphasis on *discussion* here is important. The crafting of the meeting and populating it took time and the MDBA was very actively engaged in this "conversation."

In *Victoria*, Richard Anderson (Chair, Water Committee, Victorian Farmer's Federation (VFF)) was equivocal in his views about the proposed Plan but said that it was "good from Victoria's perspective" given that northern Victoria was in "credit" where water savings were concerned. Doubt remained over how further savings would be achieved and the VFF wanted to see savings driven through environmental infrastructure projects. The tone of the Victorian conversation ebbed and flowed. There were already significant achievements in water savings being demonstrated through infrastructure and buybacks in areas like the Wimmera, and these stories were well known across the farming community. Also in Victoria, floods followed the Millennium drought across a significant part of the state and paradoxically this both exacerbated and calmed tensions. Additionally, the incoming coalition government linked water, agriculture, and environmental agencies into a single mega-agency, which helped to alleviate matters.

In *NSW*, where the town of Griffith had been the site of infamous Guide burnings, another meeting was organized in protest but it did not draw as much attention as the first rally. Demonstrating an organizational commitment to the flexibility necessary to engage the affected sectors and community members, negotiations over the Proposed Basin Plan were extended to work out NSW's position (NSWLGA, 2012). To reduce the impact of the Basin Plan in the Murrumbidgee region (and elsewhere), water-saving projects received federal funding as part of the commitment to continue access to consumptive water.

Jettisoning the larger, more formal approach, in Phase 2 the MDBA held 24 public meetings, 56 roundtable and technical meetings, 18 social and economic briefings for representatives from rural financial organizations, 5 regional briefings on water trading, 31 bilateral and work group meetings with Basin states, and a tailored Aboriginal "roadshow" in more than 30 towns (MDBA, 2012).

However, across the Basin, one particular stakeholder group remained strongly opposed to the Proposed Basin Plan. In what might initially have appeared as a paradox, environmentalists, including but not limited to peak bodies, objected on the grounds that water recovery under the Proposed Basin Plan did not go far enough in providing water for ecosystem needs and also had neglected the potential impact of climate change. Among them were the Wentworth Group and sixty scientists signing a petition.[3]

---

[2]ABC News, 4 October 2011, http://www.abc.net.au/pm/content/2011/s3332142.htm.
[3]ABC News, 4 October 2011, http://www.abc.net.au/pm/content/2011/s3332142.htm.

The MDBA received almost 12,000 submissions after the 20-week formal consultation period to the Proposed Basin Plan and over 300 changes were made to the Proposed Plan (MDBA, 2012).

The modified Phase 2 engagement process, discussions and information-sharing processes eventually reduced the scepticism about the potential benefits of the sustainable diversion limits (SDL). A pragmatic compromise was reached when the general community and state governments appeared more confident that a balance between competing interests had been found. An Authority officer now finds a key lesson as "it's 20% about the science and 80% about communication" (Personal Communication, Dr. Tony McLeod, MDBA Officer, Jun. 2015).

ABORIGINAL ENGAGEMENT

Aboriginal people have retained water relationships to the river systems they had grown up with. In her evidence to the Native Title Tribunal, a senior woman giving evidence supporting the Yorta Yorta Native Title claim in Echuca (Trish Johnson in MDBA, 2011d) made a very telling point. She said that even though she, her mother, and her siblings spent the best part of two decades fleeing the interference of the welfare system, they always maintained a link with the river system and had water relationships. Even if they slept under bridges, lived in substandard fruit pickers' housing, infrequently attended school, and never had shoes, they planted their feet close to the rivers. Her mother moved her fractured family up and down the river system, rarely leaving the Basin, and when doing so, always returning.

This is not a singular narrative, and even in brief, the water relationship of river-based people is a foundational narrative. This link to rivers was well summed up by Trish Johnson when she said: "I'm going to put this bluntly, water is our life" (MDBA, 2011d).

Additionally, Tom Trevorrow, a Narrindjeri Traditional Owner, expressed the Aboriginal view of water management as: "…our traditional management plan was—don't be greedy, don't take anything more than you need, and respect everything around you. That's the management plan—it's such a simple management plan, but so hard for people to carry out."

Across all the noise and reflection on the Phase 1 engagement process, the MDBA's elevated commitment to engaging with the Basin communities also witnessed a growing engagement with Aboriginal people, who had never stopped caring for river country in spite of the recent and well-documented history of violence and dispossession (Weir, 2007, 2011).

The predecessor of the MDBA (the Murray-Darling Basin Commission—MDBC) had worked with Aboriginal people of the Basin, particularly in spearheading cultural flows research, funding, and undertaking cultural mapping of Aboriginal use of land and water, and in building capacity (Personal Communication, Yorta Yorta senior man, 2014). In 1998, the MDBC supported the formation of the Murray-Lower Darling Rivers Indigenous Nations (MLDRIN), a confederation representing the 21 first nations in the southern part of the Murray-Darling Basin. Additionally, in 2012 the MDBA also supported the Northern Basin Aboriginal Nations (NBAN) to represent Aboriginal people in the northern Basin.

In Phase 1 there was little focused engagement with Aboriginal people. However, in Phase 2, the authors, in field work, conversations and observations, have noted that the MDBA appeared actively committed to improving relationships with Aboriginal people as evidenced by the number of internal working documents and also in the public face of the Authority.

The engagement project *A Yarn on the River—Getting Aboriginal Voices into the Basin Plan* (MDBA, 2011d) made a serious effort to incorporate Aboriginal people's voices and in doing so it reflected a "set of principles" endorsed by NBAN and MLDRIN in Jun. 2011.[4]

Over 450 submissions from Aboriginal participants in the Basin were received as a result of this project, including around 21 from Aboriginal organizations. Most organizational submissions centered around five major themes[5]:

- Surface water has been badly mismanaged and Aboriginal people desire protection of ecosystems, cultural sites, places, and landscapes; and groundwater is under threat as scientific knowledge of groundwater recharge and dynamics is poorly understood.
- The draft Plan failed to recognize the sovereign rights of Aboriginal people to water; it is unjust to prioritize historical consumptive entitlements, as this undermines Aboriginal people's right to maintain their distinctive spiritual relationship with land and water.
- Scientific analysis is conflicted and, echoing the calls for local input, Aboriginal people would have more confidence in the plan if traditional ecological knowledge were incorporated into the process.
- Recognition of a right to cultural water, referring to entitlements that are legally and beneficially owned by First Nations, of a sufficient quantity and quality to improve Aboriginal people's spiritual, cultural, environmental, social, and economic conditions.
- The need for adherence to the United Nations Declaration of the Rights of Indigenous Peoples, particularly free, prior and informed consent when negotiating with Aboriginal peoples, and culturally appropriate representation of Aboriginal people in decision making.

In addition, MLDRIN demanded the Proposed Basin Plan give highest priority for protecting end-of-system flow at the Murray Mouth, relating to ecology and sustainability of the whole river system.

The next part of the chapter examines two examples from more recent engagement by the MDBA to illustrate the step-change in engagement and discusses the three important themes that emerge.

DISCUSSION

A distinct outcome of the change that Phase 2 promoted is the manner in which the Authority sought to consult with Aboriginal people. One of the authors joined in a broad-ranging MDBA and Indigenous community information exchange session in 2014, loosely based on the cultural flows of work that was unfolding at that time at Echuca on the Murray River. The meeting, bringing together many people from many places along the southern reaches of the Basin system, had been a long time in preparation. It was jointly managed by the MDBA and Aboriginal groups, and senior Aboriginal people attended from all along the length of the River Murray, from the South Australian mouth to the border with Victoria and New South Wales, and on to Yorta Yorta country. Women and men brought their views, concerns and knowledge about the cultural importance and environmental health of the River to the discussion. Authority personnel did their homework and took great care in the presented material.

---

[4]This publication has been taken out of circulation due to the death of Aboriginal people featured in photographs.

[5]All 21 submissions from organizations were read, and these themes emerge clearly. Many of the words used here are taken from the submissions from MLDRIN, Victorian Traditional Owner Land Justice Group, and NSW and ACT Native Title Service Corporation. The authors did not have access to submissions of individuals.

The forum was not only a conversation about these important cultural and environmental matters; it also involved a field trip to infrastructure works at Gunbower Forest. This field trip led to a free-ranging and respectful conversation about the success of these works, the involvement of Aboriginal people in the development of the works program, the manner in which the works were installed, the concessions which might or might not have needed to be made, the need to take the time to have the discussions and bring a range of rather disparate people to the table (engineers and Aboriginal people had different views about priorities), and the outcomes for the environment and culture by the construction of such works.

The Echuca meeting was highly successful in bringing all these views and aspirations together in a respectful exchange of ideas. The forum was properly organized, structured and funded, and it was not rushed or unsympathetic to the need to take the time to talk about issues.

Further the MDBA, again in active conversation with Aboriginal people, has embarked on a staged commitment over time to produce a Reconciliation Plan titled *Strengthening Connections* (MDBA, 2015). In that Plan the MDBA commits to promoting employment opportunities, partnerships in educational programs about traditional knowledge relating to water, and, importantly, reporting and reviewing its consultation and reconciliation processes apart from cultural water considerations.

With respect to the irrigation community, a second large-scale meeting at Deniliquin in 2014 demonstrates the extent of the change that has taken place. While the first meeting was typified by tension and ill will, the second was organized to update the community about Basin Plan developments. One of the authors attended this meeting with the following observations.

PUBLIC MEETING, PBP DENILIQUIN 2014

The large number of attendees—up to 80—included women and a range of age groups. A person came over from South Australia to speak about the issues at the Murray mouth. Irrigators attended, clearly keen to have their say about how the Plan was adversely affecting them. It was clear that there were some very angry people in the gathering. There were no Aboriginal people present.

MDBA's CEO Rhondda Dickson spoke at the opening of the meeting and answered questions and remained at the forum. Her team included women who were administrators, community engagement specialists and scientists.

The Authority was very well organized, well prepared, familiar with the material and capable of answering all questions. There were many questions, and those asking them had agendas and interests, which they were pursuing.

The facilitator was a male, well known to the local community and clearly respected in the region.

In spite of the fact that science was being promulgated, the meeting was held in a conversational manner, jargon was not used, and scientific content was not used to exclude people from the discussion. Complex issues about engineering and the environment were aired in normal, practical and everyday language and concepts.

Those who attended were given the opportunity to speak. They were not cut off. Their questions were frequently statements of fact rather than interrogations.

Holding the meeting in the place where people worked and lived provided both advantages and challenges. It was suggested that the town of Deniliquin was dying as a result of the operations of the Plan, but as the meeting was held in the town it was quite apparent that this was not true.

There seemed to be a degree of resignation amongst the group that the Plan was very well developed and it was most unlikely there would be any retreat.

The meeting was open and public. The dialogue was respectful if sometimes heated. No one from the Authority retreated in the face of difficult questioning.

The overall impression from Deniliquin meeting was that the discussion had reached a level of maturity and that the hard work to build trust and confidence in the long lead-in to this meeting had been time well spent. It was clear that members of the public had developed relationships with Authority members. The CEO was approachable. The scientists were engaged. The community could not complain of being left out of the loop or of being ignored. Their concerns were put in the context of the Basin as a whole and while that message was difficult, if not unpalatable, it was obvious it had greater salience than might have been the case in the past.

Three dominant themes emerge from MDBA's more recent communication and engagement efforts: blending local and Basin perspectives of place, cohesive and ethical leadership and style, and social learning. These are discussed in the following sections.

BLENDING LOCAL AND BASIN PERSPECTIVES OF PLACE

The water reforms that have taken place since the mid-1990s, particularly in the southern Murray-Darling Basin, generated considerable disquiet within Basin communities. The evidence confirms that across the southern Basin there was an erosion of trust in government, which required collaborative process with better-integrated governance structures to overcome. In Phase 1, the MDBA emphasis was on the formal consultation required under the Water Act, an essentially top-down approach, perhaps informed by an understanding that engagement was needed for a "social license" for the Basin Plan to operate. But, on the other hand, local communities wanted a different approach that demanded a brand of "local" decision making.

Evans et al. (2013, p. 404) state that localism "has been seen as having the potential to strengthen democratic engagement at a time when trust in representative government, and politicians in particular, is declining." The term *localism* is almost immediately perceived by the general community as taking a bottom-up approach; however, in the academic literature, the concept is more complex with the type and the mix of localism important. Evans et al. (2013) now describe the concept as an "umbrella term" referring to "devolution of power … functions … and resources towards local democratic structures and … communities within an agreed framework of minimum standards" (p. 405).

Issues of scale need to be central in any process to drive a planning process to consider Basin-wide needs. Further efforts need to be mindful that local communities are diverse, and that the minimum standards derived from local and state-based vested interests may continue the compromises in water management, which can perpetuate the lowest common denominator results that characterized Basin actions prior to the current reforms.

Undeniably, Phase 1 engagement efforts by the MDBA were top-down and prioritized cognitive engagement, while neglecting emotional and behavioral engagement. It was only in Phase 2 of the Basin planning process that engagement considered the sociocultural context of change management. It is apparent that the modified engagement process adopted in Phase 2 resulted from consideration of the lessons of the failures and upheavals that typified the Phase 1 engagement.

After the MDBA reset the engagement efforts, conflicts were still articulated forcefully and vested interests were not mollified completely, but the tensions never played out with the same white heat evident in Phase 1. The blended approach that framed the Phase 2 engagement efforts is shown in Fig. 4. "Inclusiveness" was the basis of the process to inform, recalibrate, guide and promote action.

Subtle and overt changes of tack typified the new method of engagement. Significantly, local government (often missing, unheard, or poorly briefed in respect of Phase 1) had become a major player in the discussions in Phase 2. The differing views of Councils across the Basin were engaged, heard and considered. But, it was quickly recognized that there would be no two-way learning as the SA local government sector supported the Plan, while the NSW Councils were mostly opposed.

The MDBA faced (and continues to face) a continuing challenge to develop a Basin-wide perspective within stakeholders while maintaining a sense of importance of local places. While farmers and community members argue that they know their part of the river well, arguably Aboriginal people best articulate the sense that their local place is part of a wider and complex system.

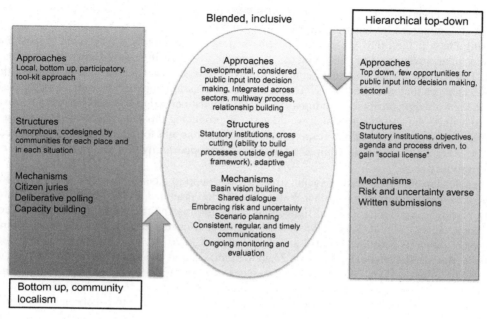

FIG. 4

Bottom-up, blended/collaborative and top-down approaches in community engagement in water planning.

Although the MBDA adopted a blended approach in Phase 2 of Basin planning, this policy change has not been clearly articulated. No statement has yet been made by the Authority clearly defining and communicating what it means by localism. Having a clear policy statement that underpins future engagement activities would help in the following ways: (a) ensure that the MDBA does not overpromise or commit beyond its capability to deliver, (b) assist in managing community expectations, (c) assist in determining the scale or intensity of engagement (e.g., CMAs as the relevant scale on ecological matters), (d) ensure that a mix of top-down and bottom-up approaches to planning and policy development occurs (such as that proposed in developing quantified environmental objectives), (e) make sure that issues of reliability and acceptability of local knowledge are considered, (f) ensure that the precedence for incorporating different types of local knowledge is reviewed, and (g) see that the feasibility of and avenues for incorporating local knowledge, and assessing and adopting appropriate structured tools to harness community input, are considered.

COLLABORATIVE AND ETHICAL LEADERSHIP

As might be expected, the new MDBA leadership that took over in Phase 2 of the Basin planning process played a major role in recalibrating the engagement process. Planning, engagement, and change management are interrelated processes requiring deliberate phases or steps, and skipping steps can be disastrous, with often the appearance of attaining a speedy outcome being just that—appearance only (Harvard Business Review, 1998). Innovation, which the Basin Plan seeks to achieve, can only be produced by a clear, strategic, organized and disciplined approach, which itself is driven by healthy, adaptive and immersed leadership (Drucker, 2006/2008).

During Phase 2, the MDBA's leadership embarked on reinventing the engagement process and its outcomes, in an atmosphere of tension across the community and within the MDBA itself and its personnel. The Knowles leadership team appears to have possessed the necessary skills and abilities to drive the changes that were necessary. To use Drucker's classic template for leadership, these skills included the following attributes: having a clear view of what needed to be done, knowing it was necessary to acquire the knowledge which would assist in producing this outcome, developing an action plan, then taking action and taking responsibility for decisions and for communicating the outcomes (Drucker, 2006/2008). Success required a focus on every opportunity and the ability to make every meeting productive.

Ethical leadership in the MDBA sought to engage and support Aboriginal views and aspirations. In early days of water reform, the National Water Commission provided leadership in this arena through funding research, organizing national Indigenous water forums and then setting up the First Peoples Water Engagement Council. This vital role needed to be continued when the National Water Commission stepped down its activities in 2012 and was abolished in 2014.

Finally, the MDBA leadership of Phase 2 had to think, say and mean "we" instead of "me" in order to create an organizational culture of inclusiveness extending into the public realm (Drucker, 2006/2008). It is reported that Knowles would say to his engagement personnel "find the diamonds in the dust," meaning they needed to identify community leaders willing to collaborate on Basin Plan outcomes.

In Phase 2, engaging the community and gathering evidence was placed at the center of the policy development process. While not strictly a democratic bottom-up approach, Knowles demonstrated and implemented a listening, networking engagement style that promised "no surprises" for the community. It was a "brave" approach in that it exposed MDBA's thinking at each step of the way to give MDBA the opportunity to seek out improvements, road test ideas and identify gaps as they worked to develop the draft plan (MDBA, 2012).

This was a key piece of learning by the MDBA and is consistent with collaborative policy making that many governments currently espouse, but find difficult to implement. Fig. 5 shows one structure for the interactions that need to occur in an effective community engagement process in policy making (Head, 2010, p. 20).

One of the key messages that was consistently sent by the community to the MDBA was that the potential social and economic impacts of the Basin Plan on local communities were important, and were missing from the assessment. This was eventually acknowledged and addressed by MDBA, although communities want this to go even further (Alston et al., 2016).

SOCIAL LEARNING

Arguably the MDBA leadership during Phase 1 was aware of the potential for ugly conflict, as security personnel were employed to attend some town-hall meetings. Recourse to such arrangements demonstrates a failure on a number of levels. Conflict management, however effective, does not equate to social learning where profound and critical learning is not just found within individuals but is socialized and permeates into organizations. Flexible, inclusive, interactive processes are fundamental and need to increasingly focus on outreach and multiple-way communication or discussion.

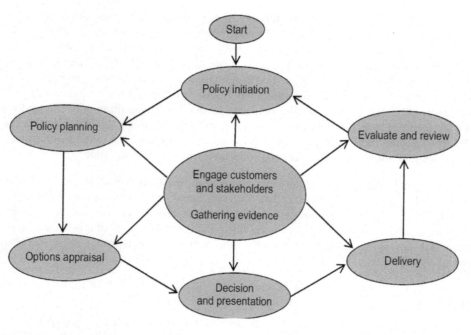

FIG. 5

Community engagement in policy making, Scottish Executive model.

From Head, B. (2010). Evidence-based policy: Principles and requirements. Strengthening evidence-based policy in the Australian federation *(pp. 13–26). Melbourne: Productivity Commission.*

In a parallel process, from 2004 to 2011 a multistage planning process to implement the Marine Life Protection Act Initiative (MLPAI) in California was taking place (Sayce et al., 2013). Kirlin et al. (2013) found the Initiative to have succeeded in overcoming scientific complexity and uncertainty and legal challenges and conflict between participants. A number of parallels exist between the MBD basin planning process and that of the MLPAI. To effectively engage with communities, whether local, scientific, or other, requires:

- Going beyond cognitive engagement. Recognizing, understanding and addressing the unique characteristics of a region or community addresses emotional and behavioral engagement.
- Time and resources spent building relationships with stakeholder groups, particularly those skeptical of the process results, in a valuable exchange of information among members of the public, staff and groups, and scientists that otherwise might not have occurred. Having top executives enter into dialogue with stakeholders in local places in the manner exemplified in the Deniliquin helps build trust.
- Customized strategies with neutral and inclusive messaging including long-term support of underrepresented groups. Particularly for the Aboriginal community, customized strategies are needed as a starting point to redress historical injustice and dispossession of land and water.

- Leaders committed to involving interested members of the public as "key communicators" (as Knowles terms it, "finding diamonds in the dust") helps relay process information to a broader audience, is effective and efficient, and helps build support for the planning process.
- Traditional and "new media" strategies are valuable for reaching a diverse and inclusive public (Sayce et al., 2013). While conventional outreach relies on websites, news releases, public meetings, brochures and posters, informational videos, reviews of public comment, an unconventional outreach strategy relies on social media, hosted community events and regional field trips such as the Echuca community information session. Johns (2014) notes that the MDBA has become increasingly engaged in the use of social media in order to improve relationships and allow for more effective communication and consultation.

Dare et al. (2014) identifies that community engagement can be effective in achieving local-scale social learning but can be limited in achieving that objective across regional and societal scales. This is perhaps the most significant challenge, as social learning across the regions and sectors remains to be developed. As adjustment mechanisms in the Basin Plan may be triggered it is becoming clear that engagement has not as yet succeeded in generating a Basin-wide narrative of sustainable water management.

CONCLUSION

This chapter began by asking whether it was possible for the MDBA to begin with an appreciation of community expectations. At the time the Basin Planning process began, it was apparent to many that Murray-Darling Basin communities were tired, disengaged, and distrusted many government water entities. Multiple conflictual positions—community, sectoral, jurisdictional and cultural—were well known.

We argue that the MDBA should have been better prepared for the backlash when the Guide was launched in 2010. The Guide was an inappropriate document to focus community engagement, and there was general rural hardship resulting from the Millennium drought (1999–2009), making it relatively easy for those opposed to the Basin Plan to promote the "narrative of destruction."

Early efforts of the MDBA focused on cognitive engagement, neglected social and economic impacts, and demonstrated a lack of awareness of multiple local and sectoral interests interacting in an uncontrolled environment. The lack of a foundational narrative for change was apparent and not enough time was given to developing the story around the need for the Basin Plan.

These efforts were rapidly recalibrated when the incendiary response to the Guide resulted in a change in the MDBA's leadership. Learning from earlier communication failures, the MDBA adopted a "no surprises" strategy that eventually led to a successful Basin Plan that combines best-available science and a consideration of social and economic impacts, while delivering a raft of environmental benefits to the Basin. Involvement of the authors in some of the subsequent meetings and discussions, namely at Deniliquin and Echuca, helps to demonstrate how the lessons had been learned.

We note that minority and vulnerable groups were included in engagement efforts. Clearly articulated Aboriginal views, which sought culturally sensitive input into decision making and a right to cultural water, were appreciated and are currently being acted upon. However, there was no articulation of a strategy to include women in the planning process, except that it is worth noting that the CEO of the MBDA in Phase 2 of the process and many of her senior executives were women and this would have lent a different dynamic to the conversation.

By the time the Basin Plan was tabled in Federal parliament in 2012, there was overall agreement on an overarching plan. Independent and multiple assessments and reviews of socioeconomic impacts and the environmental science gave communities more comfort about the methodology used in developing the Basin Plan, even though there was still no consensus. Most parties agreed that the environment was degraded and that ecosystems and communities were suffering from drought.

An outstanding challenge for the Authority in bringing the public along with it in the water reform agenda across the whole Basin is the complexity in the Basin's governance arrangements. While the boundaries of agencies' responsibilities may be clear to Commonwealth and state agencies, they remain murky to most stakeholders, thus creating uncertainty and anxiety (Alston et al., 2016). Additionally, the implications of climate change on the Basin are also a challenge, although the MBDA assert that the Basin Plan provides a sufficiently adaptive framework that allows adjustment as climate trends become more certain (Neave et al., 2015).

The MDBA is still required to develop a shared social-scientific Basin-wide narrative based on a belief that we all have a shared fate in the future of water. There will always be tensions in the interests of the individual, the collective and across cultural boundaries, as there are differences in understanding and inequities as to rights. This is particularly true in a society with a colonial history such as Australia. The MDBA is working to encourage a collective imagining of a future about which there remains conflict and it is trying to do this in the face of established processes of public administration in water management, which are undergoing fundamental change.

The lessons learned by the MDBA in the Basin planning process have application for other river basin organizations, in Australia and overseas, grappling with similar challenges. State agencies that need to deliver on water resource plans that comply with the Basin Plan will benefit from understanding community expectations for "local" input and solutions and that in a "posttruth" era, they will need to explore emotional and behavioral engagement approaches in water planning. Top-down approaches will need to be tempered by the blended attitude adopted by the MDBA.

Water policy will continue to be conflicted and even more fractious as climate change bites and floods and droughts start to occur even more often. To negotiate these events in the context of inflexible community perceptions based on personal, industry or sectoral interests is the task of future Basin managers. Future river basin policy and management organizations will need to be flexible, responsive, robust and adequately resourced to be able to address the ongoing challenges.

ACKNOWLEDGMENTS

We thank Karen Delfau for initial assistance in research, and the editors for their helpful comments. Both authors are members of the MDBA's Advisory Committee on Social, Economic and Environmental Sciences appointed in 2012 after the finalization of the Basin Plan. The contents of this chapter are accurate as of 10 Sep. 2016.

REFERENCES

Alston, M., Whittenbury, K., Western, D., & Gosling, G. (2016). Water policy, trust and governance in the Murray-Darling Basin. *Australian Geographer, 47,* 49–64.

Arnstein, S. (1971). A ladder of citizen participation in the USA. *Journal of the American Planning Association, 35,* 216–224.

Aust Govt. (2012). *Water Act 2007—Basin Plan 2012, extract for the Federal Register of Legislative instruments.* Canberra: Office of Parliamentary Counsel, Australian Government [245 pp.].

Aust Govt. (2014a). *Water Act 2007 (with amendments).* Canberra: Australian Government, Office of Parliamentary Counsel [571 pp.].

Aust Govt. (2014b). *Water recovery strategy for the Murray-Darling Basin.* Canberra: Australian Government.

BDA Group. (2010). *Review of social and economic studies in the Murray-Darling Basin: Report to the Murray Darling Basin Authority, Canberra.* http://www.mdba.gov.au/sites/default/files/archived/basinplan/834-MDB-SEA-Review-final-report-15-Mar-2010.pdf.

Biswas, A. K., & Tortajada, C. (2010). Future water governance: Problems and perspectives. *International Journal of Water Resources Development, 26,* 129–139.

Boully, L., & Maywald, K. (2011). Basin bookends, the community perspective. In D. Connell & Q. Grafton (Eds.), *Basin futures: Water reform in the Murray-Darling Basin.* Canberra: Australian National University Press.

Crase, L., Dollery, B., & Wallis, J. (2005). Community consultation in public policy: The case of the Murray-Darling Basin of Australia. *Australian Journal of Political Science, 40,* 221–237.

Crase, L., O'Keefe, S., & Dollery, B. (2009). The fluctuating political appeal of water engineering in Australia. *Water Alternatives, 2,* 440–444.

CSIRO. (2012). *Assessment of the ecological and economic benefits of environmental water in the Murray-Darling Basin: Report by CSIRO for the Murray Darling Basin Authority, Canberra.* [250 pp.].

Dare, M. L., Schirmer, J., & Vanclay, F. (2014). Community Engagement and social licence to operate. *Impact Assessment and Project Appraisal, 32*(3), 188–197. http://dx.doi.org/10.1080/14615517.2014.927108.

Dean, A. J., Lindsay, J., Fielding, K. S., & Smith, L. D. G. (2016). Fostering water sensitive citizenship: Community profiles of engagement in water-related issues. *Environmental Science & Policy, 55,* 238–247.

Drucker, P. F. (1998). *The discipline of innovation.* Cambridge, MA: Harvard Business Review on Breakthrough Thinking, Harvard University.

Drucker, P. F. (2006/2008). *Classic Drucker. Essential Wisdom of Peter Drucker from the pages of Harvard Business Review.* Boston, MA: Harvard Business School Publishing.

Deliotte Access Economics. (2012). *Benefits of the Basin Plan for the fishing industries in the Murray-Darling Basin.* Melbourne: Deliotte Access Economics. http://www.mdba.gov.au/sites/default/files/archived/basinplan/2131-BenefitsBasinPlanForFishingIndustries.pdf.

Evans, M., Marsh, D., & Stoker, G. (2013). Understanding localism. *Policy Studies, 34,* 401–407.

Evans, M., & Pratchett, L. (2013). The localism gap—The CLEAR failings of official consultation in the Murray Darling Basin. *Policy Studies, 34,* 541–558.

Frontier Economics. (2010). *Structural adjustment pressures in the irrigated agriculture sector in the Murray-Darling Basin: Report for the Murray-Darling Basin Authority, Canberra.* http://www.mdba.gov.au/files/bp-kid/392-Frontier-Economics-2010-adjustment-report-Final-for-public-release.pdf.

GHD. (2012). *Assessment of benefits of the Basin Plan for primary producers on floodplains in the Murray Darling Basin: Report to the Murray-Darling Basin Authority, Canberra.* http://www.mdba.gov.au/sites/default/files/archived/basinplan/2132-GHD_AssessmentBenefitsOfTheBasinPlanForPrimaryProducers.pd.

Head, B. (2010). *Evidence-based policy: Principles and requirements. Strengthening evidence-based policy in the Australian federation.* Melbourne: Productivity Commission [pp. 13–26].

Hildreth, P. (2011). What is localism and what implications do different models have for managing the local economy. *Local Economy, 26,* 702–714.

House of Representatives. (2011). *Of drought and flooding rains, inquiry into the impact of the Murray–Darling Basin Plan in Regional Australia.* Canberra: House of Representatives Standing Committee on Regional Australia, Australian Parliament.

IAPP. (2015). In *Core values awards, international association of public participation.* http://c.ymcdn.com/sites/www.iap2.org/resource/resmgr/Core_Values/WEB_1510_IAP2_Core_Value_Awa.pdf.

Johns, R. (2014). Community change: Water management through the use of social media, the case of Australia's Murray-Darling Basin. *Public Relations Review, 40*, 865–867.

Kirlin, K., et al. (2013). California's Marine Life Protection Act Initiative: Supporting implementations of legislation establishing a statewide network of marine protected areas. *Ocean and Coastal Management, 74*, 3–13.

Lacroix, K. E. M., & Megdal, S. B. (2016). Explore synthesize and repeat: Unraveling complex water management issues through the stakeholder engagement wheel. *Water, 8*, 118.

LGASA. (2012). *Submission on the proposed Basin Plan for the Murray-Darling Basin.* Adelaide: Local Government Association of South Australia. https://www.lga.sa.gov.au/webdata/resources/files/LGA_basinplan_submission.pdf.

Marsden Jacob Associates. (2012). *Preliminary assessment of the benefits of the Basin Plan for the recreational boating industries: Report to the Murray-Darling Basin Authority, Canberra.* http://www.mdba.gov.au/sites/default/files/archived/basinplan/2130-D1236150FinalMJAReport-RecreationalBoating.pd.

MDBA. (2010). *Guide to the proposed Basin Plan Vol 1 Overview* Canberra: Murray-Darling Basin Authority. http://www.mdba.gov.au/kid/guide/index.php.

MDBA. (2011a). *The living Murray story.* Canberra: Murray-Darling Basin Authority (Publication Number 157/11, 95 pp.).

MDBA. (2011b). *The proposed 'environmentally sustainable level of take' for surface water of the Murray-Darling Basin: Methods and outcomes.* Canberra: Murray-Darling Basin Authority (218 pp.).

MDBA. (2011c). *A yarn on the river—Getting aboriginal voices into the Basin Plan.* Canberra: Murray-Darling Basin Authority.

MDBA. (2011d). *Annual report 2010-2011.* Canberra: Murray-Darling Basin Authority.

MDBA. (2012). *Proposed Basin Plan consultation report.* Canberra: Murray-Darling Basin Authority.

MDBA. (2015). *Strengthening connections: Innovative Reconciliation Action Plan 2015-18.* Canberra: Murray-Darling Basin Authority. http://www.mdba.gov.au/sites/default/files/pubs/Strengthening-connections-report-2015-18.pdf.

Mitchell, M. (2013). From organisational learning to social learning: A tale of two organisations in the Murray–Darling Basin. *Rural Society, 22*, 230–241.

Mostert, E., Pahl-Wostl, C., Rees, Y., Searle, B. T., Tàbara, D., & Tippett, J. (2007). Social learning in European river-basin management: Barriers and fostering mechanisms from 10 river basins. *Ecology and Society, 12*, 1.

Neave, I., McLeod, A., Raisin, G., & Swirepik, J. (2015). Managing water in the MDB under a variable and changing climate. *Water (AWA), 42*, 102–107.

NSWLGA. (2012). *Submission to the Murray-Darling Basin Authority's draft Basin Plan.* Sydney: Local Government Association of NSW and Shires Association of NSW. http://www.lgnsw.org.au/files/imce-uploads/35/submission-to-mdba-draft-basin-plan.pdf.

NWC. (2004). *Intergovernmental agreement on a National Water Initiative.* Canberra: National Water Commission [39 pp.].

NWC. (2008). *Australian water markets report 2007–2008.* Canberra: National Water Commission.

Pahl-Wostl, C. (2007). A conceptual framework for analyzing adaptive capacity and multi-level learning processes in resource governance regimes. *Global Environmental Change, 19*, 354–365.

Reed, M. S. (2008). Stakeholder participation for environmental management: A literature review. *Biological Conservation, 141*(10), 2417–2431.

Ross, H., Baldwin, C., & Carter, R. W. (2016). Subtle implications: Public participation versus community engagement in environmental decision-making. *Australian Journal of Environmental Management, 23*(2), 123–129.

Sayce, K., Shuman, C., Connor, D., Reisewitz, A., Pope, E., Miller-Henson, M., et al. (2013). Beyond traditional stakeholder engagement: Public participation roles in California's statewide marine protected area planning process. *Ocean and Coastal Management, 74*, 57–66.

Tan, P. -L., Bowmer, K., & Mackenzie, J. (2012). Deliberative tools for meeting the challenges of water planning in Australia. *Journal of Hydrology, 474*, 2–10.

Thomson, I., & Boutilier, R. (2011). Social license to operate. In *SME mining engineering handbook* (pp. 1779–1796). (3rd ed.). Littleton, CO: Society of Mining Metallurgy and Exploration.

Wahlquist, A. (2011). The media and the guide to the Basin Plan. In D. Connell & R. Q. Grafton (Eds.), *Basin futures: Water reform in the Murray-Darling Basin*. Canberra: ANU Press.

Weir, J. K. (2007). *Murray River Country: An ecological dialogue with traditional owners*. Canberra: Australian National University.

Weir, J. (2011). Water planning and dispossession. In D. Connell & R. Q. Grafton (Eds.), *Basin futures: Water reform in the Murray-Darling Basin*. Canberra: ANU Press.

Wittwer, G. (2011). Confusing policy and catastrophe: Buybacks and drought in the Murray-Darling Basin. *Economic Papers*, *30*, 289–295.

Young, W. J., Bond, N., Brookes, J., Gawne, B., & Jones, G. I. (2011). *Science review of the estimation of an environmentally sustainable level of take for the Murray-Darling Basin, final report to the Murray-Darling Basin Authority: CSIRO Water for a Health Country Flagship Report to Murray-Darling Basin Authority, Canberra*.

INTEGRATED APPROACHES WITHIN WATER RESOURCE PLANNING AND MANAGEMENT IN AUSTRALIA—THEORY AND APPLICATION

C.A. Pollino*, S.H. Hamilton[†], B. Fu[‡], A.J. Jakeman[‡]

CSIRO Land and Water, Canberra, ACT, Australia Edith Cowan University, Joondalup, WA, Australia[†]
Australian National University, Canberra, ACT, Australia[‡]

CHAPTER OUTLINE

INTRODUCTION

The challenges for planning and management of water are complex and interconnected. Integrated approaches seek to break down disciplinary silos, bringing together communities to understand and explore problems, promoting inclusive solutions to complex challenges. Within Australia, this can apply to restoration of flows in the Murray-Darling Basin (MDB) or development of water resources in

Northern Australia. Both situations threaten both natural systems and market economies. Despite the broad recognition of water management being a complex problem, there are few successful examples where an integrative approach has been successfully applied, from problem definition through to decision making. In this chapter we overview recent literature on integration theory and demonstrate some of these concepts, using examples in the water resources domain from the MDB.

Integrated research seeks to cut across science disciplines to explore problems of an integrated nature, applying multidisciplinary and transdisciplinary approaches. Such approaches are central to tackling real-world, complex science problems, where solutions cannot be found via a single disciplinary approach (O'Rourke et al., 2016). Integrative research often involves actors from outside the project team (and academia) being involved in the research process (Lang et al., 2012). It is increasingly rare in environmental management and policy decision-making contexts that solutions are sought that do not require an integrated approach, be that through formal or informal processes.

The terminology for describing concepts of integration is dense and often inconsistently applied across the literature. In this chapter we have simplified the terms used. Disciplines have a defined set of tools, concepts, and theories. Knowledge is exchanged through multidisciplinary approaches, whereas transdisciplinary approaches seek to create new knowledge and theory, and take a participatory approach (Tress et al., 2005, 2007). Essentially, while integration is central to a cross-disciplinary activity, the degree of integration is reflected in the terminology. Multidisciplinary approaches typically involve a loose coupling across disciplines, whereas transdisciplinary approaches cross discipline boundaries and seek to engage and generate new knowledge across different scientific, policy, management, and civil communities (Tress et al., 2005).

The implementation of integrated approaches within the water domain spans the technically oriented physical-based models (Letcher et al., 2007), where knowledge between systems is exchanged, through to social-learning soft systems (Pahl-Wostl, 2007), where conceptualization of broader social-ecological systems is gained. Mauser et al. (2013) describes three modes of integration: *additive*, sharing information for analysis; *combinatory*, having hard integrative models; and *systematic*, a complex systems model that incorporates feedbacks.

Challenges in real-world applications of integration are the potential for irreducible uncertainty and conflicting values that contribute to increasing complexity (Funtowicz & Ravetz, 1995). Consequently, the approach taken to analyzing an integrated problem needs to reflect the diversity of objectives, the scale of problems being assessed and the desired impact in decisions and science. Issues of water resources span local-scale decisions through to exploring issues of sustainability at regional, national, and global scales. Thus, integration across these scales is a challenge. Indeed, while integrative research is conceptually attractive, implementation is challenging and progress within sustainability science is still a challenge (Kliskey et al., 2016).

THE DIMENSIONS OF INTEGRATION

In a recent paper, Hamilton et al. (2015) explored 10 dimensions of integration. Jakeman et al. (2016) went on to expand these, with a focus on integrated groundwater management. The 10 dimensions are:

1. *Issues of concern*—explore the sociotechnical problem space and the interconnectedness of issues which can lead to "wicked" problems.

2. *Governance setting*—explores the potential for intervention to influence a system and the need for scientific solutions to deliver the needs of decision makers.
3. *Stakeholders*—are impacted by, have expertise in, or have responsibility in responding to problems.
4. *Human setting*—relates to the sociocultural context of the problem, be that political, social, economical, or organizational.
5. *Natural setting*—relates to the biophysical aspects of a problem, spanning from global drivers to local outcomes.
6. *Spatial scale*—recognizes cross-scale linkages and the inherent nested scales required in representing the scale of process and information across the problem domain.
7. *Temporal scale*—distinguished from spatial scale by being intrinsically unidirectional.
8. *Disciplines*—recognizing the diversity of knowledge and expertise needed in tackling problems within complex systems, and the need to merge these different perspectives.
9. *Methods, models, other tools and data*—recognizing the plethora of approaches and data used in exploring integrated problems, and issues of compatibility and interoperability.
10. *Uncertainty*—encompasses knowledge variability, linguistic, and decision-making uncertainty.

It is argued that integrated solutions to integration problems should address each of these dimensions (Hamilton et al., 2015). The 10 dimensions are intricately linked, such that the nature of one dimension is dependent on all others. Therefore, inadequate attention to any one dimension could potentially lead to impacts of interventions being overlooked or the analysis being rendered irrelevant.

It is also possible to define the dimensions of integration as both technical and in application. Mauser et al. (2013) defined technical dimensions as: data, assumptions, and systematic analysis approaches that retain integrity and that are adaptive. These authors also went on to define dimensions of application as: the science-to-policy interface; relevance and demand; quality (transparency, replicability); cross-sectoral communication and governance.

Indeed, problems in the water resource policy and management domain require integrated solutions. However, tackling water issues as an integrated problem is not easy. The challenges in integration can limit the utility of findings and diminish the potential for impact, as is explored in the following section.

THE CHALLENGES OF INTEGRATION

Brandt et al. (2013) undertook a review of papers within the context of transdisciplinary research in sustainability science, and defined five challenges. Here we explore and expand on these challenges.

1. *Coherent framing of an issue*—this reflects the diversity of terminologies, which can hamper communication. Lack of a common terminology is one of the main obstacles in integration of disciplines (Tress et al., 2005). Collaborative research teams should overcome this by defining a "common language" and joint definitions (Lang et al., 2012). The nature of integration projects generally requires the involvement of researchers as well as stakeholders from different disciplinary backgrounds and experiences. Hence, in an integration project there are often different perspectives on what are the issues and the key questions that need investigating (Brandt et al., 2013).
2. *Best practice for integration*—this requires process and knowledge types to be clearly stated, particularly in translating theoretical conceptual frameworks and the "essentials" (such as in Hamilton et al., 2015) or evaluation criteria within the framework, through to application in

real-world contexts. This involves defining specific research questions and defining criteria to assess whether the objectives are successfully met (Lang et al., 2012).

3. *Integration models and methods*—this explores the role of frameworks and the difficulties associated with generalizations and applicability, recognizing that a diversity of methods is needed to identify, analyze and solve complex real-world problems. In applying a common framework across disciplines, some consistency in applying data, assumptions and knowledge may be necessary, which can diminish the scientific rigor within a single discipline (Mauser et al., 2013). Scientific integration enables the necessary disciplinary research questions to be defined from the problem framing and then researched by the respective discipline, at the quality required (Mauser et al., 2013).

4. *Science impact*—despite integration and transdisciplinary science being seen as essential in tackling real-world decision problems, with many approaches gaining momentum, publications fail to deliver high-impact science. Transdisciplinary research has also been found to be less likely to receive funding (Bromham et al., 2016).

5. *Lack of empowerment*—this reflects the lack of translation of the outcomes through to impact within a decision context. There is clear evidence in the literature of engagement between science and management communities, but few demonstrations that this interaction translated into changes in decision pathways.

Similarly, Tress et al. (2007) explored barriers to integration. These included lack of communication and cooperation across disciplinary boundaries; a lack of resources to conduct integrative research; lack of a conducive environment to integrative research; a lack of commitment to an integrative approach; and a lack of recognition of the science of integration within the broader science community (Tress et al., 2007). Another issue raised was the lack of a common terminology across disciplines.

One of the major hurdles in achieving integration within the water resource domain is in translating theory to practice, and this is explored further in the following section.

FROM INTEGRATION THEORY TO PRACTICE

Mauser et al. (2013) acknowledged that while there is debate about applying science within multi- and transdisciplinary contexts, pragmatic approaches are needed to address user needs and to overcome one of the key challenges, which is the barrier in translating theory into practice. These barriers span the technical through to the participatory and solution space.

The typical barriers of a technical nature can be similar to those of disciplinary research. These barriers often constitute a lack of data to support the framing and analysis of a problem, systemic uncertainties and system feedbacks that are difficult to analyze and often poorly understood, and limited analytical frameworks. Consequently, the problem is often oversimplified or is unwieldy and infinitely complex. Likewise, there are institutional barriers to the uptake of solutions. This often stems from a limited capacity, a lack of demonstrated benefits (particularly in the short term), and a limited political will to adopt a solution, particularly in the face of high uncertainties. Some of the institutional factors that can increase the success of integrative studies include: minimizing organizational barriers through organizational structures that are more suited to administering research with a disciplinary basis; reducing the time demands for integration through investing in time to understand different perspectives, individual disciplines and defining the interacting points between disciplines; and overcoming the compartmentalization of academic disciplines, where the focus of the effort and the consideration of scales can be vastly different (Kliskey et al., 2016; Tress et al., 2007).

Below we explore some of the "essentials" in overcoming these barriers:

- *Going beyond participation*—from the outset, if the objective is to move from theory to practice, the goal of taking a participatory approach should address issues of legitimacy, ownership and accountability for the problem as well as for the solution (Mauser et al., 2013). Legitimacy issues arise from a lack of representation of the team in the problem definition and solution space (Lang et al., 2012). This comes back to the issue of problem framing, where the team may have insufficient problem awareness or no agreement on the problem itself (Lang et al., 2012). Ownership of the problem and solution is needed as part of the assessment framing where rights and obligations are unclear (Lang et al., 2012; Talwar et al., 2011). Accountability is in ensuring objectives are met (such as through defining success criteria) and the actions are undertaken through a commitment to change (Talwar et al., 2011).
- *Going beyond system understanding*—if integrated assessment approaches and solutions are to be applied in practice, a commitment to adoption is needed. Many integration projects are successful in better understanding problems through framing and then translating this analysis into a framework to improve system understanding. However, to ensure adoption of the integrated solutions, the integration products need to be designed to be relevant to the breadth of stakeholders engaged, including communicating the outcomes such that they are simple enough to be readily communicated across audiences (Mauser et al., 2013).
- *Evaluating solutions to avoid surprise*—analytical tools such as models provide an opportunity to test potential changes or interventions in a system, which can be based on a set of assumptions, knowledge of the system behavior or empirical evidence. However, models are inherently uncertain and require thorough evaluation to determine how much confidence can be placed in their predictions. This evaluation should extend beyond quantitative measures of performance criteria and include a qualitative assessment of the model by experts or relevant stakeholders (Bennett et al., 2013). The reassurance of quality in tools used to support planning and management also applies to the quality of the science used. This necessitates standards of quality of science that are broadly applicable across disciplines, which may include criteria such as transparency, replicability, excellence, and data quality (Mauser et al., 2013). An assessment of uncertainty in predictions should also be undertaken. A useful approach to dealing with uncertainty of future change is *scenario analysis*, which allows policies or interventions that are robust under a range of plausible future states to be developed, rather than trying to predict the state of the future (Liu et al., 2008).
- *Demonstrating transparency*—Funtowicz and Ravetz (1993) contend that in postnormal science, uncertainty should be managed not banished, values should be made explicit not presupposed, and models should be used for interactive dialog not formalized deduction. To achieve this requires more transparency in integration processes. Different stakeholders (including experts) can have different, but equally valid, conceptions about the boundaries of the system, the framing of the issues and the formulation of the problem to be addressed (Dewulf et al., 2005). Effort should be put into defining problems with rigor and transparent (i.e., less biased) problem definition. This involves looking for the relevant questions before digging into problem solving, modeling solutions, and trying to reduce uncertainties that have little impact on the problem concerned.

An interactive, communicative and discursive approach to issue framing is required for continuous negotiation, tuning, and connecting different frames (Dewulf et al., 2004). Boundary judgements affect modeling choices and assumptions. Reflection on and appropriate choices of system boundaries are important in ensuring the effective use of integration approaches and in achieving beneficial outcomes from integration studies (Nabavi et al., 2016). Fu et al. (2017) defined two types of philosophies held by people when they consider their preferences for valuing environmental changes. The first is where "more of a good thing is better" such that benefits are desired for certain attributes in a system. The second is a "restoration philosophy" whereby a person wants to restore a modified system to its natural condition. Recognizing these philosophies has significant bearing on effectively formulating and communicating how results are communicated.

- *Being innovative*—innovation is often claimed as being the domain of science and the scientific method. However, Pahl-Wostl et al. (2013) contend that there remain "deeply embedded assumptions that physical-numerical, computational models constitute a core technology to support policy, and that quantitative data are to be prioritized relative to qualitative evidence, information and 'value-laden' judgement." Innovation in integration requires direct inclusion of human values, environmental ethics and social justice into the conventional paradigm for analysis (Pahl-Wostl et al., 2013).

- *Decision support*—Brandt et al. (2013) contend that the plurality of methods used does compromise the notion of the reproducibility that is demanded by science, increases the "costs" of method integration, and hampers communication within and outside the transdisciplinary research community. A completely reproducible, uniform approach to methods is probably neither possible nor desirable for integrated problems. Nevertheless, it would be beneficial to apply methods consistently so as to generate reproducible approaches within projects and to allow communication between different projects. Systems are needed that can exchange, replicate, adapt, and integrate knowledge. Decision support and information systems can formalize knowledge collation and communication through visualization tools.

Integration "products" can range from formal decision-based models and tools, integration frameworks, new knowledge and insights into solutions for integrative problems. Other outcomes focus on the process, and can include building capacity in collaborative engagement and learning (Kliskey et al., 2016), through to improving processes for group learning, deliberation and negotiation among experts and stakeholders, as well as building awareness across disciplinary boundaries and in stakeholder perspectives.

INTEGRATED ASSESSMENT AND INTEGRATED MODELS: DEMONSTRATIONS FROM AUSTRALIA

In this section, we overview three case studies, which have each taken a different approach to integration using modeling. All studies are done in the MDB. The first case study considers different aspects of a biophysical environment; the second explores a cross-disciplinary approach considering a biophysical and economic interaction; and the third is a multidisciplinary approach, valuing the market and nonmarket benefits of the MDB Plan (see Chapter 13).

CASE STUDY 1: DELIVERING A DECISION SUPPORT SYSTEM FOR ENVIRONMENTAL FLOWS

Inland wetlands support high biodiversity that is dependent on a regime of flows from rivers (Bunn & Arthington, 2002). Dam construction, irrigation infrastructure, water extraction from rivers and groundwater, and land-use changes across the landscape have led to significant and widespread decline in the health of wetland ecosystems (Ward & Stanford, 1995). In Australia, the recent so-called Millennium drought (1997–2009) led to record low flows in extensive areas of the MDB, severely impacting many of its wetland ecosystems. Environmental flows have been used as an important strategy to support wetland ecosystems.

To assist environmental flow management, the IBIS DSS was developed for the New South Wales state government to explore the likely outcomes of environmental water planning scenarios on ecological characteristics of floodplain wetland systems. It has three applications: Narran Lakes, Gwydir Wetlands, and Macquarie Marshes (Merritt et al., 2010). IBIS DSS integrates a dynamic, spatially distributed hydrology model (IQQM) with Bayesian network models (BNs), which represent probabilistic ecological responses (Merritt et al., 2010). The integration allows spatial and temporal representation of hydrology while utilizing the strengths of BNs, such as explicit representation of uncertainty and the capacity to use a range of data types to populate the network. The IBIS DSS has been a useful tool and has continually been used by the state government in wetland-valley scale and mid- to long-term planning in environmental flow management.

One important lesson from this integration project was a lack of consideration of multiple perspectives in problem framing. On advice from the client (NSW state government), IBIS was intended to inform medium- to longer-term strategic planning activities, with the focus at the regional rather than the local scale. Perspectives from local water planners were neither adequately elicited during problem framing nor throughout the integration process. At the end of the project, local water planners were confused about the purpose of the tool and how it would support their planning processes. Developing a shared problem framing and engaging these people throughout the process might have allayed these concerns and managed expectations more effectively.

CASE STUDY 2: UNDERSTANDING GROUNDWATER-SURFACE-WATER INTERACTIONS

Many rural areas are faced with overallocation of water resources and the associated environmental problems. An integrated model was developed by Jakeman et al. (2012) to support the resolution of the problem in the Namoi catchment in New South Wales, which suffers from overallocation of both groundwater and surface-water resources. The model integrates multiple components: (a) a hydrological model that predicts the impact of surface and groundwater extraction on surface flows, aquifer storage and discharge; (b) a social model that captures the behaviors of farmers using changes in farming systems and water use efficiency; (c) an ecological model that estimates the suitability of water conditions (e.g., flood duration and timing, interflood dry period, water quality) for four indicator plant species; (d) an economic model that captures farmers' decision regarding crop choice, area planted, and irrigation water use; and (e) a crop metamodel that estimates crop yield based on climate and irrigation application (Jakeman et al., 2012).

The model components were mostly based on existing and previously tested models, which were modified for the application. For example, the hydrological component was based on the IHACRES rainfall-runoff model, which was modified to include surface-water and groundwater interactions, a two-layer aquifer system, and upscaled to match the spatial scale of the social and economic model components. The model was used to explore trade-offs associated with different decisions, with the outcomes focusing on including short- and long-term profit, capital costs, groundwater impacts, and ecological impacts.

The model was developed collaboratively by a team with an integration focus, who worked together with research scientists from each relevant discipline (hydrology, ecology, economics, governance, and social science). This inclusive and integrative approach to model development was time-consuming; however, it provided significant value including a greater insight into the problem and a platform for discussion among stakeholder groups.

CASE STUDY 3: EVALUATING THE BENEFITS OF A WATER RESOURCE PLANNING FRAMEWORK

As with the previous case studies, this third example explores water planning within an integrated framework. This project sought to evaluate the benefits of the MDB Plan, a legislative instrument that establishes a framework for water management. The project had an interdisciplinary team, exploring benefits of the Basin Plan across physical, social and economic dimensions (Bark et al., 2016; CSIRO, 2012). One of the distinguishing features of this project relative to the others discussed previously is in the scale, with outcomes expressed at a whole-of-Basin scale.

The MDB is approximately 1 million km^2 in size and has a diversity of physical environments and socioeconomic uses and values (see also Chapter 13). The evaluation was at the whole-of-Basin scale, although some of the models that underpin the assessment are disaggregated to catchments, and in some cases, sites. The temporal scale was related to the benefits of the planning framework over a modeled period of 114 years (1895–2009), considering scenarios that represented different water planning and operating rules over this period. These were a modeled "baseline" scenario representing current infrastructure and operations, and a scenario that considered the planning and operating environment postimplementation of the Basin Plan, which subsequently changed river flows (CSIRO, 2012).

The integrated model was not coupled. There was no attempt to either have a fully integrated model sitting in a single platform or a workflow where models interacted. Rather, the outputs from one aspect were the inputs to the next. The integrative framework is reported by Bark et al. (2016). The framework used the outputs from a hydrological model as inputs into ecological models, which considered waterbirds, native fish, floodplain vegetation and, at the end of system, the Coorong and Lower Lakes. Other models considered environmental outcomes, including water-quality measures. These provided outputs that were inputs into an ecosystem services benefits assessment and a hedonic pricing assessment. These were used in an economic benefits assessment, which included avoided costs, nonmarket benefits, and market benefits.

The project team was multidisciplinary. The project required a coherent problem statement, considering the whole-of-Basin scale, for the integration framework to be successfully implemented. The project outcomes were used to inform an Impact Statement put forward to the Australian Commonwealth parliament to support the MDB Plan being adopted into legislation. The findings as an economic benefit over a Basin scale and across a modeled time period were the most challenging to communicate.

KEY LESSONS FROM CASE STUDIES

A first lesson across the case studies is the need to develop a common understanding of the problem across disciplines, as per the first of the dimensions for integration introduced previously. We wish to emphasize that, while consistently recognized, our experiences demonstrate that rarely is this well-resourced in projects, including by project funding agencies. Problem definition, when considering water as an integrated resource, is a lengthy process. It can involve lengthy stakeholder consultations, joint project team and stakeholder meetings, the need to develop a shared language to articulate the problem, and participatory conceptual modeling, to characterize the problem domain from a range of perspectives. This also requires well-developed facilitation skills, an open perspective with a focus on sharing and learning, an integration team, and an implementation plan (Tress et al., 2007). An implementation plan outlines the process for collaboration, the commitments required from researchers, and a commitment to openness to exploration to finding a fit-for-purpose solution to a problem, rather than fitting an existing solution to the problem.

The second lesson is the need to have an in-depth knowledge of stakeholders and their networks, beliefs, and concerns. As stated previously, this includes engaging those with an understanding of and experience in stakeholder processes, such as social scientists. To do this, there needs to be a joint understanding of the objectives of the participatory process from the outset, to ensure expectations are clear. These objectives can span from information exchange through to empowerment, and the stakeholder engagement process should be decided to fulfill the objectives agreed. The first and second case studies discussed previously undertook broad stakeholder consultations to gain input from those involved. The first case study had some difficulties in the consultation process due to the lack of clarity in the scale of the study and agreed focus of the scale of the application of the model in the problem definition phase.

The third lesson relates to understanding the context, including the governance settings, in which the findings are being delivered. Most often this involves developing a set of product- and relationship-based evaluation criteria from the project outset, which can help in ensuring that the analysis undertaken and the findings represent the problem with its sociopolitical context, and also in ensuring that the "product" is fit-for-purpose for the objectives, the scale of application, and the intended use. To achieve this, the policy context must be understood, through steering committees, establishing research partnerships, and formal exchanges and a regular dialog. All this requires a commitment of time and resources, and project timeframes are rarely generous enough to go beyond a steering committee arrangement.

The fourth lesson relates to scale. The challenge of how to apply integrative modeling and what are the best methods to use typically varies with scale. In our third case study, the basin scale project did not require coupling of models, but passing of outputs as inputs to the subsequent model. The other projects used models that were coupled. While we could have pursued a fully integrated model at the basin scale within a single modeling platform, this would have limited the modularity of the analysis components. All components were being developed in parallel and needed to be tested and refined independently. If the analysis framework were to be reused, the components could also be updated and replaced as needed. A fully coupled model could potentially oversimplify the problem, as well as become computationally complex.

The fifth lesson relates to packaging and communicating the product outcomes. There is still a steep learning curve in how to communicate complex, and often highly uncertain, outcomes across diverse audiences. Tools to communicate outcomes span from technical decisions-support systems with complex scenario visualizations or simulation-based analyses (as in the first two case studies) to simple visualization design tools, spatial visualization tools, and construction of narratives. There is room for innovation in developing products, to enhance adoption, to ensure a broad understanding of the integrative methods and outcomes, and in embedding outcomes within a context of an adaptive setting.

The final lesson is in recognizing that integration is expensive. It requires time and a special set of skills within a project team. Investment is needed to develop a skilled integration team that can facilitate relationship building and the transdisciplinary exchange, and understand the integration points across the disciplinary span and ensure this is reflected in the project products. The integration team must have a diverse set of skills, including those sourced from the social sciences, to monitor and evaluate progress of the project collaboration.

KEY CHALLENGES: INTEGRATIVE APPROACHES IN THE MDB

Australia is a dry continent. At national and regional scales, rainfall is unevenly distributed, being variable in both frequency and intensity, and evapotranspiration is responsible for 90% of losses in surface-water flows. This harsh climate profoundly influenced the development of urban and rural settlements, in particular in government investment in infrastructure to secure water supplies, which led to broad changes in flow regimes (see Chapters 1 and 2).

The development of irrigated agriculture in Australia has been heavily reliant on the harvesting of surface-water (and groundwater) resources, and the use of storages and delivery structures to offset the variability of the climate. Consequently, there are now a multitude of water-control devices, including locks, floodplain levee banks, and large dams, throughout Australia. Arguably, as water is a common pool resource, there is a higher potential for overexploitation, as has been observed in many water basins around the world. In the MDB, widespread development has been accompanied by a trajectory of decline in the health of the Basin's surface- and groundwater dependent ecosystems.

As a common pool resource, the benefits from water can be both public and private, benefiting both natural and market economies. Water planning and management is quintessentially an integration problem, with many of the features described by Hamilton et al. (2015). The wicked nature of water resources is represented in the implicit uncertainties, the risks, the diversity of values and perspectives on water, and the need to blend social, technical and science aspects to address water resource problems (Reed & Kasprzyk, 2009). The value of taking an integrated approach is in the diversity of knowledge brought to the problem, better quality and acceptance of the problem and possibly decisions, and increased support for adoption and implementation. Participation in decision processes can promote the development of social capital, which can lead to enabling stakeholders to more easily solve problems and address conflicts or decision points arising in the future. This is particularly pertinent under a future of global climate change.

In a recent article, Abel et al. (2016) argue that in the MDB, there is a need to shift from incremental adaptation to transformational change. They challenge the water management community to engage with the broader stakeholder community, to explore beliefs and options. While this is a desirable path, they do not discuss how to do this at a Basin scale, considering local, regional, and global drivers, and how to balance competing objectives across physical and socioeconomic contexts. In their article, Campbell et al. (2013) state that decisions are ultimately political and involve value judgements, supported by peer-reviewed technical work and sound knowledge for evidence-based decision making. The challenge is how to practically apply participatory approaches to integration, handing the diversity of beliefs and values across time and space. The MDB is a highly variable system that is interconnected, and in which decisions and changes interact across space and time.

The Australian water-reform process has demonstrated the evolution of processes for addressing water planning and management in large complex basins, exploring complex governance and physical basin scales. The participatory process and how this is handled across scales within the Basin remain a challenge.

THE FUTURE

Integrative research can contribute to better decision making in water resources planning and management. It is argued that scientists, policymakers, and the broader community need to "seize the initiative" to act together and tackle real-world problems with a set of common objectives and methods that are fit-for-purpose and reproducible (Brandt et al., 2013).

The MDB represents a challenge in how best to tackle the social-political dimensions within the biophysical constraints of the system. There is an ever-growing body of research and tools that describe the biophysical environment of the Basin. The next challenge is how to embed these within a broader integrated approach to explore water-planning options for the future, looking at outcomes across biophysical and socioeconomic disciplines.

With growing interest in the expansion of irrigation in northern Australia, there is a need for more integrative solutions that can tackle the complex decisions that will need to be made. Detailed knowledge of the biophysical systems is lacking in northern Australia, the environmental systems are of high conservation value and there are large areas of indigenous estates across the north. Consequently, the social-political context will be challenging.

Investment in integrative science requires commitment and ample time and resources to tackle complex problems. Pahl-Wostl et al. (2013) argue that there is ample opportunity for innovation in integrative research and that this research can provide solutions to meet the science challenges of the future (Funtowicz & Ravetz, 1993). Further, Laniak et al. (2013) believe that a body of expertise and experience can be gained through sponsoring, nurturing, and training in the theory and practice of integrated management.

REFERENCES

Abel, N., Wise, R., Colloff, M., Walker, B., Butler, J., Ryan, P., et al. (2016). Building resilient pathways to transformation when 'no one is in charge': Insights from Australia's Murray-Darling Basin. *Ecology & Society*, *21*(2), 23.

Bark, R. H., Colloff, M. J., MacDonald, D. H., Pollino, C. A., Jackson, S., & Crossman, N. D. (2016). Integrated valuation of ecosystem services obtained from restoring water to the environment in a major regulated river basin. *Ecosystem Services*, *22*, 381–391.

Bennett, N. D., Croke, B. F., Guariso, G., Guillaume, J. H., Hamilton, S. H., & Jakeman, A. J. (2013). Characterising performance of environmental models. *Environmental Modelling & Software*, *40*, 1–20.

Brandt, P., Ernst, A., Gralla, F., Luederitz, C., Lang, D. J., Newig, J., et al. (2013). A review of transdisciplinary research in sustainability science. *Ecological Economics*, *92*, 1–15.

Bromham, L., Dinnage, R., & Hua, X. (2016). Interdisciplinary research has consistently lower funding success. *Nature*, *534*, 684–687.

Bunn, S. E., & Arthington, A. H. (2002). Basic principles and ecological consequences of altered flow regimes for aquatic biodiversity. *Environmental Management, 30*, 492–507.

Campbell, I. C., Hart, B. T., & Barlow, C. (2013). Integrated management in large river basins: 12 lessons from the Mekong and Murray-Darling Rivers. *River Systems, 20*, 231–247.

CSIRO. (2012). *Assessment of the ecological and economic benefits of environmental water in the Murray-Darling Basin*. Canberra: Report by CSIRO for the MDBA.

Dewulf, A., Craps, M., Bouwen, R., Taillieu, T., & Pahl-Wostl, C. (2005). Integrated management of natural resources: Dealing with ambiguous issues, multiple actors and diverging frames. *Water Science & Technology, 52*, 115–124.

Dewulf, A., Craps, M., & Dercon, G. (2004). How issues get framed and reframed when different communities meet: A multi-level analysis of a collaborative soil conservation initiative in the Ecuadorian Andes. *Journal of Community & Applied Social Psychology, 14*, 177–192.

Fu, B., Dyer, F., Kravchenko, A., Dyack, B., Merritt, W., & Scarpa, R. (2017). A note on communicating environmental change for non-market valuation. *Ecological Indicators, 72*, 165–172.

Funtowicz, S. O., & Ravetz, J. R. (1993). The emergence of post-normal science. In R. Von Schomberg (Ed.), *Science, politics and morality* (pp. 85–123). Dordrecht: Springer.

Funtowicz, S. O., & Ravetz, J. R. (1995). Science for the post normal age. In L. Westra & J. Lemons (Eds.), *Perspectives on ecological integrity* (pp. 146–161). Dordrecht: Springer.

Hamilton, S. H., ElSawah, S., Guillaume, J. H., Jakeman, A. J., & Pierce, S. A. (2015). Integrated assessment and modelling: Overview and synthesis of salient dimensions. *Environmental Modelling & Software, 64*, 215–229.

Jakeman, A., Kelly, R., Ticehurst, J., Blakers, R., Croke, B., Curtis, A., et al. (2012). In: *Modelling for the complex issue of groundwater management. SIMULTECH, international conference on simulation and modeling methodologies, technologies and applications*. http://hdl.handle.net/1885/68969.

Jakeman, A. J., Barreteau, O., Hunt, R., Rinaudo, J.-D., Ross, A., Arshad, M., et al. (2016). Integrated groundwater management: An overview of concepts and challenges. In A. Jakeman, O. Barreteau, R. Hunt, J. D. Rinaudo, & A. Ross (Eds.), *Integrated groundwater management* (pp. 3–20). Switzerland: Springer International Publishing.

Kliskey, A., Alessa, L., Wandersee, S., Williams, P., Trammell, J., Powell, J., et al. (2016). A science of integration: Frameworks, processes, and products in a place-based, integrative study. *Sustainability Science, 11*, 1–11.

Lang, D. J., Wiek, A., Bergmann, M., Stauffacher, M., Martens, P., Moll, P., et al. (2012). Transdisciplinary research in sustainability science: Practice, principles, and challenges. *Sustainability Science, 7*, 25–43.

Laniak, G. F., Olchin, G., Goodall, J., Voinov, A., Hill, M., Glynn, P., et al. (2013). Integrated environmental modeling: A vision and roadmap for the future. *Environmental Modelling & Software, 39*, 3–23.

Letcher, R. A., Croke, B. F. W., & Jakeman, A. J. (2007). Integrated assessment modelling for water resource allocation and management: A generalised conceptual framework. *Environmental Modelling & Software, 22*, 733–742.

Liu, L., Gupta, H., Springer, E., & Wagener, T. (2008). Linking science with environmental decision making: Experiences from an integrated modelling approach to supporting sustainable water resources management. *Environmental Modelling & Software, 23*, 846–858.

Mauser, W., Klepper, G., Rice, M., Schmalzbauer, B. S., Hackmann, H., Leemans, R., et al. (2013). Transdisciplinary global change research: The co-creation of knowledge for sustainability. *Current Opinion in Environmental Sustainability, 5*, 420–431.

Merritt, W., Powell, S., Pollino, C., & Jakeman, T. (2010). IBIS: A decision support system for managers of environmental flows into wetlands. In N. Saintilan & I. C. Overton (Eds.), *Ecosystem response modelling in the Murray-Darling Basin* (pp. 119–136). Melbourne: CSIRO Publishing.

Nabavi, E., Daniell, K. A., & Najafi, H. (2016). Boundary matters: The potential of system dynamics to support sustainability? *Journal of Cleaner Production, 140*, 1–12.

O'Rourke, M., Crowley, S., & Gonnerman, C. (2016). On the nature of cross-disciplinary integration: A philosophical framework. *Studies in History and Philosophy of Science Part C: Studies in History and Philosophy of Biological and Biomedical Sciences, 56*, 62–70.

Pahl-Wostl, C. (2007). The implications of complexity for integrated resources management. *Environmental Modelling & Software, 22,* 561–569.

Pahl-Wostl, C., Giupponi, C., Richards, K., Binder, C., de Sherbinin, A., Sprinz, D., et al. (2013). Transition towards a new global change science: Requirements for methodologies, methods, data and knowledge. *Environmental Science & Policy, 28,* 36–47.

Reed, P. M., & Kasprzyk, J. (2009). Water resources management: The myth, the wicked, and the future. *Journal of Water Resources Planning and Management, 135,* 411–413.

Talwar, S., Wiek, A., & Robinson, J. (2011). User engagement in sustainability research. *Science and Public Policy, 38,* 379–390.

Tress, G., Tress, B., & Fry, G. (2005). Clarifying integrative research concepts in landscape ecology. *Landscape Ecology, 20,* 479–493.

Tress, G., Tress, B., & Fry, G. (2007). Analysis of the barriers to integration in landscape research projects. *Land Use Policy, 24,* 374–385.

Ward, J. V., & Stanford, J. (1995). Ecological connectivity in alluvial river ecosystems and its disruption by flow regulation. *Regulated Rivers: Research & Management, 11,* 105–119.

AUSTRALIAN CASE STUDIES

CASE STUDY 1—THE MURRAY-DARLING BASIN PLAN 13

B.T. Hart*,†, D. Davidson*,†

Water Science Pty Ltd, Echuca, VIC, Australia* Murray-Darling Basin Authority, Canberra, ACT, Australia†

CHAPTER OUTLINE

INTRODUCTION

The focus of this case study is the decision-making process associated with the setting of the sustainable diversion limits (SDLs), a critical component of the Murray-Darling Basin Plan (Basin Plan). The Basin Plan, enacted by the Australian Parliament in Nov. 2012, represents an important advance in the integrated management of the Basin's water resources and was the culmination of over 4 years of intensive work, following a long process of major water reforms in Australia that are summarized in the following paragraphs.

The water scarcity issues in the Murray-Darling Basin (MDB) became critical during the recent drought (1997–2010), and are likely to increase with the future impacts of climate change. Despite the water reforms that have been occurring over the past 20–30 years, the extended Millennium drought brought to a head the overallocation of water in the important MDB and the degradation of the Basin's water-related ecosystems due to the lack of environmental water. This resulted in the Australian parliament enacting a new Water Act in 2007 that required (among other things) the development and implementation of an integrated water resources plan for the Basin. Hart (2016) has covered the factors leading to the successful development of the Basin Plan.

Implementation of this first Basin Plan will be largely completed in 2019 (although some elements will not be completed until 2024), by which time Australia will have achieved a remarkable transformation in the way the Basin's water resources are managed, and will have achieved water reform the scale of which has not been accomplished elsewhere in the world. Hart (2015a, 2015b) has discussed the challenges in the implementation of the Basin Plan.

This chapter first describes the characteristics of the MDB and then briefly reviews the water reforms over the 30 years preceding the 2012 Basin Plan. The main features of the Basin Plan are then summarized, and the main decision-making steps in determining the SDLs for the Basin are discussed. The challenges in the implementation of the Basin Plan, a process that will occur between 2013 and 2019, and beyond, is the subject of papers by Hart (2015a, 2015b). The final section of this chapter discusses the review of the SDL set for the northern MDB, a process that was required in the Basin Plan because of the paucity of information on this region at the time the Basin Plan was developed. This discussion is focused particularly on the updated decision-making process used to review and change the northern Basin SDL.

CHARACTERISTICS OF THE MDB

The essential physical, ecological, socioeconomic and political features of the Basin are described in MDBA (2010a, 2010b, 2011a, 2011b, 2011c, 2012a) and Rogers and Ralph (2011).

The Murray-Darling Basin (catchment) has an area of around 1 million km^2 (or one-seventh the area of Australia) and is located in southeastern Australia (Fig. 1). The system is divided by climate into northern rivers (Darling system) and southern rivers (Murray system). The Darling system is more influenced by tropical weather patterns with most rainfall occurring in the summer (Dec. to Mar.), while rainfall in the Murray system is more winter-spring (Jun. to Sep.) dominated.

Runoff is very low compared with other major river systems around the world, despite the Basin covering a large area (Gordon et al., 2004). Additionally, the year-to-year variations in the Murray-Darling system are very large, ranging from 118,000 GL in 1956, a very wet year, to less than 7000 GL in 2006, a very dry year (MDBA, 2010a, 2010b). Prior to significant human changes in the Basin, about 32,000 GLy^{-1} (or 6% of average annual rainfall) occurred as runoff and flow into the Basin's rivers and streams (Grafton et al., 2014; MDBA, 2010a, 2010b).

The MDB is very important economically, being one of the most productive agricultural regions in Australia. In 2014–15, the MDB contributed 50% of Australia's irrigated production, utilizing 1.3 million ha, with a gross value of around $7.2 billion (MDBA, 2016a, 2016b).

The Basin has a large area of floodplain forests and wetlands, with 16 of the wetlands listed as internationally significant (Ramsar). It also supports a great diversity of nationally and internationally significant plants, animals and ecosystems, many of which are now threatened, vulnerable or degraded. The degradation of floodplain river red gum forests, native fish populations, water bird populations and

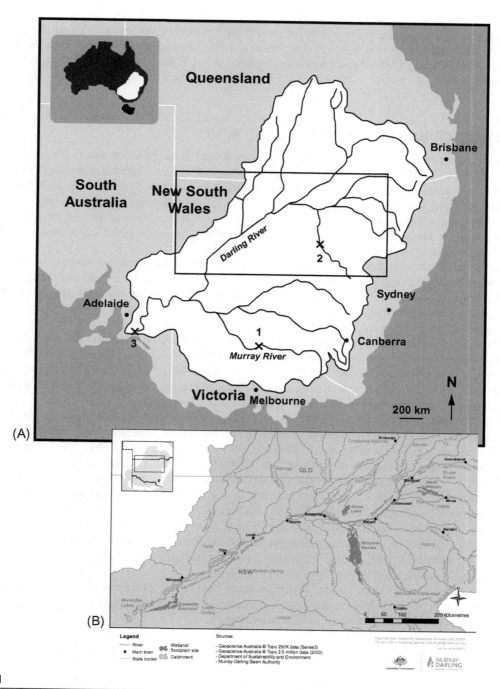

FIG. 1

(A) Location map of the Murray-Darling Basin, which encompasses parts of four Australian states (Queensland, New South Wales, Victoria, South Australia). (B) The insert shows the northern Basin in more detail (1, Barmah Forest; 2, Macquarie Marshes; 3, Lower Lakes and Coorong).

the southern Coorong Lake at the end of the system are now well documented (Colloff et al., 2015; MDBA, 2010a, 2010b, 2011a). Much of this degradation has been caused by the increasing regulation of the Murray-Darling River system, and overallocation of water for consumptive uses, over the past 100 years.

CONTEXT OF WATER REFORMS IN THE MDB

In many respects, the water planning and management system in the MDB is close to the world's best practice (NWC, 2004, 2014). Consumptive water is allocated as legal secure and tradable access entitlements or water "rights" and these water rights have been separated (unbundled) from land (see Chapter 3). Thus, water entitlements are now independent assets in the same sense as land.

Secure and tradeable water entitlements provide more confidence for those owning or investing in the entitlements, better and more compatible registration arrangements, better monitoring, reporting and accounting, and improved public access to information.

However, despite this, from the early 1990s it was recognized that there was an urgent need to reform water resource management in the Murray-Darling River system. There was general agreement that the system was overallocated, with too great a proportion of the water resource allocated for consumptive uses (mostly irrigation). This has led to a number of important changes in the management of the Basin's water resources over the past 30 years (see Chapter 1).

The first of these changes occurred in 1995, when Basin governments[1] through the Ministerial Council agreed to a cap[2] (an upper limit) on the surface water diversions in the MDB. This was in response to reports that confirmed that diversions in the Basin had grown rapidly, could grow further and that this growth had caused decline in river health. The cap placed "a line in the sand" to ensure conditions did not get worse for both the environment and the irrigators.

The Millennium drought (1997–2010) provided the crisis necessary for achieving subsequent water reforms (Connell & Grafton, 2011; Grafton et al., 2012), starting in 2004, with the Council of Australian Governments (COAG) adopting the National Water Initiative (NWI; NWC, 2004), the aim of which was to phase out overuse of water, reform the water entitlement system, and develop an active water market. The National Water Commission (NWC) was established to help achieve these reforms.

The final steps in these water reforms occurred in Jan. 2007 when the prime minister at the time (John Howard) outlined a $10 billion, 10-point National Plan for Water Security[3] to improve water efficiency and address overallocation of water in rural Australia, particularly in the MDB. Later in the same year (Aug.), the Australian Parliament passed the Water Act (2007) (Aust Govt, 2014), with Basin governments agreeing that the Australian government would take a larger coordinating role in the integrated management of the Basin's water resources.

These reforms were historic in that the Basin states ceded some of their powers to the Australian government, a significant change since, under the Australian Constitution, state governments have responsibility for managing their state's water resources.

---

[1]The Basin governments include the federal government and the governments of Victoria, New South Wales, Queensland, South Australia and the Australian Capital Territory.

[2]http://www.mdba.gov.au/what-we-do/managing-rivers/the-cap.

[3]http://pandora.nla.gov.au/pan/10052/20070321-0000/www.pm.gov.au/docs/national_plan_water_security.pdf.

This Commonwealth Water Act (2007) established a new independent Murray-Darling Basin Authority (MDBA), with a mandate to rebalance water allocations between the environment and consumptive uses, through the development and implementation of a Basin Plan. The development of this Basin Plan is now complete, having been approved by Australian Parliament in Nov. 2012 (Aust Govt, 2012), although its implementation will occur over the period 2013–19 (and beyond).

THE BASIN PLAN
THE NEED

There have been two periods in recorded history when the ecosystems of the MDB have been most severely impacted: (a) in the 1800s when land (and many wetlands) was cleared for agriculture, and (b) in the period 1930–70 when most of the dams were built and the southern river system regulated (Colloff et al., 2015).

River regulation has resulted in two major changes to the hydrology of these southern rivers: (a) most of the small to medium-sized flows are now trapped in the many dams that have been built or captured in large on-farm storages, and (b) the flow amplitudes have been largely reversed (or dampened) with the historic strong winter and spring flows now diminished, and flows in summer and autumn increased to meet the demands of irrigators and communities (Grafton et al., 2014).

These changes to the hydrology have resulted in a gradual degradation in the system's water-dependent ecosystems (MDBA, 2010a, 2010b). The most obvious examples are: the degradation (and death) of many river red gum trees, particularly in the lower reaches of the River Murray; a significant reduction in the numbers of waterbirds and native fish throughout the system; the poor ecological health of the Coorong lake system located at the end of the system; and during the Millennium drought, a significant reduction in the water levels in South Australia's Lake Alexandrina and Lake Albert, exposing acid sulfate soils, and the closure of the Murray Mouth (Brooks et al., 2009; Kingsford et al., 2009).

These issues provided the catalyst for the new Water Act (2007) and the development of the Basin Plan as discussed earlier.

KEY ELEMENTS OF THE BASIN PLAN

The Basin Plan is a high-level plan aimed at ensuring that the water resources of the MDB are managed in an integrated and sustainable way to achieve "*a healthy working Murray-Darling Basin that supports strong and vibrant communities, resilient industries, including food and fibre production, and a healthy environment*" (the Vision, Aust Govt, 2012). MDBA (2011b) defined a healthy working river as a managed river in which the natural ecosystem has been altered by the use of water for human benefit, but in which the altered system retains its ecological integrity while continuing to support strong communities and a productive economy.

Long-term average SDLs

The Basin Plan aims to achieve a healthy working Basin through the establishment of new long-term average SDLs that reflect an environmentally sustainable level of water use (or "take"). These SDLs are limits on the volumes of water that can be taken for consumptive purposes (including domestic, urban, industrial and agricultural use), and are set at both a catchment and a Basin-wide scale. The SDLs are essentially new caps on consumptive water take that need to be met as long-term averages, with annual compliance assessed by taking into consideration the particular circumstances of each water year.

The Water Act (2007) required that the SDLs should be determined through a triple bottom line approach in which the environmental requirements and social and economic outcomes were balanced (or optimized). Specifically, the Water Act required the MDBA to achieve better outcomes for the water-dependent assets and ecological functions of the Basin, and at the same time to "*promote the use and management of the Basin water resources in a way that optimises economic, social and environmental outcomes.*"[4]

The water-related environmental requirements of the Murray-Darling River system were defined through the assessment of the Environmentally Sustainable Level of Take (ESLT). This was a new concept introduced in the Water Act (2007) that required the MDBA to develop a robust and scientifically defensible method for determining the water requirements for the Basin's water-dependent ecosystems. The method used to develop this ESLT is a key component of the basin planning process, and is summarized in Hart (2016) with full details provided in MDBA (2011b) and Swirepik et al. (2016).

The decision-making process used to balance the environmental, social and economic outcomes, resulting in the Basin Plan SDLs, is covered in the next section.

The Basin Plan established a long-term SDL of 10,873 GLy^{-1}, or around a 21% reduction to the current consumptive water allocations over the Basin.[5] The SDL includes all water used for consumptive purposes, including diversions from rivers and interceptions, the latter mostly being water held in farm dams or used by commercial plantations.

Basin-wide environmental watering strategy

The introduction of SDLs and other reforms will result in more water being returned to the environment. However, this in itself is not enough to ensure the best possible environmental outcomes. The Basin Plan also requires that the MDBA develop a Basin-wide environmental watering strategy (BWS) to ensure that the size, timing and nature of river flows maximize the benefits for the environment.

The first BWS, published in Nov. 2014 (MDBA, 2014b), identifies how the desired environmental outcomes in the Basin may be achieved through better coordination and cooperation between agencies and across borders, and by using the water as wisely as possible to maintain a river system that is resilient and healthy.

The BWS also requires the MDBA to establish Basin watering priorities annually. The priorities for 2014–15 are focused around three themes: (a) connecting rivers and floodplains, (b) supporting in-stream functions, and (c) enhancing and protecting refuge habitat (MDBA, 2014c).

The states are also required to develop environmental watering plans for individual rivers and their catchments. Priorities for annual environmental watering events will be set by the relevant governments (federal and state), and will be achieved in collaboration with environmental water holders,[6] local communities and Indigenous people. The BWS will be reviewed every 5 years.

---

[4]Water Act (2007), Section 3(c).

[5]The reduction is around 25% in the highly regulated southern Basin, and around 12% in the less developed northern Basin.

[6]The Water Act (2007) established the Commonwealth Environmental Water Holder (CEWH; www.environment.gov.au/ewater) with the responsibility for effectively using the Australian government's environmental water portfolio (see Docker and Robinson, 2014). Some of the states also have established equivalent environmental water holders and an environmental water portfolio (see www.vewh.vic.gov.au for details on the Victorian Environmental Water Holder).

Water quality and salinity management plan

The Basin Plan also includes an integrated Water Quality and Salinity Management Plan that provides a Basin-wide framework of objectives and targets for ensuring Basin water is "fit for purpose," i.e., suitable for irrigation and recreational uses, maintaining aquatic ecosystems, and for drinking water after treatment. The Salinity Management Plan builds upon the very successful, collaborative Basin Salinity Management Strategy (BSMS)[7] that has tackled salinity issues across the Basin for the past 15 years (2001–15; MDBA, 2014d).

Water trading rules

A water market—the buying and selling of tradeable water rights—has been in place in the MDB for over 30 years (NWC, 2013). Most activity is in the southern connected system, with trading of both permanent and temporary water entitlements occurring (MDBA, 2014e). The introduction of a water market is generally regarded as a major success in helping to ensure that water reaches its highest value use (Grafton & Horne, 2014; Grafton et al., 2012; Horne, 2013; NWC, 2011, 2013), and was very important in assisting many irrigation businesses to survive the Millennium drought (NWC, 2013). The Basin Plan introduced new water trading rules on Jul. 1, 2014; they are aimed at ensuring all the Basin's water markets function consistently, fairly, efficiently, effectively and transparently (MDBA, 2014a).

Water resource plans

Regional water resource plans set out how water resources will be managed, usually for a 10-year period, for a water resource plan area. Thirty-nine plans are being developed by the Basin states and will need to be approved by the Commonwealth Water Minister by Jun. 2019. A number of water resource plans are already in place in the states, but these will need to be updated so they are consistent with the Basin-wide planning framework, including the SDLs. The MDBA has published a Handbook for Practitioners (MDBA, 2013) to assist states in updating their water resource plans.

Monitoring and evaluation

The Basin Plan requires that the MDBA report on the effectiveness of the Plan each year and on the impacts of the Plan every 5 years (Aust Govt, 2012). The Plan sets a series of questions to guide the evaluation reports, in particular focusing on the extent to which the objectives, targets and outcomes of the Plan have been achieved, how the Plan has contributed to changes in the environmental, social and economic conditions of the MDB, and how the Basin Plan could be improved. Importantly, these evaluation reports must also assess "to what extent has the program for monitoring and evaluating the effectiveness of the Basin Plan contributed to adaptive management and improving the available scientific knowledge of the Murray-Darling Basin" (Aust Govt, 2012, p. 149).

The MDBA have developed and published a Monitoring, Evaluation and Reporting Framework (MDBA, 2014f), and also published reports on progress in the implementation of the Basin Plan for 2013–14 (MDBA, 2015b) and 2014–15 (MDBA, 2016a). These reports focus on three key aspects of the Basin Plan: (a) better decisions at the right level, (b) productive and resilient industries and confident communities, and (c) healthy and resilient rivers, wetlands and floodplains. A more detailed report is planned for 2017.

---

[7]Updated in Nov. 2015 to Basin Salinity Management Strategy 2030 (MDBA, 2015a).

KEY FACTORS IN THE SUCCESSFUL DEVELOPMENT OF THE PLAN

Hart (2016) has provided a perspective on the key factors leading to the success of this major water reform. Three of these factors that we believe are particularly important are discussed in the following subsections.

Government commitment

The commitment of both Commonwealth and State governments to the reform process has been vital (see also Chapter 1), including:

- The ongoing bipartisan support of Australian and State and Territory governments for this reform process. This includes the establishment of the NWI in 2004, the new Commonwealth Water Act in 2007, the enactment of the Basin Plan in 2012, and the ongoing commitment of Australian governments to "implement the Basin Plan in full and on time."
- The large investment portfolio allocated by the Australian government to support the changes. Since the original commitment of $10 billion, Australian governments have now committed over $13 billion for programs in support of water reform in the MDB (originally called *Water for the Future*). Much of this funding is for investment in upgrading rural water infrastructure to improve water use efficiency on and off farm, with a share of water savings helping to bridge the gap to the SDLs under the Basin Plan. There has also been substantial investment in purchase of water entitlements for environmental use. Without these investments it is certain that the reform process would have been considerably more challenging, and probably would not have been ultimately successful.

Effective community engagement

Water policy reform is a challenging task that faces many obstacles. Daniell et al. (2014) and Daniell and Barrateau (2014) identified a number of impediments to the development and uptake of a new and innovative water policy; in particular, complex politics and powerful coalitions across multilevel governance systems and vested interests (e.g., the irrigation industry) can be important barriers to the implementation of new water policies. This makes effective engagement of the key stakeholders an absolutely essential component of the process.

The Water Act (2007) required the MDBA to develop a "proposed Basin Plan" (with appropriate consultation), make this publicly available for comment over a 16-week period, receive submissions, take note of these submissions in modifying the Plan, then submit the modified Plan to the Ministerial Council, and after addressing Ministerial Council comments, to submit the final Basin Plan to the Federal Water Minister. All of this was to happen before Jun. 2011.

With hindsight, this process was inadequate and inappropriate given the scale of the changes to be made and the unrealistic timeframes. Thus, it was hardly surprising that the earlier engagement with the community went very poorly (Hart, 2016) for at least two reasons. First, the MDBA did not fully appreciate the existing resentment in the rural community from the number of contentious reforms that state jurisdictions had instituted over the decade prior to the basin planning process. Second, the attempt to introduce an additional consultation step by preparing a "Guide to the proposed Basin Plan" in 2010 (MDBA, 2010a, 2010b) and using this as a basis for initial consultation with the community was a failure and arguably put the reform process back many months. The Guide was far too large and complicated to be an effective engagement document, and the time for community members to read and assimilate the material was far too short and there was no preexposure to the ideas.

At the commencement of 2011, however, (halfway through the 4-year period), a new chair of the MDBA was appointed and soon after a new chief executive officer. This change in leadership of the Authority saw a major change in direction, with a significantly improved engagement with Basin communities and with the state agencies.

Hart (2016) concluded that "genuine engagement with key stakeholders is essential in achieving any major reform. Effective engagement requires a long-term commitment, a genuine desire to engage, skilled and professional staff to run the engagement process, sufficient resources to sustain the process, and appropriate material to facilitate community understanding and participation."

Evidence-based decision-making

Contemporary natural resource management and policy development requires evidence-based decision making (Conroy & Peterson, 2013). This was made clear in the Water Act, which required the MDBA "to act on the basis of the best available scientific knowledge and socio-economic analysis" in developing the Basin Plan.[8] This is one of the few pieces of legislation that specifically requires this approach.

The process used by the MDBA to decide on the optimum SDL (via the establishment of the ESLT) was evidence-based, and is discussed in detail in the next section. This required the development of an iterative method to determine, at both a local catchment scale and at the Basin scale, the volume of environmental water that would result in improved environmental outcomes with minimal social and economic impacts on Basin communities. To achieve this balance required use of both the best available scientific knowledge and robust social and economic analysis.

However, as is discussed in the next section, even with a concerted effort to assemble the best available science and social and economic information, and to make it publicly available on the MDBA website, it was still necessary for the MDBA to make a number of judgments. The important point here is not that judgments were made, but that the options considered, the evidence supporting each option, and the reasons for selecting the final option were documented and made widely available.

An important consideration regarding evidence-based decision making is the question of the relevance, robustness and quality of the evidence. The MDBA adopted a general principle that all reports and research should, as far as possible, be peer reviewed, and this has been done. One important example was the peer review of the ESLT methodology done by a CSIRO-lead team (Young et al., 2011). Additionally, the MDBA has since established a high-profile Advisory Committee on Social, Economics and Environmental Sciences (ACSEES)[9] to provide advice on the relevance and quality of the environmental, social and economic evidence it seeks to use.

---

[8]Water Act (2007), Section 21, 4(b).

[9]http://www.mdba.gov.au/what-we-do/research-investigations/acsees.

KEY DECISION-MAKING STEPS IN DEVELOPING THE BASIN PLAN SDLS

In seeking to achieve better outcomes for the water-dependent assets and ecological functions of the MDB, the Water Act (2007) was groundbreaking in two respects. The Act required the MDBA in developing the Basin Plan to: (a) "promote the use and management of the Basin water resources in a way that optimises economic, social and environmental outcomes,"[10] and (b) "to act on the basis of the best available scientific knowledge and socio-economic analysis."[11] Thus, the MDBA was required to adopt a "triple bottom line" approach to the decision making and also to ensure this process was evidence-based.

The process of trading off environmental, social and economic outcomes is a challenging one. Review of decision-making literature suggests that when making complex decisions, such as establishing SDLs, a form of decision support should be used. Decision-support systems can range from guidelines to matrices to structured decision-making tools, such as multicriteria decision analysis (see Chapter 8 and Janssen, 2012). These decision-support systems differ in the amount of direction they give to the decision maker, the type of data required (qualitative or quantitative), and the use, or not, of weightings for criteria.

Following, we discuss the process used by the MDBA in developing the Basin Plan to balance the environmental, social and economic outcomes in establishing the long-term SDLs and hence the level of water recovery needed.

Additionally, in the section following this we have included aspects of the modified process developed for reassessing the SDLs for the northern Basin. Because there was less information available for the northern Basin at the time the Basin Plan was developed, it was agreed that a northern Basin review would be undertaken to improve the knowledge base for this region and to use this to revise the northern Basin SDLs specified in the Basin Plan (MDBA, 2014g). This new decision-making process is superior to that originally used in 2012 and is a good example of how process can be adapted over time on the basis of new and improved knowledge.

ENVIRONMENTAL ASPECTS

The method used to determine the environmental needs of the water-related assets of the Basin (or the ESLT) is summarized by Hart (2016), with full details provided in the MDBA (2011a) and Swirepik et al. (2016). The process required deciding, at both a local catchment scale and at the Basin scale, the volume of environmental water that would result in improved environmental outcomes (with minimal social and economic impacts on Basin communities). Briefly, the process involved the following steps.

Identify Basin-wide environmental objectives
Four high-level environmental objectives were identified to achieve a healthy working Basin (the Basin Plan vision): (a) to protect and restore water-dependent ecosystems of the Basin, (b) to protect and restore the ecosystem functions of water-dependent ecosystems, (c) to ensure that water-dependent ecosystems are resilient to risks and threats; and (d) to ensure that environmental watering is

---

[10]Water Act (2007), Section 3(c).
[11]Water Act (2007), Section 21, 4(b).

coordinated between managers of "planned"[12] environmental water, owners and managers of environmental assets, and holders of "held" environmental water.

Identify the key ecological values across the Basin and determination of their environmental water requirements

The ESLT was defined in the Water Act (2007) as the volume of water that can be taken for a water resource without compromising the key ecological assets and key ecological functions of the water resource. Five criteria were identified for assessing the key environmental assets (MDBA, 2011a; Swirepik et al., 2016). And from the over 2400 named environmental assets (and many more unnamed assets), 24 key environmental assets were used in the modeling (Swirepik et al., 2016). These included large floodplain forest areas (e.g., Barmah-Millewa forest), wetlands (e.g., Narran Lakes, Macquarie Marshes), river channels where native fish breeding and migration was important, and the Lower Lakes and Coorong at the end of the system (Fig. 1B; Swirepik et al., 2016).

To obtain the environmental water requirements, the MDBA used an integrated set of hydrological models for the Basin, linking together individual models used by states for water resource planning, which simulated flow regimes over an historical period of 114 years (1895–2009). These models can represent different river operations and water sharing arrangements, including: (a) the current (at 2009) flow regimes with all the dams and operational rules in place (baseline diversion limits), and (b) the "natural" or "without development" flows assuming the dams and river operations are not in place (without development diversion limits).

The focus of the eco-hydrological modeling was on the water requirement for water birds, native fish and floodplain vegetation. For each of the environmental assets, a set of site-specific flow indicators (SFIs) was selected, using best available scientific knowledge. These SFIs identified the optimum flow regime (threshold/volume, duration, frequency, timing) which would sustain the ecological values and functions. For example, for the Barmah Forest the SFIs required to sustain the asset are shown in Table 1.

The water requirements were then modeled to obtain the water volumes required to meet the targets and objectives. The large wetland and floodplain forest ecosystems (e.g., Macquarie Marshes, Barmah Forest—Fig. 1B), of which approximately 20 were modeled, require the largest volumes of water and essentially set the size of the environmental water needs for the Basin.

Even with this concerted effort to assemble the best available science, and to make it publicly available on the MDBA website, it was still necessary for the MDBA to make a number of judgments. Important judgments made by the MDBA were: (a) as far as possible, water-dependent assets should receive between 60% and 80% of the water they would have received before development, (b) the current areal extent of wetlands and floodplain forests would be maintained as far as possible, but not extended.

---

[12]"Held" environmental water is water that has been purchased for the purpose of achieving environmental outcomes; generally it has the same conditions of use as an irrigation water entitlement. "Planned" environmental water is water that is committed in a water resource plan for the purpose of achieving environmental outcomes. During the Millenium drought (1997–2010), there were a number of instances where state water resource plans were either suspended or the planned environmental watering requirements were put on hold.

Table 1 Examples of Site-Specific Flow Indicators (SFI) for the Murray-Darling Basin

Location	Flow Rate (ML/d)	Duration (days)	Timing	Target (%) or Max Time Between Events (y)	Reason	Without Development Frequency (%)	Baseline Frequency (%)
Barmah-Millewa Forest[a]	12,500	7–70	Jun. to Nov.	70–80 of years with an event	Maintain 100% of current freshwater meadows, moira grass plains, and red gum forests	87	50
	25,000	7–42	Jun. to Nov.	40–50	As above	66	29
	35,000	7–30	Winter/spring	33–40	Maintain 100% of current red gum forests and woodlands	39	18
	15,000	7–35	Jun. to Dec.	30	Provide for successful breeding of colonial nesting waterbirds	44	12
Bourke[b]	6000	14	Min 1 event between Jul. and Jun.	80–90	Small scale fish movement and regular access to habitat	96	66
	10,000	14	Min 1 event between Aug. and May	60–80	Fish migration	89	54
	10,000	20	Min 2 events between Aug. and May	25–35	Flow-dependent fish spawning and recruitment	42	20
	30,000	24	Any time of year	2–3 years between events	Connectivity with riparian zone including forests	1.8 years	4.1 years

Table 1 Examples of Site-Specific Flow Indicators (SFI) for the Murray-Darling Basin—cont'd

Location	Flow Rate (ML/d)	Duration (days)	Timing	Target (%) or Max Time Between Events (y)	Reason	Without Development Frequency (%)	Baseline Frequency (%)
Lower Balonne Floodplain[b]	Any flow	1	Any time of year	1.0–1.2 1 years between events	Drought refuge	0.7 years	1.2 years
	3500	14	Aug. to May	40%–60% of years with >1 event	Fish migration	30 years	68
	9200	12	Any time of year	2–3 years between events	Connectivity with riparian zone	1.3 years	5.6 years
	24,500	7	Any time of year	6–8 years between events	Connectivity with mid-floodplain zone	3.5 years	8.7 years
Narran Lakes[b]	25,000	60	Any time of year	1– 1.3 years between events	Vital breeding and nursery habitat	0.6 years	1.3 years
	50,000	90	Any time of year	1.3– 1.7 years between events	Vital breeding and nursery habitat—larger area	0.8 years	2.0 years
	154,000	90	Any time of year	4–5 years between events	Trigger large-scale waterbird breeding	2.6 years	8.3 years

[a]*MDBA (2012b)*
[b]*MDBA (2016c).*

Selection of ESLT options for assessment

Before the Basin Plan was enacted, the long-term average consumptive use of water was 13,623 GLy^{-1}. Initial MDBA modeling, and past assessments by other organizations, indicated that a Basin-wide reduction in diversions of 2800 GLy^{-1}, or around a 20% reduction, would provide a reasonable compromise between improved environmental outcomes with minimal social and economic impacts. However, there was some controversy over this assessment, with irrigators and rural communities suggesting the reduction should be less, and environmental groups and the South Australian government suggesting it should be more.

MDBA agreed to run two additional scenarios—2400 and 3200 GLy^{-1}—representing a difference of ±400 to the 2800 GLy^{-1} reduction scenario. The new modeling showed some key differences between environmental outcomes associated with the three ESLT scenarios. The most significant differences were evident for the Murray downstream of the Murrumbidgee junction, including the Coorong, Lower Lakes and Murray Mouth (Fig. 1), particularly during dry conditions. Both the 2800 and 3200 GLy^{-1} reduction in diversions options showed a greater capacity to mitigate potential drought periods when there is extreme environmental stress. Additionally, the 2400 GLy^{-1} reduction option was shown to compromise the ability to manage salinity levels within the Coorong, to maintain an open Murray Mouth, and to maintain the resilience of lower elevation parts of the lower River Murray floodplain and associated wetlands during dry periods.

SOCIAL AND ECONOMIC ASPECTS

The MDBA commissioned over 20 studies on the social and economic implications of rebalancing water use under the Basin Plan (MDBA, 2012a, 2012b). From a macroeconomic viewpoint, the economic modeling estimated that the environmental water recovery volume identified previously would have a small overall effect, reducing agricultural production by no more than 9% and the gross regional product by less than 1%. Modeling also indicated that the reduction in the aggregate value of Basin production would be more than offset by underlying economic growth over the period to 2019 (MDBA, 2012a, 2012b).

These studies did show, however, that the local effects could be significant for smaller rural communities with a high dependency on irrigated agriculture and limited capacity to adjust to the changes. The new social and economic modeling undertaken as part of the northern Basin review, and detailed in the next section, was able to predict impacts at a much finer scale, at the level of individual townships.

The social and economic assessments highlighted the fact that a range of factors influences social and economic outcomes in rural communities. The major factors include technological change, climate variability, commodity prices and the exchange rate for the Australian dollar, and demographic/generational changes. Thus, the reduction in water available for consumptive purposes proposed in the Basin Plan is just one of a range of factors that will influence social and economic outcomes, but nevertheless is an important one for vulnerable communities.

While the reduction in the consumptive water pool may have impacts on vulnerable local communities in the Basin, these impacts will be tempered to some extent by the large investment being made by the Australian government to support the changes. Since the original commitment of $A10 billion, Australian Governments have now committed over $A13 billion for programs in support of water reform in the MDB (originally called *Water for the Future*) already described.

TRIPLE BOTTOM LINE ASSESSMENT

As noted previously, the process of trading off environmental, social and economic outcomes is a challenging one, and it is generally accepted that some form of decision support should be used to assist the decision makers.

In deciding on the optimum ESLT (and hence SDL) for the original Basin Plan, the MDBA used the results of the three ESLT modeling scenarios (reduction of 2400, 2800, and 3200 GLy^{-1}), together with the social and economic studies discussed previously.

They also took into account the following:

- For each environmental asset, the threshold of 60% of the water they would have received before development was the target (unless the asset was already above the 60% threshold).
- The ESLT was set within the constraints and operative rules of the current system, which has been designed for irrigation and other water use.
- Protecting the reliability of entitlements by avoiding third-party impacts.
- Managing the "held" environmental water portfolio according to existing rules, in order to retain the productive capacity of the water-dependent enterprises.
- Taking note of the effects in rural communities of the Commonwealth government investment in water buy-backs[13] and irrigation infrastructure modernization.[14]
- Providing more time for the adjustment by allowing for a 7-year transition (2012–19) for the implementation of the Basin Plan.

The method used by the MDBA to optimize the environmental, social and economic outcomes was largely a judgment-based method utilizing the environmental, social and economic evidence available to it. No formal multicriteria analysis, benefit-cost analysis or other method was used. The process is largely documented in the report outlining the ESLT method (MDBA, 2011a). However, this is a rather large report and it would have been better if a simpler version of the optimization (trade-off) method had been published on the MDBA website.

FINAL RESULT

Through the process outlined here, the MDBA established a new long-term SDL of 10,873 GLy^{-1} for the MDB, compared with the baseline (2009) diversion limit of 13,623 GLy^{-1}. This will require that an additional 2750 GLy^{-1} be recovered for the environment or around a 21% reduction in the consumptive water allocations available in 2009. The recovered volume is made up of 2289 GLy^{-1} (25% change) in the southern Basin, 390 GLy^{-1} (12% change) in the northern Basin, and 71 GLy^{-1} in the unconnected river systems (Lachlan, Wimmera). The recovery volume for the northern Basin may change as a result of the northern Basin review, which is discussed in the next section.

As on Jun. 30, 2016, the progress towards bridging the gap was assessed as 1981 GLy^{-1}, long-term average annual yield, or around 72% of the total needed.[15]

NORTHERN BASIN REVIEW

At the time the Basin Plan was developed, it was accepted that there was less information available for the northern Basin, and as a result it was agreed that a northern Basin review would be undertaken to improve the knowledge base for this region and then to use this new knowledge to revise the northern

---

[13]http://www.environment.gov.au/water/rural-water/restoring-balance-murray-darling-basin.

[14]http://www.environment.gov.au/water/rural-water/sustainable-rural-water-use-and-infrastructure.

[15]www.environment.gov.au/topics/water/commonwealth-environmental-water-office/about-commonwealth-environmental-water.

Basin SDLs specified in the Basin Plan (MDBA, 2014g). This review provided an opportunity for the MDBA to examine new evidence and modify its decision-making process over that used in the original Basin Plan. In the following, we provide a summary of these advances. A more detailed account of the process is provided by Hart et al. (in preparation).

EVIDENCE BASE

The northern Basin review involved three components:

- Reviewing the available science (Sheldon et al., 2014) and updating of the SFIs, particularly for the Condamine-Balonne and Barwon-Darling systems (MDBA, 2016c).
- Modeling a number of possible water recovery scenarios (MDBA, 2016f).
- Undertaking a detailed social and economic study of 21 northern Basin communities (15 irrigation dependent) that involved the development of a quantitative predictive model relating water recovery to the change in area of irrigated cotton and then to the change job numbers (MDBA, 2016d).

This additional knowledge then formed the triple bottom line evidence-based—environmental, social (including cultural) and economic—for the MDBA to use in its decision making regarding whether the northern Basin SDL should be changed (i.e., should the existing proposed water recovery volume of 390 GLy^{-1} be changed).

The evidence base, summarized in the following subsections, was shared with the northern Basin communities through a series of local community meetings.

Environmental evidence

The ESLT method used in the development of the Basin Plan was also used to refine the SFIs used in the northern Basin review. The major refinements occurred in the Condamine-Balonne and Barwon-Darling systems, since these are the most overallocated (Table 2). For example, the flow in the Culgoa River [part of the Condamine-Balonne system (Fig. 1B)] is currently around 64% allocated for consumptive uses; with the potential water recovery program (specific volume yet to be decided), this would be slightly reduced to between 52% and 58% allocated.

The SFIs were targeted at the lifecycle needs of fish, waterbirds and floodplain vegetation, or the likely persistence times of refuge waterholes. The SFIs consisted of a number of components, including the volume or flow rate required, the duration of the flow, and the timing (season) of the flow event(s) (MDBA, 2016c). Each SFI has a "target frequency range,"[16] which was generally somewhere between 60% and 80% of the "without development" conditions (Table 1).

Resulting from a review of the science (Sheldon et al., 2014) and a number of other specific scientific studies (e.g., DPI 2015), some existing SFIs were modified and additional SFIs included for the Condamine-Balonne and Barwon-Darling systems (MDBA, 2016c). The original SFIs for the other northern Basin rivers were not changed. Thus, in the northern Basin review a total of 41 SFIs have been considered, 13 for Condamine-Balonne, 11 for the Barwon-Darling, and 17 for the other river basins (MDBA, 2016c).

---

[16]The lower range value is generally referred to as the "high uncertainty" or "high risk" target and the higher range value as the "low uncertainty" or "low risk" target.

Table 2 Data Showing the Volume of Water Allocated for Two Northern Basin Rivers Without Development, With Development at 2009, and Under Two Recovery Scenarios

System	WOD (GL LTA)[a]	Baseline (2009)[b]	% Allocated	280 GL Recovery	% Allocated	390 GL Recovery	% Allocated
Culgoa @ Brenda	620	224	64	263	58	295	52
Darling @Bourke	3810	2150	44	2310	40	2350	38

[a]Without development (Gigalitre long term average).
[b]Diversion limit at Jul. 2009 (taken as the baseline).

Water recovery scenarios

The MDBA assessed a number of water recovery scenarios, ranging from 278 (the current recovery at the end of 2015) to 415 GLy^{-1}, and also included the Basin Plan default recovery of 390 GLy^{-1}, the baseline scenario (situation at 2009) and the "without development" scenario (MDBA, 2016c, 2016e).

The SFIs were then assessed against each of the recovery scenarios using hydrological modeling (MDBA, 2016f). The models estimate how often each of the flow indicators happens (frequency) for each recovery scenario (using the 114-year climate record), which was then compared to the target frequency range (for details see MDBA, 2016c; Hart et al., in preparation).

The results of each scenario against each SFI were assessed in two ways: (a) as a pass or fail in meeting the high uncertainty target value, and (b) as a percentage towards achieving the target range of the SFI (MDBA, 2016c). For example, relatively few of the 9 SFIs for the Lower Balonne river were met with any of the water recovery scenarios (2/9 with 278 GLy^{-1} recovery and 4/9 with 415 GLy^{-1} recovery), a hardly surprising result given the level of overallocation even after these levels of water recovery.

Social and economic evidence

A new *hydrology-irrigated area-jobs model* was developed for the northern Basin review (KPMG 2016). The model calculates the change in area of irrigated agriculture, focused on cotton as the predominant crop, based on water recovery scenarios for 21 communities in the northern Basin. These data were then input to a new northern Basin *community model*, which generates outputs showing change in jobs numbers (across the farming sector, agricultural suppliers and the nonagricultural sector) for the 21 communities based on the water recovery scenarios (MDBA, 2016d). Fig. 2 provides a simplistic schematic of the influence of water recovery on the social and economic well-being of the 21 communities modeled, together with the environmental and floodplain grazing outcomes. The social-economic model was populated using data from 1999 to 2014, including census data (2001, 2006, 2011), and extensive data collected during fieldwork and interviews with irrigators and town businesses. The individual communities were assessed as "affected" or "nonaffected" on the basis of the loss of jobs being >5% or <5% respectively due to the particular water recovery scenario.

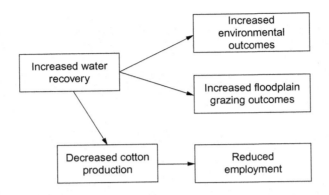

FIG. 2

Schematic showing the relationship between increased water recovery and (a) increased environmental outcomes, (b) increased outcomes for floodplain graziers (Condamine-Balonne system only), and (c) impacts on cotton production and subsequently reduced employment.

As expected the social-economic modeling predicted that the water recovery would have the greatest effect on towns in the Condamine-Balonne system. For example, decrease in total jobs for 278–390 GLy^{-1} recovery scenarios were 9%–19% for Dirranbandi and 3%–9% for St. George.

Additionally, a *hydrology-floodplain grazing production model* (MDBA, 2016b) was developed for the lower Balonne region where floodplain graziers have been significantly impacted due to upstream irrigation developments. This model uses approximately 10 years of production data collected during consultation with graziers, and a 29-year climate sequence to model production levels based on system hydrology (i.e., higher flows resulting in inundation of the floodplain increases production). The outputs from this model are reported as a percentage change in production per year (as measured between baseline and different water recovery scenarios).

Water recovery between the current 278 and 390 GLy^{-1} was predicted to result in between 2% to 5% increase in production per year.

The triple bottom line process

The MDBA developed a triple bottom line (TBL) decision-making framework incorporating "effects matrices" as the method to support their decision making in deciding on the "optimum" SDL for the northern MDB (see Table 3; MDBA, 2016b). This form of structured decision-making framework was used to promote good process in decision making rather than specifying an outcome. The framework was also aimed at generating trust and legitimacy among stakeholders through the adherence to transparency of process.

This decision-making framework did not remove the need for expert judgment and in-depth consideration of the trade-offs between water recovery options.

Fig. 3 outlines the steps involved in this TBL decision-making framework. The steps can be broken into two components: (a) planning the decision and (b) making the decision.

Table 3 Example of a Triple Bottom Line Effects Matrix Used for Northern Basin Review

Water Recovery Scenario	Environmental Effects		Social and Economic Effects		
	River channel connectivity[a]	Floodplain and wetland connectivity[b]	Effected communities[c]	Floodplain grazing[d]	Aboriginal socio-cultural values[e]
415 GL	8/19	16/24			
390 GL[f]	6/19	15/24	Dirranbandi (50%, 18%) Collarenebri (>80%, 21%) Warren (35%, 11%) St George (23%, 9%)	5.1%	
350 GL	7/19	14/24			
320 GL	7/19	15/24			
278 GL[g]	3/19	13/24	Dirranbandi (29%, 9%) Collarenebri (>80%, 21%) Warren (35%, 11%) St. George (9%, 3%)	2.0%	

[a]The number of river channel SFIs met over total number.
[b]The number of floodplain and wetland SFIs met over total number.
[c]Record for each community with >5% change in water availability (decrease in maximum area irrigated (ha), decrease in the job number in max production years).
[d]Percent increase in production (dry sheep equivalents (DSE) per hectare).
[e]Narrative description of the aboriginal socio-cultural values met with particular scenario.
[f]Basin Plan default recovery.
[g]Recovery at Jun. 2016.

Planning the decision

As shown in Fig. 3, there are four steps in this component:

- *Define the problem:* For the northern Basin review, the decision revolved around whether Basin Plan settings (the SDL volume of 390 GL for the northern Basin) should be changed (and if so by how much) and also providing information to inform water recovery and other management actions.
- *Objectives:* Explicit objectives are required to focus the decision makers' need to answer the question "what are we trying to achieve?" when contemplating options and trade-offs. In this case the objectives (listed as outcomes for the Basin Plan) are: (a) communities with sufficient and reliable water supplies that are fit for a range of intended purposes, including domestic, recreational and cultural use, (b) productive and resilient water-dependent industries, and communities with

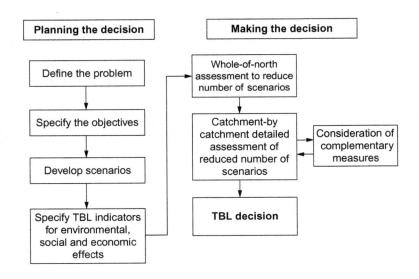

FIG. 3

Triple bottom line decision-making process used by the MDBA in the northern Basin review (MDBA, 2016b). "Complementary measures" are ways to provide additional environmental benefits often without additional water (e.g., protection of environmental water, building of fishways).

confidence in their long-term future, and (c) healthy and resilient ecosystems with rivers and creeks regularly connected to their floodplains and ultimately the ocean.

- *Alternative scenarios:* A range of alternative water recovery scenarios were investigated (see the preceding).
- *TBL indicators*: Environment and social-economic indicators were defined as described above and the effect of the various water recovery scenarios on each indicator was modeled. The environmental indicators were the 41 SFIs, and the social and economic indicators were: area of irrigated cotton production lost and loss of jobs, in both the agricultural and nonagricultural sectors for each of the 21 townships.

Making the decision

The decision-making space confronting the MDBA was complex, consisting of at least 5 possible water recovery scenarios, 41 environmental indicators (SFI) and 2 social and economic indicators (productive area lost, jobs lost) for each of the 21 local communities surveyed. For this reason the MDBA sought to focus on a smaller number of scenarios by first considering the whole of the northern Basin to shortlist the scenarios, and then to consider in detail the TBL effects of these scenarios (Fig. 3). Priority was given to the Condamine-Balonne and Barwon-Darling catchments in this second step. Effects matrices were used in both steps. An example of an effects matrix is given in Table 3.

The Authority identified the trade-offs between scenarios in terms of environmental benefits and social and economic impacts, and also assessed the opportunities for "complementary measures"[17] to provide additional environmental benefits (perhaps with less water recovery). Stakeholder perspectives on the water recovery scenarios were also considered.

Based on the assessment of scenarios in the two steps, the Authority were able to then use their judgment (backed by the evidence) to recommend SDL volumes for the northern catchments (local and shared reduction).

They recommended that the recovery volume be reduced from 390 to 320 GLy^{-1} (MDBA, 2016b).

All technical, hydrological and economic modeling reports are available on the MDBA website. To ensure the credibility and robustness of this evidence, all this work has been peer-reviewed, and these peer reviews are also on the Authority's website (e.g., Bewsher, 2016; Blackwell et al., 2016). Additionally, the "effects tables," accompanied by supporting narrative summarizing the differences between each scenario for the environmental, economic and social indicators, are also available publicly. The narratives attempted to articulate what is gained or lost between the reduction scenarios, and what trade-offs were made.

CONCLUSIONS

In this chapter we first provided an analysis of the development of the Murray-Darling Basin Plan. The key elements of the Basin Plan—long-term SDLs, basin-wide watering strategies, water quality and salinity management plans, new water trading rules, regional water resource plans and monitoring and evaluation—are discussed. Then we commented on what we believe are the key factors leading to the successful completion of the Basin Plan—government commitment to the reform process, an effective community engagement process and the application of an effective evidence-based decision-making process. Finally, and in line with the focus of this book, we discussed the key decision-making steps adopted in developing the SDLs, the key policy initiative that underpins the Basin Plan.

We also provided an assessment of the updated decision-making process used to review and recommended change (from the current 390 to 320 GLy^{-1}) the volume to be recovered for the northern Murray-Darling Basin. A review of the northern Basin was agreed upon in 2012 at the time the Basin Plan was enacted, because there was less knowledge of this system. The review resulted in updated environmental indicators (SFIs), modeling of a range of water recovery scenarios, and the development and application of a new quantitative social-economic model that related water recovery volumes to change in area of irrigated cotton and then to change in job numbers in 21 local communities. This evidence base was shared with the northern Basin communities through a series of local community meetings.

This evidence on the relationships between the water recovery scenarios and the environmental, social and economic indicators was then considered by the MDBA using "effects matrices" to assist the decision-making process.

---

[17]These include such actions as adequate shepherding on environmental water down a river, and building of fishways on weirs to allow fish passage.

REFERENCES

Aust Govt. (2012). *Water Act 2007—Basin Plan 2012, extract for the federal register of legislative instruments.* [28 November 2012]. Canberra: Australian Government, Office of Parliamentary Counsel. 245 p.

Aust Govt. (2014). *Water Act 2007 (with amendments).* Canberra: Australian Government, Office of Parliamentary Counsel. 571 p.

Bewsher (2016). *Review of the Hydrological Modelling Frameworks used to inform potential Basin Plan Amendments (draft).* Report prepared by Bewsher Consulting for Murray–Darling Basin Authority, Canberra, September (82 p.).

Blackwell, B., McFarlane, J., & Stayner, R. (2016). *An independent review of the social and economic modelling inputs to the northern Basin review.* Report prepared by University of New England for the Murray–Darling Basin Authority, Canberra, October (43 p.).

Brooks, J. D., Aldridge, K. T., Deegan, B. M., Geddes, M. C., Gillanders, B., & Paton, D. C. (2009). *An ecosystem assessment framework to guide the management of the Coorong.* Adelaide: University of Adelaide.

Colloff, M. J., Lavorel, S., Wise, R. M., Dunlop, M., Overton, I. C., & Williams, K. J. (2015). Adaptation services of floodplains and wetlands under transformational climate change. *Ecological Applications, 26*(4), 1003–1017. http://dx.doi.org/10.1890/15-0848.

Connell, D., & Grafton, R. Q. (Eds.), (2011). *Basin futures: Water reform in the Murray-Darling Basin.* Canberra: ANU Press.

Conroy, M. J., & Peterson, J. T. (2013). *Decision making in natural resource management: A structured adaptive approach.* Stafford: Wiley-Blackwell.

Daniell, K. A., & Barrateau, O. (2014). Water governance across competing scales: Coupling land and water management. *Journal of Hydrology, 519,* 2367–2380.

Daniell, K. A., Coomes, P. J., & White, L. (2014). Politics of innovation in multi-level water governance systems. *Journal of Hydrology, 519,* 2415–2435.

DPI. (2015). *Fish and flows in the northern Basin: Responses of fish to changes in flow in the Northern Murray-Darling Basin.* Report by NSW Department of Primary Industries for the Murray-Darling Basin Authority, Canberra, August (90 p.).

Docker, B., & Robinson, I. (2014). Environmental water management in Australia: Experience from the Murray-Darling Basin. *International Journal of Water Resources Development, 30*(1), 164–177.

Gordon, N. D., McMahon, T. A., Finlayson, B. L., Gippel, C. J., & Nathan, R. J. (2004). *Stream hydrology—An introduction for ecologists* (2nd ed.). New York: John Wiley & Sons Ltd. 448 p.

Grafton, R. Q., & Horne, J. (2014). Water markets in the Murray-Darling Basin. *Agricultural Water Management, 145,* 61–71.

Grafton, R. Q., Libecap, G. D., Edwards, E. C., O'Brien, R. J., & Landry, C. (2012). Comparative assessment of water markets: Insights from the Murray-Darling Basin of Australia and the Western USA. *Water Policy, 14,* 175–193.

Grafton, R. Q., Pittock, J., Williams, J., Jiang, Q., Possingham, H., & Quiggin, J. (2014). Water planning and hydro-climatic change in the Murray-Darling Basin, Australia. *Ambio, 43,* 1082–1092.

Hart, B. T. (2015a). The Australian Murray-Darling Basin Plan: Challenges in its implementation (part 1). *International Journal of Water Resources Development.* http://dx.doi.org/10.1080/07900627.2015.1083847.

Hart, B. T. (2015b). The Australian Murray-Darling Basin Plan: Challenges in its implementation (part 2). *International Journal of Water Resources Development.* http://dx.doi.org/10.1080/07900627.2015.1084494.

Hart, B. T. (2016). The Australian Murray-Darling Basin Plan: Factors leading to its successful development. *Ecohydrology & Hydrobiology* (in press).

Hart, B. T., Williams, B., Gole, K., Spencer, B., & James, D. (in preparation). Triple bottom line decision-making process adopted for the Northern Murray-Darling Basin review. *Australasian Journal of Water Resources.*

Horne, J. (2013). Economic approaches to water management in Australia. *International Journal of Water Resources Development, 29,* 526–543.

Janssen, R. (2012). *Multi-objective decision support for environmental management.* New York, NY: Springer Science & Business Media. 233 p.

Kingsford, R. T., Fairweather, P. G., Geddes, M. C., Lester, R. E., Sammut, J., & Walker, K. F. (2009). *Engineering a crisis in a Ramsar wetland: The Coorong, lower lakes and Murray mouth.* Sydney: Australian Wetlands and Rivers Centre, University of NSW.

KPMG. (2016). *Northern Basin community modelling: Economic assessment of water recovery scenarios (draft).* Report by KPMG for Murray-Darling Basin Authority, Canberra, August (70 p.).

MDBA. (2010a). *Guide to the proposed Basin Plan. Vol 1 overview.* Canberra: Murray-Darling Basin Authority. October 2010. http://www.mdba.gov.au/kid/guide/index.php.

MDBA. (2010b). *Guide to the proposed Basin Plan: Volume 2 technical background, Part II.* Canberra: Murray-Darling Basin Authority. p. 614.

MDBA. (2011a). *The proposed 'environmentally sustainable level of take' for surface water of the Murray-Darling Basin: Methods and outcomes.* Canberra: Murray-Darling Basin Authority. 218 p.

MDBA. (2011b). *The living Murray story.* Publication number 157/11 (95 p.) Canberra: Murray-Darling Basin Authority. http://www.mdba.gov.au/media-pubs/publications/living-murray-story.

MDBA. (2011c). *Delivering a healthy working basin—About the Draft Basin Plan.* Canberra: Murray-Darling Basin Authority. http://www.mdba.gov.au/media-pubs/publications/delivering-healthy-working-basin. November.

MDBA. (2012a). *The socio-economic implications of the proposed Basin Plan.* Canberra: Murray-Darling Basin Authority. 35 p. http://www.mdba.gov.au/what-we-do/basin-plan/development/socio-economic-reports.

MDBA. (2012b). *Assessment of environmental water requirements for the proposed Basin Plan: Barmah–Millewa Forest.* MDBA publication no: 19/12 (27 p.) Canberra: Murray-Darling Basin Authority.

MDBA. (2013). *Handbook for practitioners: Water resource plan requirements.* Canberra: Murray-Darling Basin Authority. October (108 p.) http://www.mdba.gov.au/what-we-do/basin-plan/development/consultation-report/ch09.

MDBA. (2014a). *Guidelines for water trading rules.* Murray–Darling Basin Authority. http://www.mdba.gov.au/what-we-do/managing-rivers/water-trade/trading-rules/guidelines-for-the-water-trading-rules.

MDBA. (2014b). *Basin-wide environmental watering strategy.* Publication no. 20/14 (125 p.) Canberra: Murray–Darling Basin Authority. http://www.mdba.gov.au/sites/default/files/pubs/Final-BWS-Nov14.pdf.

MDBA. (2014c). *Overview of the 2014–15 basin annual environmental watering priorities.* Canberra: Murray–Darling Basin Authority. 6 p.

MDBA. (2014d). *Basin salinity management strategy: 2012-2013 summary.* MDBA publication no. 02/14 (p. 125) Canberra: Murray–Darling Basin Authority.

MDBA. (2014e). *Water markets.* Canberra: Murray–Darling Basin Authority. http://www.mdba.gov.au/what-we-do/managing-rivers/water-trade.

MDBA (2014f). *Murray-Darling Basin water reforms: Framework for evaluating progress.* Technical report 09/14. Canberra: Murray Darling Basin Authority.

MDBA. (2014g). *Northern Basin review.* MDBA publication 49/14 (8 p.) Canberra: Murray-Darling Basin Authority.

MDBA. (2015a). *Basin salinity management strategy 2030 (BMS 2030).* MDBA publication no. 21/15 (30 p.) Canberra: Murray–Darling Basin Authority.

MDBA. (2015b). *Towards a healthy, working Murray-darling basin: Basin plan annual report 2013-14.* MDBA publication no. 46/14 (34 p.) Canberra: Murray-Darling Basin Authority.

MDBA. (2016a). *Towards a healthy, working Murray-Darling Basin: Basin plan annual report 2014-15.* MDBA publication no. 23/15 (47 p.) Canberra: Murray-Darling Basin Authority.

MDBA. (2016b). *The Northern Basin Review: Understanding the economic, social and environmental outcomes from water recovery in the northern basin*: (p. 50). Report 39/16. Canberra: Murray-Darling Basin Authority.

MDBA. (2016c). *Environmental outcomes of the northern Basin review*. Canberra: Murray-Darling Basin Authority. 149 p.

MDBA. (2016d). *Northern Basin review—Technical overview of the socio-economic analysis: Interim report*. Canberra, October: Murray-Darling Basin Authority. 160 p.

MDBA. (2016e). *Lower balonne floodplain grazing model report (draft)*. Canberra: Murray-Darling Basin Authority. 43 p.

MDBA. (2016f). *Hydrological modelling for the northern Basin review*. Canberra: Murray-Darling Basin Authority (in preparation).

NWC. (2004). *Intergovernmental agreement on a National Water Initiative*. Canberra: National Water Commission. 39 p.

NWC. (2011). *Water markets in Australia: A short history*. Canberra: National Water Commission.

NWC. (2013). *Australian water markets report 2012-13*. Canberra: National Water Commission. http://www.nwc.gov.au/publications/topic/water-industry/australian-water-markets-report-2012-13.

NWC. (2014). *National water initiative—Ten-year anniversary*. Canberra: National Water Commission. http://nwc.gov.au/nwi/nwi-10-year-anniversary.

Rogers, K., & Ralph, T. J. (2011). *Floodplain wetland biota in the Murray-Darling Basin*. Collingwood: CSIRO Publishing.

Sheldon, F., Balcombe, S., & Capon, S. (2014). *Reviewing the scientific basis of environmental water requirements in the Condamine-Balonne and Barwon-Darling: A synthesis*. Report by Griffith University to the Murray-Darling Basin Authority, Canberra (22 p.).

Swirepik, J. L., Burns, I. C., Dyer, F. J., Neave, I. A., O'Brien, M. G., Pryde, G. M., et al. (2016). Science informed policy: Establishing environmental water needs for Australia's largest and most developed river basin. *River Research and Applications, 32*(6), 1153–1165. http://dx.doi.org/10.1002/rra.2975. 24.

Young, W. J., Bond, N., Brookes, J., Gawne, B., & Jones, G. J. (2011). *Science review of the estimation of an environmentally sustainable levels of take for the Murray-Darling Basin*. Final report to MDBA, Water for a Healthy Country, CSIRO, November, Canberra (40 p.).

THE VICTORIAN WATERWAY MANAGEMENT STRATEGY

S.E. Loo, A. Clarke

Department of Environment, Land, Water and Planning

CHAPTER OUTLINE

INTRODUCTION

This chapter details the decision-making process used to develop the most recent state policy for waterway management in Victoria. Waterways (rivers, estuaries, wetlands, and their floodplains) have many environmental, social, cultural, and economic values that underpin community wellbeing, productive economies, livable cities and regional towns, as well as supporting diverse plants and animals. Additionally, many Aboriginal values of cultural, social, spiritual, and customary significance are located on or near waterways. Managing waterways is a long-term task that requires an understanding that many of the things we value (e.g., high-quality drinking water and recreational opportunities) depend on the environmental condition of waterways.

The *Victorian River Health Strategy* (VRHS) released in 2002 was the first state-wide policy framework for communities to work in partnership with the Victorian government to manage and restore the state's rivers over the long term (DNRE, 2002). Prior to the development of this policy framework, river health management in Victoria was hampered by:

- Absence of clear principles and objectives for making decisions about river protection and restoration.
- Not having clear criteria for assigning priorities to management activities.
- Inadequate policy direction on important management issues impacting river health.

A review of the policy framework in 2010 concluded that the VRHS was a significant achievement for river management in Victoria (DSE, 2010). New policy challenges, areas for improvement and key lessons were identified and used to inform the development of the next-generation strategy, the *Victorian Waterway Management Strategy* (the Strategy) (DEPI, 2013). The key policy advances that have been made and factors that were vital for successful policy development and implementation of this Strategy are covered in this chapter.

BUILDING ON STRONG FOUNDATIONS

Victoria has a long history of recognizing the importance of its waterways and catchments and working to improve their condition through:

- An integrated catchment management approach, facilitated through legislation that establishes 10 catchment regions, a catchment management authority (CMA) for each and the preparation of 10 Regional Catchment Strategies (RCSs).[1]
- State policy framework for river health, provided originally by the VRHS.
- A regional plan for river health in each of the 10 catchment regions (e.g., regional river health strategies—RRHSs).[1]
- A regional service delivery model for implementation and strong community partnership.
- Funding from the Environmental Contribution Levy (ECL),[2] which is a legally required contribution from water corporations for the purposes of funding initiatives that seek to promote the sustainable management of water or to address adverse environmental impacts of water extraction and consumption.

---

[1]http://www.delwp.vic.gov.au/water/governing-water-resources/catchment-management-authorities.
[2]http://www.depi.vic.gov.au/water/governing-water-resources/environmental-contribution.

- The environmental water reserve, a legally recognized amount of water set aside to meet environmental needs (including environmental water entitlements).
- The establishment of the Victorian Environmental Water Holder (VEWH), an independent statutory body responsible for making decisions on the most efficient and effective use of Victoria's environmental water entitlements.[3]
- Four regional sustainable water strategies (SWSs), which outline policies and actions to ensure sustainable water management over a 50-year period.[4]
- State environment protection policies to protect the quality of surface water and groundwater.[5]
- Commitments to work within the Murray-Darling Basin Plan's salinity and water quality targets, and to implement the Basin Plan (see Chapter 13).
- Management plans for each of Victoria's 11 Ramsar wetlands.
- Community-based natural resource management activities, such as Landcare[6] and Waterwatch.[7]

Victoria's clear management arrangements, along with direct government investment into waterway health activities, has delivered many achievements, including:

- Over 13,800 km of vegetation works, improving the health and resilience of priority waterways since 2010.
- A total of 280 sites where works have been undertaken to improve in-stream health since 2010.
- Approximately 650 GL (long-term average) of environmental water entitlements now (early 2016) held by the VEWH.
- The Commonwealth Environmental Water Holder (CEWH) held approximately 621 GL (long-term average) of environmental water entitlements in northern Victoria, including those that contribute to Victoria's recovery target under the Basin Plan.
- Environmental water was provided to 382 rivers and wetlands between 2011–12 and 2014–15 and significant environmental works and measures were undertaken to improve environmental water use efficiency (DELWP, 2016).

PURPOSE OF THE VICTORIAN WATERWAY MANAGEMENT STRATEGY

Building on the success of the foundational policy (VRHS), the Strategy provides the framework for government, in partnership with the community, to maintain or improve the condition of rivers, estuaries and wetlands so that they can support agreed environmental, social, cultural, and economic values. The framework is based on regional planning processes and decision making within the broader system of integrated catchment management in Victoria. The Strategy addresses the community expectations and obligations for waterway management expressed in the Victorian *Water Act 1989*,

---

[3]http://www.depi.vic.gov.au/water/governing-water-resources/victorian-environmental-water-holder.
[4]http://www.depi.vic.gov.au/water/governing-water-resources/sustainable-water-strategies.
[5]http://www.delwp.vic.gov.au/water/rivers,-estuaries-and-wetlands/state-environment-protection-policy-waters-review.
[6]http://www.delwp.vic.gov.au/environment-and-wildlife/landcare.
[7]http://www.depi.vic.gov.au/water/using-water-wisely/education/waterwatch-victoria-program.

Catchment and Land Protection Act 1994, other relevant state and national legislation and policies, and international agreements.

The Strategy updates and replaces the policy framework of the 2002 VRHS and, for the first time, provides a single management framework in Victoria for rivers, estuaries, wetlands and their connection to the marine receiving environment.

The Strategy is intended for state government, waterway managers, land managers, local government, other regional agencies and authorities, management partners, Traditional Owners and landholders or community groups involved in waterway management or activities that may affect waterway condition. The Strategy establishes:

- A vision, guiding principles and management approach.
- A transparent, integrated waterway management framework that: facilitates regional decision making with community input, sits within an integrated catchment management context, and comprehensively integrates waterway management activities.
- An adaptive management framework and flexible approach to manage through the challenges of drought, flood, bushfire, and the potential impacts of climate change.
- Aspirational targets that summarize key regional management activities over the next 8 years that aim to maintain or improve the condition of waterways.
- Clear policy directions for waterway management issues.
- Specific actions that will deliver more effective and efficient management of waterways.

The vision for Victoria's waterways is that *Victoria's rivers, estuaries and wetlands are healthy and well-managed, and support environmental, social, cultural and economic values that are able to be enjoyed by all communities.*

MANAGEMENT ARRANGEMENTS IN VICTORIA

Institutional arrangements for waterway management in Victoria exist at both the state and regional levels. These arrangements, often supported by legislation, define the roles and responsibilities for key agencies involved in waterway management. Having clearly defined management arrangements has been vital to the effectiveness of waterway management in Victoria.

The *Catchment and Land Protection Act 1994* establishes 10 catchment regions (Fig. 1), each with a CMA to coordinate integrated management of land, water, and biodiversity.[1] Catchment management authorities also have specific responsibilities for waterway management under the *Water Act 1989*, except in the Port Phillip and Westernport region, where a water corporation (Melbourne Water) has the waterway management responsibilities.[1]

THE VICTORIAN WATERWAY MANAGEMENT PROGRAM

The Strategy is implemented through the Victorian Waterway Management Program (the Program) (DEPI, 2013). The Victorian Government, through the Department of Environment, Land, Water and Planning, is primarily responsible for oversight of the Program and establishing the state policy

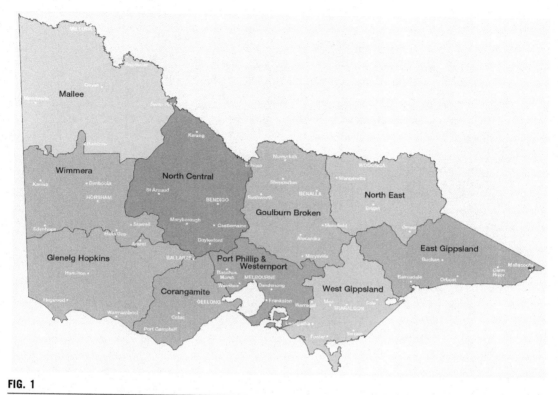

FIG. 1

The 10 catchment management regions of Victoria.

framework for waterway management. The "waterway managers" (i.e., nine regional CMAs and Melbourne Water in the metropolitan region) lead the regional implementation of the Program. The Program is based on an 8-year adaptive management cycle where learning occurs at all stages and is used to update and improve the program in subsequent cycles. The Program involves the following stages and components (Fig. 2):

- *Strategy and planning*—state-wide policy framework and targets, planning for waterway management through Regional Waterway Strategies (RWSs) with priorities and regional targets.
- *Implementation and monitoring*—government and other investment in regional priorities, implementation of priority management activities, monitoring of management activities (intervention monitoring), and long-term resource condition assessment.
- *Evaluation and reporting*—management reporting, resource condition reporting, program evaluation, and improvement.

Community participation and research and innovation occur across all parts of the Program.

FIG. 2

The 8-year adaptive management cycle of the Victorian Waterway Management Program.

KEY ADVANCEMENTS ACHIEVED IN THE POLICY FRAMEWORK

The review of the VRHS highlighted several key improvements required to progress the state policy framework (DSE, 2010). The Strategy provides an updated framework for managing waterways in Victoria that incorporates lessons learned over the decade from 2002 to 2012. Descriptions of the key advancements to the policy framework are detailed in the following sections.

MANAGING ALL WATERWAYS WITHIN A SINGLE FRAMEWORK

Victoria has a long history of integrated catchment management, which recognizes that improvements in the environmental condition of land, water, and biodiversity cannot be achieved in isolation from each other. Similarly, rivers should not be managed in isolation from the management of their catchment, including wetlands, estuaries, and marine receiving waters. To improve integration, the policy framework for rivers outlined in the VRHS was expanded and updated to incorporate the management of estuaries and wetlands and consideration of impacts on marine receiving waters, providing a single management framework for all waterways in Victoria.

SETTING PRACTICAL OBJECTIVES FOR WATERWAYS

The environmental condition of waterways in Victoria varies from excellent through to very poor, reflecting the fact that some waterways are largely unmodified but others are intensively used and highly modified. The environmental condition of a waterway has a strong influence on the types of values present. A waterway with excellent environmental condition often has many natural values. Waterways in moderate environmental condition may have few natural values, but high social and recreational values.

The Strategy describes four types of waterways, characterized by the environmental condition and the level of modification (Fig. 3). These four types are not necessarily mutually exclusive and do not have clearly defined boundaries between them. Different objectives apply to the management of the

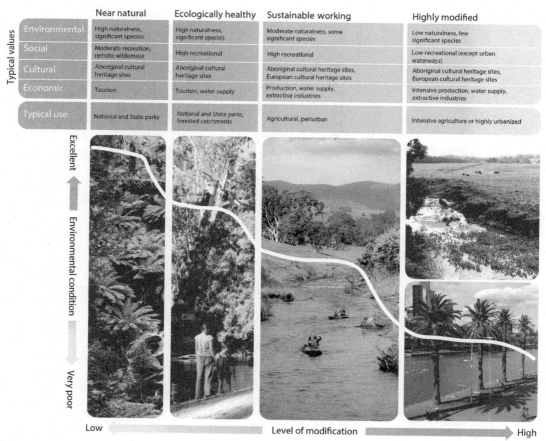

FIG. 3

Types of waterways depending on their values, condition, and typical uses.

four different types of waterways. For example, in managing an *ecologically healthy waterway* the objective may be to maintain its good condition. The objective for a priority *sustainable working waterway* may be to improve its condition, and for a priority *highly modified waterway* it may be to prevent further decline or to have amenity features such as shade, bike paths and no visible litter. For priority *urban waterways*, a large amount of management effort may be needed just to maintain current condition and prevent further decline. A key advantage of this approach is the clear recognition that there are different acceptable endpoints for waterways and that the return to a "reference" condition for all waterways is neither feasible nor desirable by communities in some cases.

The overarching management objective of the Strategy is to maintain or improve the environmental condition of Victoria's waterways to support environmental, social, cultural, and economic values. Management activities focus on priority waterways to provide public benefits and target key drivers of environmental condition that support the multiple values of waterways (Fig. 4). Waterway managers

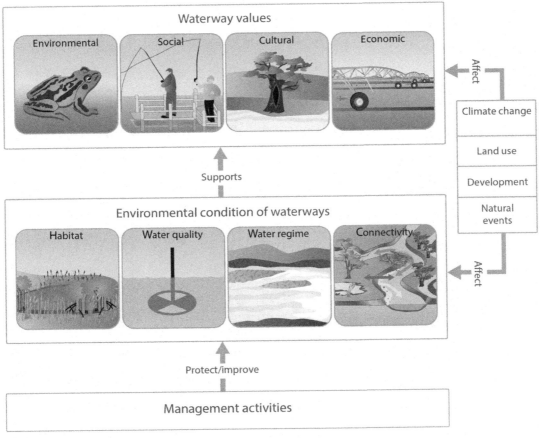

FIG. 4

The drivers of environmental condition that support the values of waterways.

and an expert panel have determined an agreed list of environmental, social, cultural, and economic values for consideration in regional waterway management programs (DEPI, 2013). Data on the values relevant for each major waterway come from the scientific experts, government departments, regional agencies, and local knowledge.

Catchment management authorities set waterway objectives in partnership with local communities, Traditional Owners and other management partners every 8 years. Objectives are informed by community values and uses of waterways, science and traditional ecological knowledge, and also the need to consider what can realistically be achieved within the context of historical catchment land use. Waterway objectives are formally assessed every 8 years, in consultation with the community, as part of the development of RWSs. Changes in climate, water availability, land use, and population are considered as part of the review process.

OUTLINING A PROGRAM LOGIC AND DEVELOPING TARGETS

Program logic is an approach to planning that uses a diagram to demonstrate the rationale for a program and express how change is expected to occur (Roughley, 2009). The Strategy includes a simplified program logic for the Victorian Waterway Management Program (Fig. 5) that describes how each year specific management activities (outputs) are delivered by regional agencies across Victoria in order to achieve particular management outcomes. Over the 8-year planning period, these outputs and

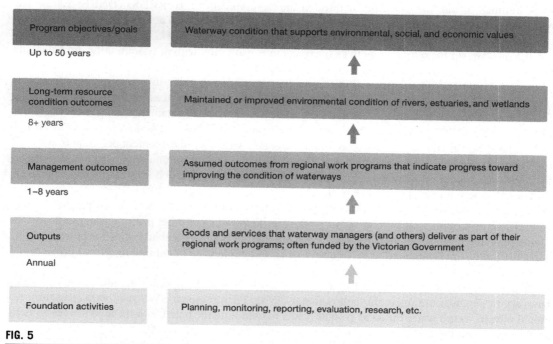

FIG. 5

The simplified program logic for the Victorian Waterway Management Program.

outcomes collectively contribute to either maintaining or improving the environmental condition of waterways.

Targets for the Strategy were developed on the basis of the program logic structure, but were limited by available sources of data that can be used to confidently report against each target. Targets are set at the long-term resource condition outcome and management outcome level.

An expert scientific panel provided advice on the scientific aspects of the strategy. The panel advised that the information required to accurately predict specific (quantitative) changes in waterway condition at the state-wide level as a result of management activities was currently not available. Current knowledge is only likely to be sufficient to support general, descriptive (qualitative) predictions about the directional changes in condition (e.g., declining, maintaining or improving) that are expected as a result of regional work programs across Victoria. The focus for improving knowledge to enable quantitative assessments of condition change needs to be at the regional level. The use of evidence-based models, research and monitoring will ensure that knowledge and confidence about predicting the effect of management activities is improved over time. For example, the Victorian Environmental Flows Monitoring and Assessment Program (VEFMAP) was established to assess the long-term ecological responses to environmental watering in priority regulated rivers across Victoria.[8]

STRENGTHENING THE REGIONAL PLANNING FRAMEWORK

Regional planning processes for waterway management were established as part of the policy framework outlined in the VRHS, to occur through the development of 10 RRHSs (DNRE, 2002). The RRHSs were umbrella documents that coordinated all other river-related action plans and were developed by CMAs and Melbourne Water between 2004 and 2006. The RRHS development process provided a mechanism for setting objectives for rivers in consultation with the community and balancing the environmental, social, and economic values of rivers.

A list of priority management activities was then developed to guide investment in river management in the region over a 6-year period. The RRHSs were the cornerstone of the regional planning framework for waterway management and have since been replaced by statutory plans under the *Water Act 1989*, called RWSs. Following the release of the Strategy in 2013, 10 RWSs were developed by late 2014. The RWSs are the new centerpiece of an integrated waterway management planning framework for Victoria's rivers, estuaries and wetlands. For some issues (e.g., environmental water management) the RWSs are supported by management plans that provide more detailed information on objectives, management activities and targets (Fig. 6).

The purpose of the RWSs is to: identify high-value waterways (based on environmental, social, cultural, and economic values and using a set of principles), determine priority waterways for the 8-year planning period, include a regional work program of management activities for priority waterways (including environmental water management), and guide investment into multiyear projects and annual work programs.

The RWSs were developed in accordance with regional decisions about water allocation and water recovery targets that are made through regional SWSs. Each regional SWS sets out a long-term regional plan to secure water for local growth, while maintaining the balance of the area's water system and safeguarding the future of its waterways and groundwater (DSE, 2009).

---

[8]http://www.delwp.vic.gov.au/__data/assets/pdf_file/0011/342569/VEFMAP-Fact-sheet-1-Benefits-of-ewater.pdf.

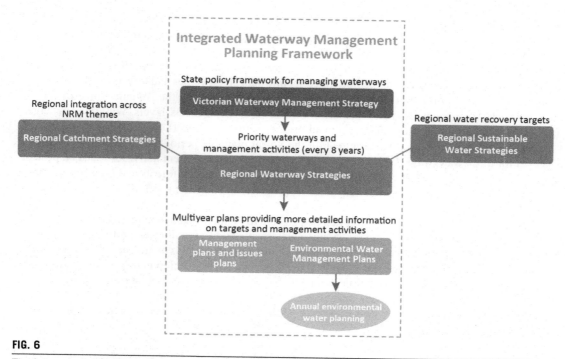

FIG. 6

The integrated waterway management planning framework.

The RWSs were developed in consultation with regional agencies and boards, Traditional Owners, the regional community and other key stakeholders, and endorsed by the relevant Ministers for water and for environment and climate change.

There are many other strategies and plans that do not have waterway management as their primary focus, but are considered in regional waterway management planning. These include state strategies (such as the Victorian Coastal Strategy (VCC, 2008), protecting Victoria's biodiversity[9]), plans for the management of public land (such as forests and parks plans, fire plans, and regional coastal plans) and other relevant longer-term strategies such as regional growth plans.

At a regional level, the RCSs are the primary integrated strategies for managing land, water, and biodiversity in Victoria. The RCSs identify priority areas and a program of measures to protect and manage those places. The long-term objectives and priorities in the RCS that relate to waterways are implemented through the RWSs.

AN IMPROVED REGIONAL PRIORITY-SETTING PROCESS

Government investment must be directed to activities that provide the most efficient and effective long-term improvements in waterway condition and the greatest community benefits. Hence, priority setting is a core activity in planning. In Victoria, a transparent process for setting priorities using the

---

[9]http://www.delwp.vic.gov.au/environment-and-wildlife/biodiversity.

asset-based approach (that considers the environmental, social, cultural, and economic values of waterways and the level of risk to those values) is implemented at the regional level through the development of the RWSs.

Supporting work has been undertaken by the Victorian government and waterway managers to develop a transparent and consistent method for collecting data about values and threats and giving each a simple score. This information was housed in a central database, called the Aquatic Value Identification and Risk Assessment (AVIRA) database, which also determines risk levels. Waterway managers use this information to assist with the regional priority-setting process (DEPI, 2013).

Once priority waterways are determined for the 8-year planning period, management activities are selected for those waterways. Depending on the level of risk, management activities focus on protecting the current environmental condition of a waterway or aim to reduce threats and improve condition over time. Logic models are used to help select appropriate management activities, based on the best available knowledge.

Where multiple values exist, it will often be possible to identify management activities that will be beneficial for all values (e.g., reducing threats to waterway condition will often protect environmental and social values). Where values potentially conflict (e.g., managing threatened native fish and recreational fishing for salmonids within a single waterway), the RWS development process provides the mechanism for regional decision making about which values will be managed for (with reference to the regional goals and consultation with stakeholders).

In summary, the regional priority-setting process that was implemented through the development of the RWSs was as follows:

- Developed regional goals for waterway management.
- Identified high-value waterways (based on environmental, social, cultural, and economic values).
- Filtered the high-value waterways and selected those that aligned with the regional goals.
- Identified threats to the values of those waterways and assessed the level of risk.
- Determined priority waterways for the 8-year planning period.
- Identified high-level management activities and assessed their feasibility and cost effectiveness.
- Selected priority management activities to form the regional work program for the 8-year planning period.

The majority of Victorian government investment in regional waterway management is directed to management activities on priority waterways. However, there are also circumstances when investment can occur on nonpriority waterways; for example, it may be necessary to undertake work in upstream areas to reduce threats to downstream priority waterways, such as erosion control works in highly modified waterways to protect downstream wetlands. Investment may also be required to protect public infrastructure, provide important connectivity between priority waterways or to support dedicated community groups who are actively working to improve the environmental condition of their local waterway.

Finally, existing regulatory controls apply across all waterways in Victoria and work that is required to comply with any legal or statutory requirements must be undertaken. Waterways that are not a priority in the RWSs may still be a priority for local communities, who can apply for grants from local government, the Australian government and private or philanthropic donors to undertake management activities.

Community input is a critical part of the regional priority-setting process and there are multiple opportunities for this to occur. Local communities are an important source of information about many of the social values that waterways provide. Waterway managers involve communities in developing management objectives for waterways in the region during the development of the RWSs. The feasibility of many management activities depends on consultation with and agreement of local landholders. Regional communities also have the opportunity to comment on the overall regional waterway program when the draft RWSs are released for public consultation.

USING ENVIRONMENTAL WATER EFFICIENTLY AND EFFECTIVELY

Environmental water is a key component of integrated waterway management. Victoria has a clear framework for making decisions about water allocation and entitlements (see Chapter 3). This includes environmental entitlements, which provide for a share of the water resource specifically for the environment. The Victorian government has established robust arrangements for the planning and use of environmental entitlements. In 2011, the VEWH, an independent statutory authority, was established to hold and manage environmental entitlements. The VEWH works with waterway managers to ensure environmental water entitlements are used to achieve the best environmental outcomes.

Environmental water has mostly been recovered through investment in on-farm and regional irrigation infrastructure to reduce operational losses, and purchases (buy-back) from farmers (see Chapter 13). Most water purchases from farmers took place between 2009 and 2012 when the "restoring the balance" in the Murray-Darling Basin was at its peak.

The framework for making decisions about water allocation and entitlements was strengthened through the development of four regional SWSs to plan for long-term water security across the state (DSE, 2009). The key focus for waterway management now is to ensure that the water set aside for the environment is used as efficiently and effectively as possible and that its management is comprehensively integrated with other management activities. Efficient use of environmental water may also mean that less water recovery is required, which will help to reduce social and economic impacts on agricultural production and regional communities.

Some river systems have a high level of interaction between surface water and ground water. As a result, increased groundwater extraction can reduce streamflow and therefore affect groundwater-dependent ecosystems (GDEs) and the reliability of water for existing surface water users. Where a GDE relies on groundwater that is clearly connected to surface water, the management approach needs to be integrated. The Upper Ovens River Water Supply Protection Area was the first in Victoria to have an integrated surface water-groundwater management plan.

For environmental water management, the Strategy:

- Reaffirms Victoria's approach to managing environmental water, in recognition of the advances that have been made in recovering water for the environment and establishing robust frameworks for its management.
- Clarifies the roles, responsibilities and relationships between waterway managers, the Victorian government, the VEWH and the CEWH, as well as Victoria's interaction with the Murray-Darling Basin Authority and Murray-Darling Basin Plan.
- Provides policy direction to address complex environmental water management issues, such as considering social and cultural benefits that are supported by environmental water.

- Describes tools to achieve more efficient and effective use of existing environmental water.
- Outlines actions to reduce risks to environmental water availability.
- Identifies and manages GDEs.
- Supports research and monitoring to improve knowledge about the ecological outcomes of environmental water use.

RECOGNIZING WATERWAYS NEED ONGOING MANAGEMENT

While the protection of Victoria's few waterways that remain in excellent condition is vital, it must be recognized that the majority of on-ground works and active environmental water management will occur in systems where some decline in condition has occurred. On-ground works include activities to improve habitat, water quality, water regimes and connectivity, such as riparian fencing and revegetation, erosion control, provision of fish passage, in-stream habitat improvements and the delivery of environmental water where available. It is the waterways where trade-offs between human use and environmental condition have occurred that require high levels of ongoing management in order to prevent further degradation or improve their environmental condition where possible. Maintenance activities are critical to ensure investment in on-ground works is protected over time. To formally recognize this, waterway managers are now able to include maintenance activities as a part of funding applications.

STRENGTHENING COMMUNITY PARTNERSHIPS IN WATERWAY MANAGEMENT

Waterway management in Victoria is undertaken as a partnership between government, landholders, and the community. Identifying the opportunities for communities to participate in planning, implementing, and monitoring regional work programs strengthened the partnership approach. Community ownership and care of waterways are recognized as being critical to maintaining waterway condition. Stewardship is encouraged and community engagement programs build on the community's appreciation of the values that waterways provide. To assist in capacity building, better access to knowledge has been provided by making data available publicly and through education programs.

MANAGING THROUGH EXTREME EVENTS: BUSHFIRES AND FLOODS

The policy framework outlined in the VRHS did not have the flexibility to consider how priorities and management activities could be modified in response to extreme events, such as bushfires and floods. The new strategy better integrates flood management with waterway management; improves the consideration of waterway health in bushfire management; ensures that waterway values and assets are included in bushfire and flood rapid risk assessment processes; and provides direction for reviewing and reprioritizing resource allocation after an extreme event. Urgent works may be required after extreme flood or bushfire events to address immediate risks. These may include clearing of flood debris in waterways to protect public infrastructure, stabilizing waterways affected by erosion, addressing threats to water quality (such as overflows from sewage works or dead stock in waterways), erecting temporary silt barriers after a bushfire or relocating threatened species.

PLANNING FOR THE IMPACTS OF CLIMATE CHANGE—A SEASONALLY ADAPTIVE APPROACH

The climatic conditions experienced in Victoria over the past 20 years have included a severe 13-year drought, but also some of the worst floods in the state's history (DEPI, 2013). Under a medium climate change scenario, it is predicted that average annual streamflow will reduce by approximately 50 per cent in some catchments by 2065 (Porter et al. 2016). Higher air temperatures, reduced streamflows and more extreme events such as flood and bushfire could have short and longer-term impacts on water temperature, turbidity and frequency and severity of algal blooms. Ecological impacts are also likely (DELWP 2016). It is important to plan for a full range of possible future climate scenarios and natural climate variability.

Rather than relying only on emergency management during these extremes of climatic conditions, a flexible management approach allowing annual implementation of management activities to be adapted to reflect the prevailing seasonal conditions, and a process to change management objectives, has been adopted.

The northern region sustainable water strategy (DSE, 2009) established the "seasonally adaptive approach" as a flexible way to manage rivers and wetlands. This approach takes into account recent climate history, climate outlook and available environmental water. In drought periods, the focus is on avoiding catastrophic events (such as major fish death events) and protecting drought refuges so that plants and animals can survive and begin recolonization when conditions improve. In wet periods, the focus is on providing a water regime to restore values that were not maintained in drier periods, such as major bird breeding events.

The seasonally adaptive approach is implemented through annual works and watering plans. This involves:

- Recognizing the long-term objectives and outcomes outlined in the RWSs and regional SWSs.
- Setting short-term management aims through annual planning processes that reflect whether the current conditions are drought, dry, average, or wet.
- Adapting management activities to prevailing climate conditions in any year.
- Monitoring and drought, flood, or bushfire response planning in all years.
- Improving community awareness of the need to adapt management actions depending on current climatic conditions.

PROCESS TO CHANGE MANAGEMENT OBJECTIVES

With limited funding and major challenges (such as the potential impacts of climate change and population growth), natural resource managers need to be prepared to accept that changes to the environmental condition of waterways will occur. Some values may not persist into the future. In particular, values may change as the local climate or pattern of land use changes (e.g., movement of populations of some species and communities). In some cases, values may be lost altogether.

The Strategy outlines a process to change management objectives when there is real and defensible information to indicate that this is necessary. It is important to emphasize that this does not imply that management is "giving up" on valuable river, estuary, or wetlands systems. Instead, management will be conducted with the recognition that not every current value within every system can be maintained given the potential impacts of climate change and land use change.

Knowledge needs to be improved and the condition of waterways closely monitored to assess if management objectives for a waterway have been met at the end of each 8-year planning period. If the objectives have not been met and the scientific evidence exists to show that the values either have been, or most likely will be, lost some decades into the future, then the management objectives may need to be changed in consultation with the community. This occurs as part of the development of the RWS every 8 years.

UPDATED ADAPTIVE MANAGEMENT APPROACH AND NEW INFORMATION

The Victorian Waterway Management Program recognizes that new knowledge must be periodically used to adapt planning, management, and monitoring practices. "Best practice" is better thought of as "current best practice," as new and improved methods are continuously being developed. Our understanding of ecosystem processes and how they respond to management activities and a changing climate is also being constantly improved as new research is undertaken.

The Program operates on an 8-year adaptive management cycle that follows six steps:

- Management objectives and targets are set in state and regional strategies.
- Logic models that describe the known (or assumed) relationships between threats, management activities and their expected outcomes are used to identify management options.
- Management activities are implemented to a known standard (if a standard has been developed).
- The response to the management activity is strategically monitored for a subset of sites.

Management practices are adapted accordingly, based on monitoring information or research. It is important to understand the assumptions and uncertainty in the logic models that are used to identify the appropriate management activities. The Program has focused on documenting where there is high, medium, or low confidence in the models, based on whether the model has been built on a theoretical understanding of a system (low confidence), a single published experiment (medium confidence), or multiple published studies (high confidence). Research priorities are often directed to where the uncertainty is high and significant funding into the management activity occurs. Many models are qualitative and there is the desire to move to more quantitative models, but this will take time and resources. Nevertheless, such research is recognized as critical to improve the rigor of the regional target-setting process.

The Program monitors both broad scale, long-term environmental conditions and changes in response to management activities at sites where work has been undertaken. Victoria has established three index programs to measure the long-term resource condition of rivers, wetlands, and estuaries (DEPI, 2013). Monitoring of the effectiveness of management activities at sites of environmental works is undertaken by CMAs or by strategic research programs (e.g., the Riparian Intervention Monitoring Program and the VEFMAP) (DEPI, 2013).

Universities and research institutions also provide assistance in filling information gaps and guidance on the most appropriate natural resource management tools, methods or approaches to use. They also provide scientific evidence about the condition of natural assets.

NEW MANAGEMENT ISSUES ADDRESSED

Key policy components of the VRHS were management of riparian land, water quality, and river channels. Each of these important management areas had refreshed policy directions outlined in the Strategy. Additionally, several new policy areas were addressed in the Strategy that was not covered in the

VRHS. These included wetland management, estuary management, involvement of Traditional Owners, urban waterways, invasive species, and recreational use.

For *wetlands*, the Strategy comprehensively integrated the management of wetlands with the management of rivers and estuaries. It also committed to maintaining the values of Ramsar sites, monitoring their ecological character and providing clarity regarding the listing of new Ramsar sites. The Strategy outlined how public land managers and waterway managers could work together more closely to manage wetlands in parks and reserves, and also outlined arrangements for improving the long-term security of wetlands on public land reserves in fragmented landscapes. Importantly, the Strategy made a commitment to providing assistance for landholders to improve the conservation and sustainable use of wetlands on private land, which has guided investment into this area.

For *estuaries*, the Strategy distinguished these ecosystems and their management requirements as unique, rather than as just the last reach of the river. Organizational responsibility for estuary entrance management was clarified and formalized. Improved strategic planning of management activities to maintain or improve the environmental condition of estuaries was outlined, as well as the need to enhance knowledge to support improved estuary management. The Strategy also strengthened programs to increase community awareness of, and involvement in, estuary management, such as EstuaryWatch.

The VRHS had limited consideration of *Traditional Owner rights and interests* in waterway management. Involving Traditional Owners in the development of state policy was challenging, since Traditional Owner groups generally only speak for their particular Country. Specific policies and actions were developed for the new strategy to ensure the involvement of Victoria's Traditional Owners in waterway management. As a result of community feedback on the draft Strategy (DSE, 2013), cultural values of waterways were separated from social values to demonstrate the importance of Aboriginal values in their own right. Traditional Owners are involved in the development of RWSs and a state guidance note was developed to support this process, in consultation with relevant representatives. The RWSs must identify Aboriginal values associated with waterways in their region and incorporate traditional knowledge where culturally appropriate and desired by Traditional Owner groups. Several research projects are underway to investigate methods for identifying Aboriginal values associated with waterways and how they can be better incorporated into regional planning processes. Opportunities for education, training, and capacity building are provided through scholarships to university graduate programs in waterway management and natural and cultural resource management (DEPI, 2013).

A relatively new policy focus for the Victorian government is the focus on greater *community wellbeing and livability* by maintaining or improving the condition of waterways in urban areas. This policy focus is best articulated in the new *Water for Victoria* policy document, which contains directions for urban waterway management (DELWP, 2016).

For *invasive species management* in waterways, the Strategy established a risk-based approach that takes into account both current and future risk for invasive species management in waterways. It also identified priorities for improved knowledge, surveillance and community and industry awareness of the impacts and vectors of spread of invasive species.

The *recreational benefits* of waterways have long been recognized. The Strategy encourages and supports community involvement in waterway planning for, and management of, recreational use of waterways. It acknowledges the importance of waterways for recreation and promotes sustainable recreational use. Importantly, policy to manage risks associated with recreational use is provided through guidelines, protocols, and education.

KEY SUCCESS FACTORS

A whole-of-government approach is required if long-term waterway health policy and programs are to be delivered over successive governments. Strong relationships with central agencies (such as the Victorian Department of Premier and Cabinet and the Department of Treasury and Finance), the EPA Victoria and Fisheries Victoria were vital. A partnership approach with regional service providers (the waterway managers) has also been fundamental to the success of the Program. The department has worked closely with waterway managers to determine how policies for on-ground actions would be implemented for nearly two decades in a collaborative, adaptive management approach. Community participation in waterway management is also essential, as government cannot do the job of improving waterway health on its own.

Clear direction from relevant ministers is vital to successful policy development and implementation. Each different minister for water and minister for environment generally seek different policy outcomes for their government. Strong working relationships with ministers and their ministerial advisors/office staff are essential to ensure early and clear direction on what the government wants to achieve and the preferred policy options for doing this.

A secure funding source for implementation has been perhaps the most critical factor to the program's success. The ECL was created in 2004 as the key funding mechanism for the Victorian government's *Our Water Our Future* policy (DSE, 2004). An amendment to the *Water Industry Act 1994* enabled the ECL to be collected from Victorian water corporations and passed on to water customers. This has meant that since 2004–05, the community has been paying for the costs of sustainable water-related initiatives through their water bills. This initiative is part of a continuing shift towards full-cost recovery for water planning and management activities, as required under the National Water Initiative (NWC, 2004).

The institutional roles and responsibilities for waterway management in Victoria, defined in legislation, have meant that the service delivery model is clear and the scope of responsibilities is defined. For example, waterway managers have specific functions for waterway management, to both "develop and implement plans and programs" and "carry out works and activities." This contrasts with other natural resource management areas in Victoria that do not have clearly defined service delivery roles in legislation or strategy.

CONCLUSIONS AND FUTURE CHALLENGES

While the Strategy made numerous advancements to the policy framework for waterway management in Victoria, continuous improvement will always be required. An independent review of the Strategy and the RWSs will be undertaken in 2020, which will inform the next round of strategy development. The strong foundations of the policy framework need to be maintained and the key success factors considered.

In 2016, the *Water for Victoria* policy document (DELWP, 2016) sought to strengthen the framework further. It contains policy to focus waterway health investment into large-scale, long-term on-ground work programs, further support recreational values of water, and better recognize Aboriginal values of water, and also policy to deliver livability outcomes in urban areas and improve the health of urban waterways.

Future challenges include the following:

- The effectiveness of our catchment management system could be further improved with a review of the current establishing legislation and a dedicated multiyear funding stream for implementation of RCSs. Additional on-ground work and extension activities are required to better support the legislated objective of maintaining and enhancing long-term land productivity while also conserving the environment.
- We need to increase understanding that the task of improving environmental condition is long term. Given the scale of the issues across the state, it will take several decades to complete restoration work and then ongoing maintenance is required, as for any community asset.
- Short-term management funding and programs have not adequately recognized the significant time lag between undertaking management activities and being able to see demonstrated improvements in environmental conditions.
- With the VEWH holding more than 650 GL of water recovered for the environment, it is now vital to ensure we are making the most efficient and effective use of this significant asset and achieving the best possible environmental outcomes in the face of climate change.
- Climate change will impact on future environmental water availability and the ecological objectives that are able to be achieved. This needs to be better incorporated into our planning processes as knowledge improves.
- Regional waterways in catchments with major population growth face a significant challenge to protect waterway values or manage the transformation of these waterway systems as urban and periurban development occurs.
- We need improved information and knowledge about catchments and waterways to inform adaptive management and enable continuous improvement of our management practices.
- Given the significant government investment being made in catchment and waterway health, we must be able to clearly demonstrate the outcomes that are achieved.

REFERENCES

DELWP. (2016). *Water for Victoria—Water Plan*. Melbourne: Department of Environment, Land, Water and Planning.

DEPI. (2013). *Victorian Waterway Management Strategy*. Melbourne: Department of Environment and Primary Industries.

DNRE. (2002). *Victorian River Health Strategy*. Melbourne: Department of Natural Resources and Environment.

DSE. (2004). *Securing our water future together: Victorian government white paper—Our Water Our Future*. Melbourne: Department of Sustainability and Environment.

DSE. (2009). *Northern Region Sustainable Water Strategy*. Melbourne: Department of Sustainability and Environment.

DSE. (2010). *Victorian River Health Strategy (2002)—A Review*. Melbourne: Department of Sustainability and Environment.

DSE. (2013). *Community feedback: Draft Victorian Waterway Management Strategy*. Melbourne: Department of Sustainability and Environment.

NWC. (2004). *Intergovernmental Agreement on a National Water Initiative*. Canberra: National Water Commission. 39 p.

Potter, N. J., Chiew, F. S. H., Zheng, H., Ekström, M., & Zhang, L. (2016). *Hydroclimate projections for Victoria at 2040 and 2065*. Australia: CSIRO.

Roughley, A. (2009). *Developing and using program logic in natural resource management—User guide*. Canberra: Australian Government.

Victorian Coastal Council. (2008). *Victorian Coastal Strategy*. Melbourne: Victorian Coastal Council.

MANAGING WATER QUALITY FOR THE GREAT BARRIER REEF

15

R. Eberhard*, J. Brodie†, J. Waterhouse†

Eberhard Consulting, Brisbane, QLD, Australia* James Cook University, Townsville, QLD, Australia†

CHAPTER OUTLINE

INTRODUCTION

Australia's Great Barrier Reef (GBR) is truly one of the world's natural wonders. Widely billed as the world's largest, most spectacular and best-protected coral reef system, the GBR encompasses a wide range of globally significant biodiversity and cultural values and directly supports fishing and tourism

Decision Making in Water Resources Policy and Management. http://dx.doi.org/10.1016/B978-0-12-810523-8.00016-1

industries. Protected as a national marine park and World Heritage Area, the GBR is nonetheless threatened by climate change, overfishing and catchment runoff (Brodie & Waterhouse, 2012). In this chapter, we focus on efforts to manage the diffuse sources of water-quality pollutants from the catchment affecting the GBR.

In recent decades, the health of GBR ecosystems has continued to decline, driven by poor water quality largely derived from catchment runoff, exacerbated by extreme climate events (cyclones, coral bleaching) (Brodie et al., 2013b). Climate change is already affecting the GBR, and will have "far-reaching consequences in the decades to come" (GBRMPA, 2014).

Elevated levels of fine sediment, nutrients and pesticides are exported to the GBR lagoon from agricultural landscapes during high rainfall events (Kroon, 2012). Since 2003, the Australian and Queensland governments have developed successive bilateral plans to accelerate the adoption of improved agricultural practices. Significant but limited progress against targets for the uptake of new practices and modeled water-quality benefits have been achieved primarily through voluntary mechanisms (Aust. Govt. & Qld. Govt., 2016a).

This chapter reviews the concerted efforts by the Australian and Queensland governments to improve water quality over the last 15 years. The four phases of bilateral water-quality planning and programs, and associated scientifically robust targets and reporting systems, are discussed. Then the key decision-making features (bilateral arrangements, regional water-quality improvement plans, water-quality targets, stakeholder engagement and science communication and brokerage) are analyzed. Finally, the key achievements and future challenges in managing water quality in the GBR are discussed.

As a case study in water-quality management, the GBR demonstrates the successful use of targeted incentives to drive improved water quality across large-scale and diverse agricultural landscapes. However, the GBR also provides a salutary lesson about the limits of suasive approaches to a narrowly defined problem and the need for an integrated approach using multiple policy levers to effectively address multiple and cumulative threats.

WATER-QUALITY IMPACTS ON THE GBR
THE GBR LAGOON AND CATCHMENT

The GBR stretches 2300 km from the Torres Strait in the far northeastern tip of Australia, down the coast to Bundaberg and Hervey Bay. In the north, the reefs are found close to the coast (within 100 km), while the outer reefs are found further offshore (up to 200 km) in the south (Fig. 1).

The GBR World Heritage Area supports 2900 coral reefs, 1050 islands and 46,000 km^2 of seagrass beds within its 348,000 km^2 area. The diversity of ecosystems found along the length, depth and across the shelf of the GBR makes it arguably the most significant marine protected area on the planet (Day & Dobbs, 2013). Iconic species include whales, dolphins, turtles, sharks and dugong, as well as the coral reefs themselves.

In 1975 the GBR was protected under the *Great Barrier Reef Marine Park Act 1975*. World Heritage listing was afforded in 1981 on the basis of the GBR's "Outstanding Universal Value," specifically "having superlative natural phenomena and areas of exceptional natural beauty; it being an outstanding example of major stages in the Earth's evolutionary history; it representing significant

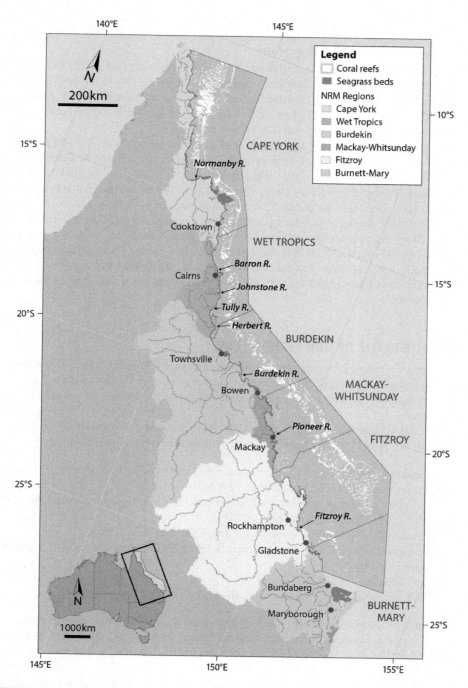

FIG. 1

Map of the Great Barrier Reef and its catchment, showing the six Natural Resource Management regions.

Map prepared by D. Tracey, TropWATER James Cook University (Lønborg et al., 2016).

ongoing ecological and biological processes and Traditional Owners' interaction with the natural environment; and ... containing the most important and significant natural habitats for in situ conservation of biological diversity" (GBRMPA, 2014).

The direct and indirect economic contribution of the GBR was estimated at $5.7 billion per annum in 2012 and just under 69,000 jobs (Deloitte Access Economics, 2013). Over 90% of this value is generated within the tourism sector, and the balance from recreation, commercial fishing, science and research. More recently, Stoeckl et al. (2014) have estimated the value of ecosystem services provided by the GBR as between $15 and $20 billion per annum.

Thirty-five major basins (424,000 km^2) drain into the GBR lagoon (Fig. 1). The landscapes and communities of the GBR are diverse. Two very large catchments (the Fitzroy and Burdekin) account for most of the GBR catchment area (70% of the area but only 28% of the flow) (Waterhouse et al., 2016a). Rangeland grazing in these very large, flat, semiarid catchments (rainfall less than 1500 mm/y) delivers high sediment loads to the GBR receiving waters in high-flow events. In contrast, the small, steep and humid tropical catchments of the Wet Tropics and Mackay-Whitsunday regions, as well as the Mary catchment, are dominated by intensive agriculture, including sugarcane and high-value horticultural crops such as bananas and tropical tree crops. These small regions with high rainfall and runoff (e.g., the Wet Tropics has only 5% of the GBR catchment area but 33% of the average annual flow) contribute high levels of dissolved nutrients and pesticides to the GBR (Waters et al., 2014).

THE CURRENT STATUS OF GBR ECOSYSTEMS

The GBR is facing the cumulative pressures of climate change, overfishing, catchment runoff, coastal development and mines and energy developments (Hughes et al., 2015). The combined effects of historical and contemporary impacts have resulted in significant deterioration in a wide range of GBR species and habitats.

Across the midshelf and offshore areas of the GBR it is estimated that coral cover has declined 51% (1985–2012) (De'ath et al., 2012). There are many causes of coral decline, including:

- Smothering and reduced light availability on inner-shelf reefs caused by fine sediment in catchment runoff (Fabricius et al., 2014; Fabricius et al., 2016);
- Predation by Crown of Thorns Starfish (CoTS) outbreaks triggered by nutrient-enriched catchment runoff (Brodie et al., 2005; Fabricius et al., 2010);
- Competition with algae (De'ath & Fabricius, 2010; Fabricius et al., 2005);
- Coral bleaching and other climate-change impacts (Hoegh-Guldberg et al., 2007; Hughes et al., 2015; Wooldridge, 2009);
- Coral disease (Haapkylä et al., 2011); and
- Direct and indirect impacts of cyclones (De'ath et al., 2012).

Seagrass has experienced similar, though less well-documented, declines in extent and abundance from storms, floods and the impacts of coastal and catchment agriculture, urban and port land uses (McKenzie et al., 2015). Many of the GBR's iconic species have seriously declined, including some species of marine mammals, turtles, sharks and seabirds. For example, dugong populations have declined from an estimated 72,000 in the 1960s to about 4200 in the 1990s (Marsh et al., 2005).

The Australian government's 2015 report to the UNESCO World Heritage Committee concluded that: "the overall outlook for the Reef is poor, has worsened since 2009, and is expected to further deteriorate in the future; and that greater reductions of all threats at all levels—Reef-wide, regional and local—are required" (Aust. Govt., 2015b). The Australian and Queensland governments provided a further update on progress in implementing the Reef 2050 Plan in Dec. 2016 (Aust. Govt. & Qld. Govt., 2016d).

The impacts of growing human pressures are exacerbated by the impacts of climate change. Coral reefs are threatened globally, with climate-change predictions suggesting that conditions in the next 50 years will exceed anything experienced by corals in the last half a million years (Hoegh-Guldberg et al., 2007; Hughes et al., 2003). Extreme sea surface temperatures in 2016 resulted in an unprecedented global coral bleaching event (Normile, 2016). Effects were severe, with preliminary results suggesting an average 22% coral mortality across the GBR. Mortality was most severe (average 50% mortality) in the northern section (Cairns to Cape York; Fig. 1) (GBRMPA, 2016). This northern part of the GBR (Princess Charlotte Bay to the Torres Strait) was generally considered to be in relatively pristine condition (Brodie et al., 2012; Brodie & Pearson, 2016; De'ath et al., 2012).

Researchers continue to call for renewed efforts to manage water quality to improve the resilience of GBR to multiple stressors while global climate agreements are enacted and implemented (Brodie & Pearson, 2016; Brodie & Waterhouse, 2012; Dale et al., 2016; Kroon et al., 2016). The declining health of the GBR requires urgent action to maintain its resilience in the medium term, since any mitigation of climate change will take time.

CATCHMENT WATER QUALITY PRESSURES

In the period since European agricultural practices were introduced into the GBR catchments (around 1830 in the south and later in the northern areas), loads of sediment, nitrogen, phosphorus and pesticides transported by rivers have increased greatly. For example, for the whole of GBR, modeled estimates of annual average total suspended sediment (TSS), total phosphorus (TP), and total nitrogen exported loads have increased by 2.9-, 2.3- and 1.8-fold, respectively, from predevelopment loads (Waters et al., 2014). These changes can be detected using historic data such as coral coring (e.g., Lewis et al., 2007), sediment coring (Clark et al., 2016; Reymond et al., 2013; Roff et al., 2012) and geochemical records (Lewis et al., 2014b; Waters et al., 2014).

The mean annual TSS load to the GBR is estimated to be 8.5 million t/y. The major sources of TSS to the GBR are from the large dry tropics grazing-dominated Burdekin and Fitzroy Basins, which contribute over 70% of the modeled TSS load. Erosion from hillslopes, streambanks, and gullies contribute to the TSS discharge with major variations in the relative contributions between individual catchments (Bartley et al., 2014; Wilkinson et al., 2013). While erosion from cropping areas and urban development on steep lands are much smaller in area, they may pose a threat to local coastal reefs and seagrass areas.

The estimated mean annual nitrogen load comprises 10,500 t/y dissolved inorganic nitrogen (DIN), 14,300 t/y of dissolved organic nitrogen, and 11,800 t/y of particulate nitrogen (PN). By far the largest increases since European settlement have occurred in the DIN load due to fertilizer use and the PN load due to increased erosion (Waterhouse et al., 2012). Similarly, the estimated mean annual TP load comprises 1100 t/y of dissolved inorganic phosphorus (DIP), 600 t/y of dissolved organic phosphorus, and

4500 t/y of particulate phosphorus (PP). The largest increase has been in PP due to increased erosion. The largest sources of nitrogen and phosphorus are from the Wet Tropics and Burdekin regions (Waters et al., 2014).

At least 16 t/y of the photosystem-II (PSII) inhibiting herbicides atrazine, ametryn, hexazinone, diuron, simazine, and tebuthiuron (for which reliable monitoring information exists) are discharged to the GBR (Waters et al., 2014). Unlike nutrients and sediments, there is no "natural" level of pesticides. Atrazine, ametryn, hexazinone and diuron originate predominantly from the sugarcane industry (Waterhouse et al., 2012), with atrazine also being used in grains cropping, and tebuthiuron originating from the beef grazing industry (Waters et al., 2014).

Practices to reduce erosion in grazing lands rely on maintaining ground cover and biomass of pastures, especially during the dry season and droughts (Thorburn et al., 2013). Increasing the levels of ground cover helps reduce runoff and prevents or reduces further hillslope and channel erosion. More recently, it has become clear that gully networks caused by livestock grazing are much more important sources of sediment than previously thought (Olley et al., 2013). Improved grazing practices alone may not be adequate to reduce gully erosion and specific interventions to stabilize and revegetate gullies may be necessary (Bartley et al., 2014).

Substantial amounts of nitrogen fertilizer are applied to high-value crops such as sugarcane in the GBR catchments with large losses of DIN to surface-water runoff and deep drainage (Thorburn et al., 2013). Large reductions in the amounts of applied nitrogen per hectare will be required to reach the pollutant load-reduction targets set in current government policies (see "Modeling Water Quality Targets" section) (Brodie et al., 2017).

Practices to effectively reduce the levels of nitrogenous fertilizers applied, while maintaining profitability, are not well understood in the sugarcane industry. New "solutions," such as slow-release fertilizers, are being tested (Di Bella et al., 2013).

Knowledge of management practices that dramatically reduce pesticide losses is well advanced, particularly in sugarcane (Devlin et al., 2015). Techniques such as banded spraying, where only 10% of the normal amount of pesticide used in full-field spraying is used, can reduce losses of herbicides by 50%–90% (Davis & Pradolin, 2016; Oliver et al., 2014). However, any changes to the mix of herbicides used may have unintended negative ecological effects (Davis et al., 2016).

IMPACTS ON FRESHWATER, COASTAL, AND MARINE ECOSYSTEMS

Land-derived pollutants disperse offshore during wet season high-flow events (Devlin et al., 2012) and pose a range of threats to valuable GBR ecosystems, including coral reefs, seagrass meadows, and associated species such as dugongs and turtles (Schaffelke et al., 2013). Excessive nutrient discharges lead to increased survival of CoTS larvae and is associated with increased outbreaks of CoTS populations, which are major coral predators (Fabricius et al., 2010). Elevated nutrients can also lead to excessive algal growth at the expense of coral diversity (De'ath & Fabricius, 2010) and can increase the susceptibility of corals to bleaching (Wooldridge, 2009, 2017a) and disease (Haapkylä et al., 2011).

Fine sediment can reduce the growth of seagrass and corals by making the water turbid and reducing light penetration (Collier et al., 2012; Fabricius et al., 2014, 2016). For example, long periods of low light, driven by frequent river discharges, has led to the loss of seagrass in Cleveland Bay (adjacent to

Townsville) during the period 2009–13 (Petus et al., 2014), with subsequent effects on the mortality of dugongs (Wooldridge, 2017b).

Pesticides are used for the control of weeds (herbicides) and insects (insecticides); however, they may not discriminate between pest and nonpest species. When released into waterways, they pose a risk to freshwater and some inshore and coastal habitats (Lewis et al., 2014a; O'Brien et al., 2016).

Increased runoff of sediment, nutrients and pesticides from agricultural and urban land uses may also be impacting freshwater and estuarine ecosystems, but there is less evidence of these effects than in the more-studied marine ecosystems (Brodie & Waterhouse, 2016; Davis et al., 2016; Waterhouse et al., 2016a).

GOVERNANCE OF THE GBR

The state of Queensland retains primary authority for the management of land and water resources of the GBR, while the Australian government administers the GBR Marine Park and World Heritage Area. Legislative boundaries between the Queensland and Australian governments overlap in the tidal and inshore areas (Hockings et al., 2014) and the day-to-day operations of the marine park are jointly managed.

Over the last 15 years, the Australian government has played an increasingly influential role through funding national Natural Resource Management (NRM) programs (Robins & de Loë, 2009; Robins & Kanowski, 2011). In Queensland, regional NRM organizations are nonstatutory bodies governed by community-based boards (Robins & Dovers, 2007b). The mandate of these organizations, funding levels and investment priorities have changed over time (Vella & Sipe, 2014; Vella et al., 2015), but in general they broker the delivery of voluntary resource management programs on behalf of the Queensland and Australian governments.

The following sections describe the evolution of reef policy and programs from the 1980s to the present day—outlining the key decisions, governance arrangements, policy initiatives, and processes to engage stakeholders in these decisions.

PHASE 1: SCIENCE TO CONSENSUS (1980–2002)

The Great Barrier Reef Marine Park was proclaimed under the *Great Barrier Reef Marine Park Act 1975*, in conjunction with establishment of the Great Barrier Reef Marine Park Authority (GBRMPA). Intergovernmental coordination between the Queensland and Australian governments for the management of the GBR was formalized under the "Emerald Agreement," which established the GBR Ministerial Council in 1979 (GBRMPA, 2005).

Scientific evidence of the water-quality impacts and risks to the GBR built through the 1980s to 2000s (Haynes & Michalek-Wagner, 2000), but it was through the GBR Ministerial Council that scientific argument finally attracted policy attention in 2000. Following a Ministerial Council meeting, the first set of catchment-based pollutant load-reduction targets was developed by the GBRMPA (Brodie et al., 2001b) and this was quickly followed by two key reports commissioned to support the development of a bilateral Reef Water Quality Protection Plan. The Baker Report (Baker, 2003) confirmed the impacts and risks of declining water quality and agricultural industries as the primary source, and the Productivity Commission Report (Productivity Commission, 2003) evaluated a number of policy options to address water-quality issues.

At the same time, the national NRM program was undergoing a significant shift, from the first phase of the Natural Heritage Trust (NHT) that resourced farmers and community groups directly, to the second phase (NHT2) that established regional organizations to oversee NRM programs under a national framework and bilateral arrangements (Robins & Dovers, 2007a; Vella et al., 2015). Regional NRM organizations were charged with developing and implementing NRM plans based on "best-available" science and strong community engagement.

During this phase, the need for actions to address water-quality risks to the GBR was strongly advocated by marine scientists and GBMRPA. While new catchment modeling capacity was employed to develop targets (Brodie et al., 2001a), monitoring systems and evidence was drawn almost exclusively from the coastal and marine environment. No formal stakeholder engagement mechanisms were established under bilateral arrangements for GBR water-quality policy at this stage, but regional NRM organizations and national NRM funding programs were anticipated to be the vehicle for delivery of "on-ground" policy actions. The Queensland government, with responsibility for land management in the GBR catchments, was a relatively passive partner in these early developments.

PHASE 2: REEF PLAN AND REEF PARTNERSHIP (2003–07)

The first Reef Water Quality Protection Plan (Reef Plan) was launched in 2003 with the goal of "halting and reversing the decline in water quality entering the Reef within 10 years" (Qld. Govt. & Aust. Govt., 2003). Activities within the Plan were largely existing government initiatives, and there were no formal stakeholder engagement structures (Aust. Govt. & Qld. Govt., 2005).

The key additional activity initiated at this stage was the development of regional Water Quality Improvement Plans (WQIPs) by regional NRM organizations, funded by the Australian government under the National Water Quality Management Strategy (see "Regional Water Quality Improvement Plans" section). The WQIPs involved a substantial investment in science, particularly in water-quality modeling and investigation of the water-quality impacts of agriculture at various scale, and involved local and regional industry representatives, as well as government officers (Taylor & Lawrence, 2012; Taylor et al., 2010).

The first generation of regional WQIPs developed a framework for the water-quality benefits of different (improved) agricultural practices in major industries (sugar, horticulture, and grazing). This framework developed into the "ABCD" management practices framework, with variations for different industries and regions (Vella et al., 2011) (refer to "Science Communication and Brokerage" section).

Another key development was the establishment of the Reef Water Quality Partnership. This emerged from a coordination arrangement across regional WQIPs, supported by regional NRM staff and contractors. The Reef Partnership was adopted within the formal Reef Plan governance arrangements for 2 years (2006–08), where it facilitated policy dialog between regional NRM organizations and Queensland and Australian governments, and was supported by scientific and technical advisory groups.

The WQIPs made a compelling case for investment in improved agricultural practices. Pilot NRM programs demonstrated the use of the ABCD framework as a metric to guide the prioritization of program grants to farmers for the adoption of improved practices with demonstrated water-quality benefit. The regional WQIPs had built regional expectations for action among stakeholders, including industry groups.

The Reef Partnership facilitated the drafting of a "Scientific Consensus Statement" and the design of an integrated monitoring program that included property, catchment and marine components. A major conclusion of the Scientific Consensus Statement was that "current management interventions were not effectively solving the problem" (Brodie et al., 2008).

A policy window was created by growing stakeholder frustration with the Reef Plan, the lack of investment in WQIP implementation, and the looming Federal election and expected change in government (2006–07). Actively encouraged by Australian government officers, regional NRM and agricultural industry organizations jointly prepared a business case for investment that was advocated to both the government and the opposition.

PHASE 3: REEF RESCUE AND REGULATION (2008–13)

Following the election in 2007, the Australian government announced a new plan: Reef Rescue, which was a $200-million grants program to accelerate the adoption of improved agricultural practices. Reef Rescue was based on the proposal put forward by regional NRM and agricultural industry organizations, who then became the delivery agents and formalized their collaboration as the Reef Alliance.

The national NRM program underwent significant reform in 2008, moving from the NHT program, which funded regional NRM priorities, to a competitive funding model to deliver national NRM priorities "Caring for our Country" (Robins & Kanowski, 2011).

At the same time, the Queensland government regulated agricultural practices in high-priority GBR catchments under the *Great Barrier Reef Protection Amendment Act 2009*. Reef Plan itself was revised in 2009, with the additional long-term goal that "by 2020 the quality of water entering the Reef from adjacent catchments has no detrimental impact on the health and resilience of the Great Barrier Reef" (Qld. Govt. & Aust. Govt., 2009). The revised Reef Plan committed to water quality and land-management practice targets, based upon the WQIPs and the business case put forward by NRM and agricultural industry organizations. A formal stakeholder advisory committee replaced the Reef Water Quality Partnership.

A more substantial monitoring program was established, including catchment monitoring and modeling, and the first GBR water-quality Report Card was released in 2011 (Aust. Govt. & Qld. Govt., 2011).

About this time, the rapid expansion in coal and coal seam gas industries triggered significant new and expanded port development proposals, with risks from dredging and dredge spoil disposal, as well as longer-term risks from the associated greenhouse gas emissions. At the same time, new governments in Queensland (2012) and then federally (2013) were winding back a wide range of environmental controls ("green tape reduction"), including a "nonenforcement" policy for the recently introduced agricultural pollution regulations.

The proposed port developments received wide media coverage. A reactive monitoring mission by UNESCO/IUCN was triggered when the Australian government approved a large liquefied natural gas hub in central Queensland without notifying the World Heritage Committee (McGrath, 2012). In response, the Australian and Queensland governments prepared strategic assessments of development in the marine and catchment areas of the GBR, respectively, under section 146 of the *Environment Protection and Biodiversity Conservation Act 1999*. Ultimately, the World Heritage Committee chose not to list the GBR as "World Heritage in Danger" but maintains a close interest. The Australian government currently provides annual reports to the committee on efforts to maintain the "Outstanding Universal Values" that define the GBR's World Heritage status (Aust. Govt. & Qld. Govt., 2016b, 2016d).

Early results of the new programs looked promising. In the first 2 years of the Reef Rescue program, 34% of sugarcane farmers, 25% of horticulture farmers, and 17% of graziers adopted improved practices with a modeled reduction of 6% of nutrient, 7% of sediment, and 15% of pesticide loads (Aust. Govt. & Qld. Govt., 2013).

This third phase of GBR management was characterized first by an increased commitment by both governments to drive change in agricultural practices, adopting many of the strategies proposed by the Reef Alliance and supported by regional WQIPs. A mix of regulation and incentives were used, and clear evidence showed that suasive measures were supporting accelerated uptake of improved management practices. But then changes in government at both levels adjusted policy settings. The resources and energy boom put new pressures on the GBR and triggered the engagement of the World Heritage Committee.

In 2013, an updated Scientific Consensus Statement was released, and Reef Plan was revised (Brodie et al., 2013b; Qld. Govt. & Aust. Govt., 2013). Targets were updated based on revised modeling of the adoption of best-management practices (BMP) in agricultural industries, predominantly grazing and sugarcane.

PHASE 4: RESPONDING TO UNESCO (2014–CURRENT)

In 2015, the Queensland and Australian governments developed the *Reef 2050 Long-Term Sustainability Plan* (Reef 2050 Plan, Aust. Govt., 2015a) with a broader remit to address all threats to the reef including climate change, land-use change and direct-use impacts. The Reef 2050 Plan remains the guiding document for addressing water-quality issues. The Reef 2050 Plan was presented to the World Heritage Committee in 2015 (Aust. Govt., 2015b). The Australian Academy of Science issued a scathing response, claiming that "the draft plan is inadequate" and "fails to effectively address any of these pressures" (AAS 2014).

In 2014, the Australian government established the Reef Trust with a $140-million commitment. The Reef Trust provides investment for the implementation of some aspects of the Reef 2050 Plan. Rather than programmatic funding that had previously provided long-term (5-year) investment certainty, the Reef Trust has engaged with a variety of new project investments, including market-based instruments for agricultural practice change.

Another change of government in Queensland in 2015 saw renewed political commitment to protecting the GBR. The first "Minister for the GBR" was established, making changes to ports policy (including a ban on sea dumping of capital dredge waste), additional resource commitments ($100 million), and the review and reinstatement of recent legislative changes to the *Water Act 2000*, *Vegetation Management Act 1999* and *Coastal Protection and Management Act 1995*. The Queensland government convened a high-level GBR Water Science Taskforce to determine the best approach to achieving revised pollutant load reduction targets (up to 80% nitrogen reduction and 50% sediment reduction in key catchments by 2025). The Taskforce recommended a suite of measures, including extension, education, incentives, and regulation, and noted the need for substantial and enduring investment and governance reform (GBRWST 2016). The Queensland government accepted in principle all of the Taskforce's recommendations.

This period (Phase 4) has been characterized by a substantial reengagement in GBR policy by the new Queensland government, but changes in Australian government funding arrangements have created competition and uncertainty for delivery agents (the NRM organizations). The World Heritage Committee continues to scrutinize efforts to protect the reef.

Currently (Dec. 2016), the Scientific Consensus Statement, pollutant load reduction targets and regional management priorities are being updated in anticipation of a revision of the Reef Plan in 2017.

FEATURES OF THE DECISION-MAKING PROCESS
BILATERAL ARRANGEMENTS

Formal bilateral arrangements for GBR policy have existed since the Ministerial Council was established in 1979 (and continues as the Ministerial Forum). Intergovernmental officers' committees have supported three rounds of the bilateral Reef Plan. Yet the degree of policy coordination between the two governments has varied significantly over that period, reflecting different policy approaches and the relationship between individual Ministers. The Australian and Queensland governments still make their investment decisions independently, for political reasons as well as for policy outcomes. Good bilateral arrangements have included the commitment to coinvestment in, for example, monitoring and reporting programs, while poor coordination is evidenced by unilateral announcements, such as the Queensland government's announcement of the GBR regulations for agricultural industries in 2009.

Bilateral arrangements are also critical for effective regional NRM arrangements. The national NRM program has gone through several iterations of change. The regional model was established under the second phase of the Natural Heritage Trust, and partnership between the Australian and State and Territory governments was instrumental for the establishment of effective regional arrangements (Keogh et al., 2006; Turnbull, 2005). The shift to reduced core funding and more competitive funding arrangements under the new Caring for our Country program (Robins & Kanowski, 2011) has impacted delivery capacity through greater uncertainty, reducing the mandate of regional plans, and weaker bilateral arrangements (Robins & Kanowski, 2011; Senate Standing Committee, 2010; Vella & Sipe, 2014; Vella et al., 2015).

REGIONAL WATER QUALITY IMPROVEMENT PLANS

Regional WQIPs were first developed between 2006 and 2008, and have since been updated to produce a complete set of WQIPs for the six NRM regions (BMRG, 2015; Cape York NRM & SCYC, 2016; FBA, 2015; Folkers et al., 2015; NQ Dry Tropics, 2016; Terrain NRM, 2015). A WQIP identifies the main issues impacting waterways and the coastal and marine environment from land-based activities, and prioritizes management actions to halt or reverse the trend of declining water quality within a region. In relation to water-quality impacts on the GBR lagoon, the main management focus relates to activities associated with sugarcane, grazing and, to a lesser extent, horticulture, grain crops and urban areas. The scope of the plans also recognizes the role of coastal ecosystems in linking catchment and marine ecosystems, and regulating hydrology and pollutants.

The WQIPs are consistent with the *Framework for Marine and Estuarine Water Quality Protection* (Aust. Govt., 2002) and apply the framework described in the National Water Quality Management Strategy (ANZECC, 2000). In Queensland, this is linked through the *Environmental Protection Act 1994*, which is the main legislation for water quality in freshwater, estuarine and marine areas. Under the Act, WQIPs may be accredited as Healthy Waters Management Plans (HWMP),[1] which set water-quality targets based on environmental values and water-quality objectives in the WQIPs. The water-quality targets are then used under the *Environmental Protection Regulation 2008* for Environmentally Relevant Activities, such as industry, agriculture, aquaculture, mining, and waste disposal.

Environmental values and water-quality objectives are identified through community consultation (e.g., Kerr, 2013; Terrain NRM, 2012) and include values for primary industries, recreational and esthetics, drinking water, industrial, cultural and spiritual and aquatic ecosystems. It is the role of the WQIP to identify where these are most relevant for protecting or improving the water quality of the region.

The WQIPs have been led by the regional NRM organizations and developed with a range of stakeholder partners and technical experts to guide management in these areas. The WQIPs have been successful in providing a technical knowledge base for decision making, and guidance on management and delivery options at a basin scale. The plans also identify the relative costs and benefits of investing in alternative management options for the first time, allowing managers to compare a range of mechanisms for improving water quality, although it is recognized that detailed social and economic data are severely limited in many of the GBR catchments.

Despite the fact that the WQIPs provide the foundation documents for action to manage water-quality impacts on the reef, their implementation is not directly funded. Rather, the WQIPs indirectly guide policy priorities and investments in three ways:

- Water-quality objectives scheduled under the *Environmental Protection (Water) Policy 2009* support the regulation of Environmentally Relevant Activities under the *Environmental Protection Act 1994*;
- GBR water-quality policy, plans and investments are "informed" by the WQIPs—for example, the target-setting processes at the GBR scale build on the work developed through the WQIPs;
- When regional NRM organizations are preparing funding proposals, they refer to WQIPs for guidance, but these proposals are typically highly constrained by investor specifications.

The lack of a coherent pathway for WQIP implementation remains a weakness of GBR water-quality governance, and undermines the collaborations that support WQIP development.

---

[1]The HWMP guidelines are available from the Department of Environment and Heritage Protection website at http://www.ehp.qld.gov.au/water/policy/water_quality_improvement_plans.html.

MODELING WATER QUALITY TARGETS

The impact of land-sourced pollutants varies throughout the GBR World Heritage Area with the degree of exposure a function of the distance from the coast and river mouth, the magnitude of river discharges, wind and ocean current directions, the varying mobility of pollutant types, and of course the different catchment characteristics and land uses (Brodie et al., 2013a). Thus, coral reefs, seagrass meadows, and other important ecosystems have different exposure to land-sourced pollutants, with important implications for the prioritization of management interventions.

Multiple criteria analysis has been used to prioritize pollutants, regions and land uses for water-quality programs in the GBR (e.g., Barson et al., 2014; Brodie et al., 2013a; Cotsell et al., 2009; Greiner et al., 2005; Waterhouse et al., 2012). While these analyses have been useful for prioritization of management response and investment under Reef Plan and Reef Rescue, more sophisticated analyses are now needed to more confidently prioritize between pollutants across the individual catchments. Importantly, these analyses are still unable to adequately incorporate social, economic, and institutional factors.

The WQIPs are underpinned by supporting technical studies that guide spatial prioritization. In most regions, this comprises an assessment of the relative risk of degraded water quality to regional coastal and marine ecosystems linked to catchment pollutant sources, analysis and evaluation of financial-economic data for agricultural production systems and assessment of priority management actions. In some regions, an integrated assessment of the benefits and costs of achieving water-quality targets has used the Investment Framework for Environmental Resources (INFFER) analysis (Pannell et al., 2009; Park & Roberts, 2014; Roberts et al., 2016). The outcomes of these studies were then used to identify priority pollutants, locations and land uses for targeting investment and management (e.g., Waterhouse et al., 2016b).

The monitoring and modeling of pollutant loads form an integral part of these assessments by providing information on priority pollutants, pollutant sources and smaller-scale generation "hot spots." A combined monitoring and modeling approach is used to estimate and report end-of-catchment pollutant loads because of the limited capacity to effectively monitor the massive GBR catchment directly.

This combined approach has allowed decision makers to access relatively consistent information over large geographical and temporal scales, although there is always some skepticism about the validity of modeled results. Modeled end-of-catchment pollutant load estimates are generated from the "Source Catchments" modeling framework for key pollutants: TSS, DIN, PN, DIP, PP, and PSII herbicides (e.g., Waters et al., 2014), incorporating new information on paddock-scale modeling of TSS, forms of N and P, and PSII herbicides, plus remote-sensed spatial and temporal inputs including groundcover. This provides a consistent set of end-of-catchment pollutant loads for each of the 35 GBR basins (Waters et al., 2014). A fixed climate period (1986–2014) has been used for all model runs to normalize for climate variability and provide a consistent representation of predevelopment and anthropogenic generated catchment loads. This period is taken to represent an "average" climatic period rather than the extremes such as those recorded in the period 2008–13.

Confidence in the modeling is increasing each year due to improvements, including improved gully mapping, hydrological calibration and additional load validation (Pers. Comm., Cameron Dougall, Department of Natural Resources and Mines, Nov. 2016). The available monitoring data in the GBR catchments shows mixed correlation with the modeled end-of-catchment data, but is comparable in many locations.

STAKEHOLDER ENGAGEMENT

Stakeholder engagement and collaboration has been an enduring feature of GBR water-quality policy development and implementation at a number of levels (Hockings et al., 2014). Key stakeholders have worked well with governments through both formal and informal engagement processes (Robinson et al., 2010). Regional NRM organizations, agricultural industry organizations and the conservation sector have built an alliance that has successfully advocated for policy change and delivered programs collaboratively (Eberhard, 2010).

In addition, WQIPs have engaged local communities and researchers in identifying local priorities and developing strategies (Kroon et al., 2009; Peterson et al., 2010; Robinson et al., 2011) and data sharing and reporting (Eberhard et al., 2013).

Good stakeholder collaboration can build trust, provide critical local knowledge, leverage resources, help to share risk, and build implementation capacity (Hill et al., 2015). But stakeholder participation at both local and policy levels is not without its issues (Taylor & Lawrence, 2012; Taylor & van Grieken, 2015). For example, the role of science and local knowledge is often unclear (Bohnet, 2014; Kroon et al., 2009; Lebel et al., 2006), aspects of the planning processes may be highly contested between stakeholders (Potts et al., 2016; Taylor et al., 2010), and some stakeholder groups may be poorly engaged (e.g., indigenous Traditional Owners).

SCIENCE COMMUNICATION AND BROKERAGE

The science of water-quality impacts on the GBR and management responses is complex, multidisciplinary and still evolving. This needs to be communicated to policy makers, farmers and the wider community in a way that balances clarity with appropriate acknowledgment of complexity and uncertainty. Two strategies employed to good effect in the GBR are worth highlighting—science consensus statements, and the ABCD framework for characterizing the effectiveness of management practices.

For science to be accepted, it needs to be perceived as credible (of an appropriate standard) and legitimate (perceived as fair and unbiased) (Cash et al., 2002). Credibility is associated with scientific consensus, and a formal process of developing and documenting a Scientific Consensus Statement on Water Quality in the GBR is now in its third iteration (Brodie et al., 2008; Brodie et al., 2013b). The Consensus Statements are used as a reference document by policy makers and stakeholders, and are an established part of the review process that updates Reef Plan.

The ABCD management practices framework is a framework that articulates different standards of practices for key industries (sugarcane, grazing and horticulture) as they relate to water-quality risk. The legitimacy of the framework comes from its development by scientists and farmers in each region. It has provided a common language between investors, program managers, industry groups and farmers who use it in different ways at different scales (Vella et al., 2011). For a grower, the framework provides a benchmark of industry standards that can show where improvements might be made. For NRM organizations running grants programs, it informs decisions about priority projects. For policymakers, it provides a framework for understanding the role of different policy levers and measuring performance at the level of practice change.

Both the Scientific Consensus Statements and the ABCD management practice framework communicate current scientific knowledge in ways that increase its legitimacy and credibility, and therefore acceptance, by multiple audiences.

ACHIEVEMENTS
PROGRESS AGAINST TARGETS

The Reef Report Card 2015 (Aust. Govt. & Qld. Govt., 2016c) demonstrates the limited success of efforts to facilitate the adoption of improved (lower water-quality risk) land-management practices over 6 years:

- Graziers manage 31.1 million ha of land and over 100,000 km of stream bank in the GBR catchments. An average of 32% of this area is now farmed using improved BMP pasture, streambank and gully management practices. However, against the GBR target of 90% BMP, this scored D (poor) on a scale from E very poor, to A very good.
- Sugarcane is grown by 3777 enterprises on 400,000 ha of land in the GBR. Over 6 years, 32% of this area implemented BMP for sediment, nutrients and/or pesticides. Against a target of 90% BMP uptake, this scored D, or poor.
- Higher rates of adoption were achieved in horticulture (47% of the area, scored as C, moderate) and grains (57% of the area, also scored as C, moderate).

Limited commensurate data is available on actual improvements in water quality. Modeled pollutant load reductions provide estimates of the benefits achieved by improved management practices (Waters et al., 2014). Measurable load reductions lag behind practice change due to the time for practice to take effect, such as the time to clear out sediment bedloads, which will vary with catchment, position in catchment, stream power and flow events.

Modeled reductions in anthropogenic pollutant loads achieved at the end of catchments due to change in practices recorded between 2009 and 2015 include:

- An 18% reduction in DIN against a target of 50% by 2018—score E, very poor;
- A 12% reduction in suspended sediment against a target of 20%—score C moderate,
- A 34% reduction in pesticides against a target of 60%—score C, moderate.

The overall condition of the inshore marine environment (water quality, seagrass, and coral) remains poor, and has not changed since the 2011 Report Card.

Given the estimated investment of around $700 million (Brodie & Pearson, 2016), progress appears to be slow, albeit against the massive scale of the GBR catchment. Modeling has also shown that even complete adoption of existing industry "best management practices" is not expected to achieve sediment and nutrient targets (Thorburn & Wilkinson, 2012; Waters et al., 2014). Policy makers therefore need to consider how to further accelerate adoption of improved practices, how to develop the next generation of improved practices, and what other actions can contribute to improving water quality, while waiting for global action on carbon emissions.

The following sections explore some of the options and constraints faced by policy makers in the GBR.

ONGOING CHALLENGES
Delivery capacity for incentive programs

The behavioral change programs to improve GBR water quality seek to accelerate the adoption of agricultural practices with lower water-quality risks in high-priority areas. Regional NRM organizations and industry partners have delivered education and awareness-raising programs, agricultural extension, grants, and more recently, market-based instruments. This is consistent with a more general shift of agricultural extension services from state agencies to the private sector and NRM organizations (Marsh & Pannell, 2000).

Changing NRM programs at both the Australian and Queensland government level have seen a decline in base funding and a shift to more competitive funding arrangements (Robins & Kanowski, 2011; Vella et al., 2015). The adoption of competitive funding models in the GBR, and NRM in general, has increased competition, and the GBR water-quality programs dominate NRM budgets in those regions. In the pursuit of more business-like efficiencies, the new programs have increased transaction costs, created additional uncertainty, reduced mandate, reduced staffing levels and impacted relations with local communities (Dale et al., 2013; Robins & Kanowski, 2011; Tennent & Lockie, 2013).

Yet we know that the suasive policy instruments employed by current GBR water-quality programs (e.g., grants, education, extension, market-based instruments) rely on human capacity such as local technical expertise and trusted relationships for the successful engagement of farmers (Emtage & Herbohn, 2012; Taylor & van Grieken, 2015). Spatial priorities across the GBR have been periodically reviewed through the life of Reef Plan, and these priorities directly inform investment amounts. The overall fragility of the regional NRM model, and the variable and uncertain funding of reef programs, directly impacts the human and institutional capacity of organizations charged with delivery of these reforms.

Regulation and land-use change

While incentive-based programs have accelerated the adoption of improved agricultural management, this is not sufficient to achieve water-quality targets (Brodie & Pearson, 2016; Kroon et al., 2016). The GBR lagoon enjoys stringent legal protection (Baxter, 2006), but controls on catchment-based activities are more complex.

Regulating agricultural runoff is challenged by the capacity to measure performance at the scale of management (attribution) and the assessment of cumulative effects (Gardner & Waschka, 2012). The Queensland government's *Great Barrier Reef Protection Amendment Act 2009* provides a risk-based approach to regulating practices in large grazing and sugarcane enterprises in priority catchments. However, changing levels of government commitment to compliance have reduced the effectiveness of these regulations. Anecdotally, stakeholders reported that regulation has driven engagement in incentive-based programs (the threat of regulation), or triggered disengagement (frustration and confusion with government programs).

Similarly, a review of pesticide regulation in relation to the GBR found that the response by Australia's national pesticide regulator has been "ad hoc, case-by-case, very slow, and ineffective" (King et al., 2013).

Additional regulatory mechanisms could be used to support water-quality improvement in the GBR, including local and Queensland government planning instruments, and vegetation and water resource management legislation. The potential perverse impacts of other policies, such as agricultural intensification, drought relief, and water resource development, are also important.

Good policy practice suggests a mix of regulatory instruments and close engagement with the agricultural community to ensure regulations achieve desired outcomes (Cherry et al., 2008; Connor et al., 2009; Stark & Richards, 2008; Van Grinsven et al., 2012). This approach is strongly supported by the regional WQIPs and was recommended by the GBR Water Science Taskforce

(GBRWST, 2016). Ultimately, however, the effectiveness of any regulation is dependent on the political will to develop regulation, enforce compliance, and evaluate effectiveness.

Coastal development and EIS processes

The recent mining boom in Queensland highlighted the potential for other industries (coal and coal seam gas) to significantly impact the GBR. Brodie (2014) estimated that the planned expansion of major ports at Cairns, Townsville, Abbott Point, Hay Point, and Gladstone would generate approximately 150 million tons of sediment over a 10-year period. The water-quality impacts of the dredge spoil are expected to be offset, but Brodie's calculations suggest that this will be both impossible and unrealistically expensive. Changes in government policy have since constrained the proposals for new port developments, but existing ports continue to expand, and pressures to add new ports are likely to return in the future.

The risks of major coastal development projects are illustrated by the significant disease and mortality event of fish and other sea life that occurred in Gladstone Harbour in 2011. Inadequate monitoring prevented the cause of the event from being clearly determined—a major dredging program, a poorly constructed and leaky bund wall and recent flood event were all likely contributors. The review of practices during construction and dredging in Gladstone Harbour revealed that "aspects of the design and construction of the bund wall were not consistent with industry best practice" and that "While considerable water quality monitoring was undertaken in Gladstone Harbour, the location of monitoring sites was inadequate to provide attribution of changes in turbidity during construction and dredging operations" (Aust. Govt., 2014).

Development approvals for major infrastructure projects involve the preparation of Environmental Impact Statements (EIS) by proponents, with assessment by the Queensland and Australian governments. Yet EIS requirements are neither rigorous nor standardized (Sheaves et al., 2015). Of particular concern is the inadequacy of cumulative impact assessment, which, although required as part of an EIS, is not well understood or adequately assessed (Hockings et al., 2014). A number of authors have called for independent, scientific review of EIS documentation and monitoring design (Grech et al., 2013; Hughes et al., 2015; Sheaves et al., 2015), and propose an independent, collaborative, and transparent program to provide more comprehensive and ongoing cumulative impact assessment (Grech et al., 2016; Jacobs, 2014).

Climate change

The GBR Outlook Report 2014 (GBRMPA, 2014) defines climate-change impacts as the single largest threat to the GBR ecosystem. Better water quality improves the resilience of the GBR ecosystem to cyclone damage, CoTS outbreaks, and the impacts of increased temperature and ocean acidification. For example, by 2018 ocean temperature increases are expected to cause coral bleaching events twice per decade, and 2035 has been used as a critical climate-change timeline within regional planning (Terrain NRM, 2015). Beyond 2035, the influence of water-quality improvement in the context of other drivers of GBR health, such as climate change, is difficult to predict. However, in light of the severe bleaching occurring on the GBR in 2016 (GBRMPA, 2016; Normile, 2016) the time frame to provide better water quality to support GBR resilience has been recently revised from 2035 to 2025 (GBRWST, 2016; NQ Dry Tropics, 2016).

CONCLUSIONS

The outlook for the GBR is dire. The most recent (2016) bleaching event demonstrates the speed and scale at which climate change can significantly impact the condition of the GBR. On the one hand, compared to many environmental issues in Australia, the GBR is a story of successful collaboration linking on-ground partnerships, strategic alliances between NRM, agriculture and conservation, and bilateral commitment to saving this global icon. But it is now clear that, despite the success of scaled up voluntary measures, with strong scientific support to increase the adoption of agricultural best management practices, this will not be sufficient to achieve the pollutant load reductions required to support the GBR in the face of climate change.

Governments have failed to take a stronger approach and use the other policy levers available to them, such as regulation and land-use change. There has been no strategic response to increasing coastal development pressures and a flawed development assessment system. There are many existing and complementary policy measures in land, water and vegetation management and coastal planning that have not been employed. The core bilateral planning documents remain weak, without strong mandate or sufficient investment. Climate change remains the "elephant in the room"—without rapid global action to reduce greenhouse gas emissions, the reef will continue to rapidly degrade.

REFERENCES

AAS. (2014). *Response to the draft Reef 2050 long-term sustainability plan.* Canberra: Australian Academy of Science.

ANZECC. (2000). *Australian water quality guidelines for fresh and marine waters.* Canberra: Australian and New Zealand Environment and Conservation Council.

Aust. Govt. (2002). *The framework for marine and estuarine water quality protection: A reference document.* Canberra: Australian Government.

Aust. Govt. (2014). *Independent review of the bund wall at the port of Gladstone report on findings—April 2014.* Canberra: Australian Government.

Aust. Govt. (2015a). *Reef 2050 long-term sustainability plan.* Canberra: Australian Government.

Aust. Govt. (2015b). *State party report on the state of conservation of the Great Barrier Reef World Heritage Area (Australia)—2015.* Canberra: Australian Government.

Aust. Govt. & Qld. Govt. (2005). *Implementation of the reef water quality protection plan: Report to the Prime Minister and the Premier of Queensland 2005. Progress to date, challenges and future directions.* Canberra: Australian and Queensland Governments.

Aust. Govt. & Qld. Govt. (2011). *Great Barrier Reef first report card—2009 baseline.* Brisbane: Australian and Queensland Governments.

Aust. Govt. & Qld. Govt. (2013). *Great Barrier Reef report card 2011, reef water quality protection plan.* Brisbane: Australian and Queensland Governments.

Aust. Govt. & Qld. Govt. (2016a). *Great Barrier Reef report card 2015, reef water quality protection plan.* Brisbane: Australian and Queensland Governments.

Aust. Govt. & Qld. Govt. (2016b). *Reef 2050 plan—Annual report and implementation strategy.* Canberra: Australian and Queensland Governments.

Aust. Govt. & Qld. Govt. (2016c). *Great Barrier Reef report card 2015.* Brisbane: Australian and Queensland Governments.

Aust. Govt. & Qld. Govt. (2016d). Reef 2050 plan—Update on progress. In: *Presented to the UNESCO World Heritage Centre and IUCN*. Canberra: Australian and Queensland Governments.

Baker, J. (2003). *A report on the study of land-sourced pollutants and their impacts on water quality in and adjacent to the Great Barrier Reef*. Brisbane: Queensland Government.

Barson, M., Randall, L., Gale, K., & Leslie, R. (2014). *Reef water quality protection plan 2013—Prioritisation project report*. Canberra: Australian Government.

Bartley, R., Corfield, J. P., Hawdon, A. A., Kinsey-Henderson, A. E., Abbott, B. N., Wilkinson, S. N., et al. (2014). Can changes to pasture management reduce runoff and sediment loss to the Great Barrier Reef? The results of a 10-year study in the Burdekin catchment, Australia. *Rangeland Journal, 36*, 67–84.

Baxter, T. (2006). Legal protection for the Great Barrier Reef World Heritage Area. *Macquarie Journal of International and Comparative Environmental Law, 3*, 67–82.

BMRG. (2015). *Water quality improvement plan for the Burnett Mary*. Bundaberg: Burnett Mary Regional Group.

Bohnet, I. C. (2014). Lessons learned from public participation in water quality improvement planning: A study from Australia. *Society & Natural Resources, 28*, 180–196.

Brodie, J. (2014). Dredging the Great Barrier Reef: Use and misuse of science. *Estuarine, Coastal and Shelf Science, 142*, 1–3.

Brodie, J., Binney, J., Fabricius, K., Gordon, I., Hoegh-Guldberg, O., Hunter, H., et al. (2008). *Scientific consensus statement on water quality in the Great Barrier Reef*. Brisbane: Queensland Government.

Brodie, J., Christie, C., Devlin, M., Haynes, D., Morris, S., Ramsay, M., et al. (2001a). Catchment management and the Great Barrier Reef. *Water Science and Technology, 43*, 203–211.

Brodie, J., Furnas, M., Ghonim, S., Haynes, D., Mitchell, A., Morris, S., et al. (2001b). *Great Barrier Reef catchment water quality action plan*. Townsville: Great Barrier Reef Marine Park Authority.

Brodie, J., Fabricius, K., De'ath, G., & Okaji, K. (2005). Are increased nutrient inputs responsible for more outbreaks of crown-of-thorns starfish? An appraisal of the evidence. *Marine Pollution Bulletin, 51*, 266–278.

Brodie, J. E., Kroon, F. J., Schaffelke, B., Wolanski, E. C., Lewis, S. E., Devlin, M. J., et al. (2012). Terrestrial pollutant runoff to the Great Barrier Reef: An update of issues, priorities and management responses. *Marine Pollution Bulletin, 65*, 81–100.

Brodie, J. E., Lewis, S. E., Collier, C. J., Wooldridge, S., Bainbridge, Z. T., Waterhouse, J., et al. (2017). Setting ecologically relevant targets for river pollutant loads to meet marine water quality requirements for the Great Barrier Reef, Australia: A preliminary methodology and analysis. *Ocean & Coastal Management*, [in press].

Brodie, J., & Pearson, R. G. (2016). Ecosystem health of the Great Barrier Reef—Time for effective management action based on evidence. *Estuarine, Coastal and Shelf Science, 183*(B), 438–451.

Brodie, J., & Waterhouse, J. (2012). A critical review of environmental management of the 'not so Great' Barrier Reef. *Estuarine, Coastal and Shelf Science, 104–105*, 1–22.

Brodie, J., & Waterhouse, J. (2016). The Great Barrier Reef (Australia): A multi-ecosystem wetland with a multiple use management regime. In C. M. Finlayson, G. R. Milton, R. C. Prentice, & N. C. Davidson (Eds.), *The wetland book: II: Distribution, description and conservation* (pp. 1–14). Dordrecht: Springer.

Brodie, J. E., Waterhouse, J., Maynard, J. A., Bennett, J., Furnas, M. J., Devlin, M. J., et al. (2013a). *Assessment of the relative risk of water quality to ecosystems of the Great Barrier Reef: Report 13/28 by TropWATER (James Cook University) to the QLD Department of the Environment and Heritage Protection, Townsville.* .

Brodie, J., Waterhouse, J., Schaffelke, B., Kroon, F., Fabricius, K., Lewis, S., et al. (2013b). *2013 scientific consensus statement: Land use impacts on Great Barrier Reef water quality and ecosystem condition*. Brisbane: Queensland Government.

Cape York NRM & SCYC. (2016). *Eastern Cape York water quality improvement plan*. Cooktown: Cape York Natural Resource Management and South Cape York Catchments.

Cash, D., Clark, W. C., Alcock, F., Dickson, N. M., Eckley, N., & Jäger, J. (2002). *Salience, credibility, legitimacy and boundaries: Linking research, assessment and decision making: RWP02-046*. Cambridge, MA: John F. Kennedy School of Government, Harvard University.

Cherry, K., Shepherd, M., Withers, P., & Mooney, S. (2008). Assessing the effectiveness of actions to mitigate nutrient loss from agriculture: A review of methods. *Science of the Total Environment, 406*, 1–23.

Clark, T. R., Leonard, N. D., Zhao, J. -X., Brodie, J., McCook, L. J., Wachenfeld, D. R., et al. (2016). Historical photographs revisited: A case study for dating and characterizing recent loss of coral cover on the inshore Great Barrier Reef. *Scientific Reports, 6*, 19285.

Collier, C., Waycott, M., & McKenzie, L. (2012). Light thresholds derived from seagrass loss in the coastal zone of the northern Great Barrier Reef, Australia. *Ecological Indicators, 23*, 211–219.

Connor, J. D., MacDonald, D. H., Morrison, M., & Cast, A. (2009). Evaluating policy options for managing diffuse source water quality in Lake Taupo, New Zealand. *The Environmentalist, 29*, 348–359.

Cotsell, P., Gale, K., Hajkowicz, S., Lesslie, R., Marshall, N., & Randall, L. (2009). *Use of a multiple criteria analysis (MCA) process to inform reef rescue regional allocations*. Townsville: Reef and Rainforest Research Centre Limited.

Dale, A., McKee, J., Vella, K., & Potts, R. (2013). Carbon, biodiversity and regional natural resource planning: Towards high impact next generation plans. *Australian Planner, 50*, 328–339.

Dale, A., Vella, K., Pressey, R. L., Brodie, J., Gooch, M., Potts, R., et al. (2016). Risk analysis of the governance system affecting outcomes in the Great Barrier Reef. *Global Environmental Change, 183*(3), 712–721.

Davis, A. M., Pearson, R. G., Brodie, J. E., & Butler, B. (2016). Review and conceptual models of agricultural impacts and water quality in waterways of the Great Barrier Reef catchment area. *Marine and Freshwater Research, 68*, 1–19.

Davis, A. M., & Pradolin, J. (2016). Precision herbicide application technologies to decrease herbicide losses in furrow irrigation outflows in a northeastern Australian cropping system. *Journal of Agricultural and Food Chemistry, 64*, 4021–4028.

Day, J. C., & Dobbs, K. (2013). Effective governance of a large and complex cross-jurisdictional marine protected area: Australia's Great Barrier Reef. *Marine Policy, 41*, 14–24.

De'ath, G., Fabricius, K. E., Sweatman, H., & Puotinen, M. (2012). The 27-year decline of coral cover on the Great Barrier Reef and its causes. *Proceedings of the National Academy of Sciences, 109*, 17995–17999.

De'ath, G., & Fabricius, K. (2010). Water quality as a regional driver of coral biodiversity and macroalgae on the Great Barrier Reef. *Ecological Applications, 20*, 840–850.

Deloitte Access Economics. (2013). *Economic contribution of the Great Barrier Reef: Report to Great Barrier Reef Marine Park Authority, Townsville*. .

Devlin, M., Lewis, S., Davis, A., Smith, R., Negri, A., Thompson, M., et al. (2015). *Advancing our understanding of the source, management, transport and impacts of pesticides on the Great Barrier Reef 2011–2015: A report by James Cook University for the Queensland Department of Environment and Heritage Protection, Townsville (134 pp.)*.

Devlin, M., McKinna, L., Alvarez-Romero, J., Petus, C., Abott, B., Harkness, P., et al. (2012). Mapping the pollutants in surface riverine flood plume waters in the Great Barrier Reef, Australia. *Marine Pollution Bulletin, 65*, 224–235.

Di Bella, L., Stacey, S., Benson, A., Royle, A., & Holzberger, G. (2013). An assessment of controlled release fertiliser in the Herbert cane growing region. *International Sugar Journal, 115*, 784–788.

Eberhard, R. (2010). *Reef alliance health check 2010: A rapid evaluation of the current partnership arrangements and future prospects*. Brisbane: Eberhard Consulting.

Eberhard, R., Johnston, N., & Everingham, J. A. (2013). A collaborative approach to address the cumulative impacts of mine-water discharge: Negotiating a cross-sectoral waterway partnership in the Bowen Basin, Australia. *Resources Policy, 38*, 678–687.

Emtage, N., & Herbohn, J. (2012). Implications of landholders' management goals, use of information and trust of others for the adoption of recommended practices in the Wet Tropics region of Australia. *Landscape and Urban Planning, 107*, 351–360.

Fabricius, K., De'ath, G., McCook, L., Turak, E., & Williams, D. M. (2005). Changes in algal, coral and fish assemblages along water quality gradients on the inshore Great Barrier Reef. *Marine Pollution Bulletin, 51*, 384–398.

Fabricius, K., Logan, M., Weeks, S., & Brodie, J. (2014). The effects of river run-off on water clarity across the central Great Barrier Reef. *Marine Pollution Bulletin, 84*, 191–200.

Fabricius, K., Logan, M., Weeks, S., Lewis, S., & Brodie, J. (2016). Changes in water clarity in response to river discharges on the Great Barrier Reef continental shelf: 2002–2013. *Estuarine, Coastal and Shelf Science, 173*, A1–A15.

Fabricius, K., Okaji, K., & De'ath, G. (2010). Three lines of evidence to link outbreaks of the crown-of-thorns seastar *Acanthaster planci* to the release of larval food limitation. *Coral Reefs, 29*, 593–605.

FBA. (2015). *Fitzroy water quality improvement plan.* Rockhampton: Fitzroy Basin Association.*http://www.fba. org.au/wqip*.

Folkers, A., Rohde, K., Delaney, K., & Flett, I. (2015). *Mackay Whitsunday Water Quality Improvement Plan 2014–2021.* Canberra: Reef Catchments and Australian Government.

Gardner, A., & Waschka, M. (2012). Using regulation to tackle the challenge of diffuse water pollution and its impact on the Great Barrier Reef: The challenge of managing diffuse source pollution from agriculture: GBR case study and Queensland's reef protection legislation: An analysis. *Australasian Journal of Natural Resources Law and Policy, 15*, 109.

GBRMPA. (2005). *Great Barrier Reef Marine Park heritage strategy.* Townsville: Great Barrier Marine Park Authority.

GBRMPA. (2014). *Great Barrier Reef outlook report 2014.* Townsville: Great Barrier Reef Marine Park Authority.

GBRMPA. (2016). *Coral bleaching.* Townsville: Great Barrier Marine Park Authority.

GBRWST. (2016). *Great Barrier Reef water science taskforce final report.* Brisbane: Great Barrier Reef Water Science Taskforce.

Grech, A., Bos, M., Brodie, J., Coles, R., Dale, A., Gilbert, R., et al. (2013). Guiding principles for the improved governance of port and shipping impacts in the Great Barrier Reef. *Marine Pollution Bulletin, 75*, 8–20.

Grech, A., Pressey, R. L., & Day, J. C. (2016). Coal, cumulative impacts, and the Great Barrier Reef. *Conservation Letters, 9*, 200–207.

Greiner, R., Herr, A., Brodie, J., & Haynes, D. (2005). A multi-criteria approach to Great Barrier Reef catchment (Queensland, Australia) diffuse-source pollution problem. *Marine Pollution Bulletin, 51*, 128–137.

Haapkylä, J., Unsworth, R. K., Flavell, M., Bourne, D. G., Schaffelke, B., & Willis, B. L. (2011). Seasonal rainfall and runoff promote coral disease on an inshore reef. *PLoS One, 6*, e16893.

Haynes, D., & Michalek-Wagner, K. (2000). Water quality in the Great Barrier Reef World Heritage Area: Past perspectives, current issues and new research directions. *Marine Pollution Bulletin, 41*, 428–434.

Hill, R., Davies, J., Bohnet, I., Robinson, C., Maclean, K., & Pert, P. (2015). Collaboration mobilises institutions with scale-dependent comparative advantage in landscape-scale biodiversity conservation. *Environmental Science & Policy, 51*, 267–277.

Hockings, M., Leverington, A., Trinder, C., & Polglaze, J. (2014). *Independent assessment of management effectiveness for the Great Barrier Reef outlook report 2014.* Townsville: Great Barrier Reef Marine Park Authority.

Hoegh-Guldberg, O., Mumby, P. J., Hooten, A. J., Steneck, R. S., Greenfield, P., Gomez, E., et al. (2007). Coral reefs under rapid climate change and ocean acidification. *Science, 318*(5857), 1737–1742.

Hughes, T. P., Baird, A. H., Bellwood, D. R., Card, M., Connolly, S. R., Folke, C., et al. (2003). Climate change, human impacts, and the resilience of coral reefs. *Science, 301*(5635), 929–933.

Hughes, T. P., Day, J. C., & Brodie, J. (2015). Securing the future of the Great Barrier Reef. *Nature Climate Change, 5*, 508–511.

Jacobs. (2014). *Institutional and legal mechanisms that provide coordinated planning, protection and management of the Great Barrier Reef World Heritage Area*. Brisbane: Jacobs.

Keogh, K., Chant, D., & Frazer, B. (2006). *Review of arrangements for regional delivery of natural resource management programs*. Canberra: Commonwealth of Australia.

Kerr, R. (2013). *Community draft environmental values for the waters of the Burdekin dry tropics*. Townsville: NQ Dry Tropics.

King, J., Alexander, F., & Brodie, J. (2013). Regulation of pesticides in Australia: The Great Barrier Reef as a case study for evaluating effectiveness. *Agriculture, Ecosystems & Environment, 180*, 54–67.

Kroon, F. J. (2012). Towards ecologically relevant targets for river pollutant loads to the Great Barrier Reef. *Marine Pollution Bulletin, 65*, 261–266.

Kroon, F. J., Robinson, C. J., & Dale, A. P. (2009). Integrating knowledge to inform water quality planning in the Tully-Murray basin, Australia. *Marine and Freshwater Research, 60*, 1183.

Kroon, F. J., Thorburn, P., Schaffelke, B., & Whitten, S. (2016). Towards protecting the Great Barrier Reef from land-based pollution. *Global Change Biology, 22*, 1985–2002.

Lebel, L., Anderies, J. M., Campbell, B., Folke, C., Hatfield-Dodds, S., Hughes, T. P., et al. (2006). Governance and the capacity to manage resilience in regional social-ecological systems. *Ecology and Society, 11*, 19.

Lewis, S., Smith, R., O'Brien, D., Warne, M. S. J., Negri, A., Petus, C., et al. (2014a). *Assessment of the relative risk of water quality to ecosystems of the Great Barrier Reef: Supporting studies: A report by James Cook University to the QLD Department of the Environment and Heritage Protection, Townsville*. .

Lewis, S. E., Olley, J., Furuichi, T., Sharma, A., & Burton, J. (2014b). Complex sediment deposition history on a wide continental shelf: Implications for the calculation of accumulation rates on the Great Barrier Reef. *Earth and Planetary Science Letters, 393*, 146–158.

Lewis, S. E., Shields, G. A., Kamber, B. S., & Lough, J. M. (2007). A multi-trace element coral record of land-use changes in the Burdekin River catchment, NE Australia. *Palaeogeography, Palaeoclimatology, Palaeoecology, 246*, 471–487.

Lønborg, C., Devlin, M., Waterhouse, J., Brinkman, R., Costello, P., da Silva, E., et al. (2016). *Marine monitoring program: Annual report for inshore water quality monitoring: A report by Australian Institute of Marine Science and JCU TropWATER for the Great Barrier Reef Marine Park Authority, Townsville*. .

Marsh, H., De'Ath, G., Gribble, N., & Lane, B. (2005). Historical marine population estimates: Triggers or targets for conservation? The dugong case study. *Ecological Applications, 15*, 481–492.

Marsh, S. P., & Pannell, D. (2000). Agricultural extension policy in Australia: The good, the bad and the misguided. *Australian Journal of Agricultural and Resource Economics, 44*, 605–627.

McGrath, C. (2012). UNESCO/IUCN reactive monitoring mission report on the Great Barrier Reef. *Australian Environment Review, 27*, 253–257.

McKenzie, L., Collier, C., Langlois, L., Yoshida, R., Smith, N., Takahashi, M., et al. (2015). *Marine monitoring program: Inshore seagrass, annual report for the sampling period June 2013 to May 2014*. Townsville: Great Barrier Reef Marine Park Authority.

Normile, D. (2016). El Niño's warmth devastating reefs worldwide. *Science, 352*(6281), 15–16.

NQ Dry Tropics. (2016). *Burdekin region Water Quality Improvement Plan. Better water quality for the Burdekin*. Townsville: NQ Dry Tropics.

O'Brien, D., Lewis, S., Davis, A., Gallen, C., Smith, R., Turner, R., et al. (2016). Spatial and temporal variability in pesticide exposure downstream of a heavily irrigated cropping area: Application of different monitoring techniques. *Journal of Agricultural and Food Chemistry, 64*, 3975–3989.

Oliver, D. P., Anderson, J. S., Davis, A., Lewis, S., Brodie, J., & Kookana, R. (2014). Banded applications are highly effective in minimising herbicide migration from furrow-irrigated sugar cane. *Science of the Total Environment, 466*, 841–848.

Olley, J., Brooks, A., Spencer, J., Pietsch, T., & Borombovits, D. (2013). Subsoil erosion dominates the supply of fine sediment to rivers draining into Princess Charlotte Bay, Australia. *Journal of Environmental Radioactivity, 124*, 121–129.

Pannell, D. J., Roberts, A. M., Park, G., Curatolo, A., Marsh, S., & Alexander, J. (2009). *INFFER (Investment Framework for Environmental Resources), INFFER working paper 0901*. Perth: University of Western Australia.

Park, G., & Roberts, A. (2014). *Wet tropics WQIP—INFFER assessment*. Newstead, Australia: Natural Decisions.

Peterson, A., Walker, M., Maher, M., Hoverman, S., & Eberhard, R. (2010). New regionalism and planning for water quality improvement in the Great Barrier Reef, Australia. *Geographical Research, 48*, 297–313.

Petus, C., Collier, C., Devlin, M., Rasheed, M., & McKenna, S. (2014). Using MODIS data for understanding changes in seagrass meadow health: A case study in the Great Barrier Reef (Australia). *Marine Environmental Research, 98*, 68–85.

Potts, R., Vella, K., Dale, A., & Sipe, N. (2016). Evaluating governance arrangements and decision making for natural resource management planning: An empirical application of the governance systems analysis framework. *Society & Natural Resources, 29*, 1325–1341.

Productivity Commission. (2003). *Industries, land use and water quality in the Great Barrier Reef catchment*. Canberra: Commonwealth of Australia.

Qld. Govt. & Aust. Govt. (2003). *Reef water quality protection plan for catchments adjacent to the Great Barrier Reef world heritage*. Brisbane: Queensland and Australian Governments.

Qld. Govt. & Aust. Govt. (2009). *Reef water quality protection plan for the Great Barrier Reef World Heritage area and adjacent catchments*. Brisbane: Queensland Department of Premier and Cabinet.

Qld. Govt. & Aust. Govt. (2013). *Reef water quality protection plan 2013: Securing the health and resilience of the Great Barrier Reef World Heritage Area and adjacent catchments*. Brisbane: Queensland and Australian Governments.

Reymond, C. E., Roff, G., Chivas, A. R., Zhao, J. -X., & Pandolfi, J. M. (2013). Millennium-scale records of benthic foraminiferal communities from the central Great Barrier Reef reveal spatial differences and temporal consistency. *Palaeogeography, Palaeoclimatology, Palaeoecology, 374*, 52–61.

Roberts, A., Park, G., & Dickson, M. (2016). *NQ dry tropics WQIP—INFFER assessment*. Newstead, Australia: Natural Decisions.

Robins, L., & de Loë, R. (2009). Decentralised governance for natural resource management: Capacity challenges in Australia and Canada. In M. Lane, C. J. Robinson, & B. Taylor (Eds.), *Contested country: Local and regional natural resources management in Australia* (pp. 179–197). Collingwood: CSIRO.

Robins, L., & Dovers, S. (2007a). Community-based NRM boards of management: Are they up to the task? *Journal of Environmental Management, 14*, 111–122.

Robins, L., & Dovers, S. (2007b). NRM regions in Australia: The "haves" and the "have nots". *Geographical Research, 45*, 273–290.

Robins, L., & Kanowski, P. (2011). 'Crying for our country': Eight ways in which 'caring for our country' has undermined Australia's regional model for Natural Resource Management. *Australasian Journal of Environment, 18*, 88–108.

Robinson, C. J., Eberhard, R., Wallington, T., & Lane, M. (2010). *Using knowledge to make collaborative policy-level decisions in Australia's Great Barrier Reef: CSIRO Water for a Healthy Country Technical Report, Brisbane*. http://www/.clw.csiro.au/publications/waterforahealthycountry/2010/wfhc-GBR-collaborative-decisions.pdf [Accessed 4 August 2010].

Robinson, C., Taylor, B., & Vella, K. (2011). *An assessment of integrated regional planning arrangements for water quality improvement in the Wet Tropics Region: Marine and Tropical Sciences Research Facility (MTSRF) transition project final report, Townsville.* .

Roff, G., Clark, T. R., Reymond, C. E., Zhao, J. -X., Feng, Y., McCook, L. J., et al. (2012). Palaeoecological evidence of a historical collapse of corals at Pelorus Island, inshore Great Barrier Reef, following European settlement. *Proceedings of the Royal Society B: Biological Sciences, 280*(1750), 2012–2100.

Schaffelke, B., Anthony, K., Blake, J., Brodie, J., Collier, C., Devlin, M., et al. (2013). *Marine and coastal ecosystem impacts. Synthesis of evidence to support the reef water quality scientific consensus statement 2013*. Brisbane: Department of the Premier and Cabinet, Queensland Government.

Senate Standing Committee. (2010). *Evaluation of current governance arrangements to support regional investment under the NHT and NAP*. Canberra: Senate Standing Committee on Rural and Regional Affairs and Transport Commonwealth of Australia.

Sheaves, M., Coles, R., Dale, P., Grech, A., Pressey, R. L., & Waltham, N. J. (2015). Enhancing the value and validity of EIA—Serious science to protect Australia's Great Barrier Reef. *Conservation Letters, 9,* 377–383.

Stark, C. H., & Richards, K. G. (2008). The continuing challenge of nitrogen loss to the environment: Environmental consequences and mitigation strategies. *Dynamic Soil, Dynamic Plant, 2,* 41–55.

Stoeckl, N., Farr, M., Larson, S., Adams, V. M., Kubiszewski, I., Esparon, M., et al. (2014). A new approach to the problem of overlapping values: A case study in Australia's Great Barrier Reef. *Ecosystem Services, 10,* 61–78.

Taylor, B. M., & Lawrence, G. A. (2012). Agri-political organisations in environmental governance: Recalcitrant participants or active partners? *Journal of Environmental Policy & Planning, 14,* 337–359.

Taylor, B., & van Grieken, M. (2015). Local institutions and farmer participation in agri-environmental schemes. *Journal of Rural Studies, 37,* 10–19.

Taylor, B., Wallington, T., & Robinson, C. J. (2010). *Risk and social theory in environmental management.* Melbourne: CSIRO Publishing [pp. 161–174].

Tennent, R., & Lockie, S. (2013). Vale Landcare: The rise and decline of community-based Natural Resource Management in rural Australia. *Journal of Environmental Planning and Management, 56,* 572–587.

Terrain NRM. (2012). *Environmental values for the wet tropics basins.* Innisfail, Australia: Terrain NRM.

Terrain NRM. (2015). *Wet tropics Water Quality Improvement Plan 2015–2020.* Innisfail: Terrain NRM.

Thorburn, P. J., & Wilkinson, S. N. (2012). Conceptual frameworks for estimating the water quality benefits of improved agricultural management practices in large catchments. *Agriculture, Ecosystems & Environment, 180,* 192–209.

Thorburn, P., Wilkinson, S., & Silburn, D. (2013). Water quality in agricultural lands draining to the Great Barrier Reef: A review of causes, management and priorities. *Agriculture, Ecosystems & Environment, 180,* 4–20.

Turnbull, W. (2005). *Evaluation of current governance arrangements to support regional investment under the NHT and NAP.* Canberra: Departments of Environment, Water, Heritage and the Arts, and Agriculture Forestry and Fisheries.

Van Grinsven, H., Ten Berge, H., Dalgaard, T., Fraters, B., Durand, P., Hart, A., et al. (2012). Management, regulation and environmental impacts of nitrogen fertilization in Northwestern Europe under the nitrates directive: A benchmark study. *Biogeosciences, 9,* 5143–5160.

Vella, K., Dale, A., Reghenzani, J., Sing, N., & Parker, D. (2011). An approach for adaptive and integrated agricultural planning to deal with uncertainty in a Great Barrier Reef Catchment. *Australian Planner, 51,* 243–259.

Vella, K., & Sipe, N. (2014). The evolving landscape of natural resource planning and governance in Australia. In: *Association of Collegiate Schools of Planning Conference, Philadelphia, Pennsylvania, 30 October–2 November 2014.*

Vella, K., Sipe, N., Dale, A., & Taylor, B. (2015). Not learning from the past: Adaptive governance challenges for Australian natural resource management. *Geographical Research, 53,* 379–392.

Waterhouse, J., Brodie, J., Lewis, S., & Audas, D. M. (2016a). Land-sea connectivity, ecohydrology and holistic management of the Great Barrier Reef and its catchments: Time for a change. *Ecohydrology & Hydrobiology, 16,* 45–57.

Waterhouse, J., Duncanson, P., Attard, S., Bartley, R., Bristow, K., Brodie, J., et al. (2016b). *Spatial prioritisation and extension of information to manage sediment, nutrient and pesticide runoff in the Burdekin NRM region: A report by NQ Dry Tropics for the QLD Department of Environment and Heritage Protection, Townsville.*

Waterhouse, J., Brodie, J., Lewis, S., & Mitchell, A. (2012). Quantifying the sources of pollutants in the Great Barrier Reef catchments and the relative risk to reef ecosystems. *Marine Pollution Bulletin, 65,* 394–406.

Waters, D. K., Carroll, C., Ellis, R., Hateley, L., McCloskey, G. L., Packett, R., et al. (2014). *Modelling reductions of pollutant loads due to improved management practices in the Great Barrier Reef catchments—Whole of GBR: Technical report (Vol. 1, p. 120).* Toowoomba: QLD Department of Natural Resources and Mines.

Wilkinson, S. N., Hancock, G. J., Bartley, R., Hawdon, A. A., & Keen, R. J. (2013). Using sediment tracing to assess processes and spatial patterns of erosion in grazed rangelands, Burdekin River basin, Australia. *Agriculture, Ecosystems & Environment, 180,* 90–102.

Wooldridge, S. A. (2009). Water quality and coral bleaching thresholds: Formalising the linkage for the inshore reefs of the Great Barrier Reef, Australia. *Marine Pollution Bulletin, 58,* 745–751.

Wooldridge, S. A. (2017a). Excess seawater nutrients, enlarged algal symbiont densities and bleaching sensitive reef locations: 1. Identifying thresholds of concern for the Great Barrier Reef, Australia. *Marine Pollution Bulletin,* [in press].

Wooldridge, S. A. (2017b). Preventable fine sediment export from the Burdekin River catchment reduces coastal seagrass abundance and increases dugong mortality within the Townsville region of the Great Barrier Reef, Australia. *Marine Pollution Bulletin, 114,* 671–678.

TRANSFER OF THE KNOWLEDGE— INTERNATIONAL CASE STUDIES

KNOWLEDGE TRANSFER IN INTERNATIONAL WATER RESOURCE MANAGEMENT—SIX CHALLENGES

16

I. Campbell*, C. Barlow†

Rhithroecology Pty Ltd, South Balckburn, VIC, Australia Australian Centre for International Agricultural Research, Canberra, ACT, Australia†

CHAPTER OUTLINE

INTRODUCTION

In this chapter we crystallize key challenges we have encountered in using the understanding and expertise developed in Australia to assist water-resource managers in developing regions, the lower Mekong Basin in southeast Asia and elsewhere. Much of this is based on our experience of working in a multijurisdictional agency, the Mekong River Commission, as well as being engaged in and funding many research and aid projects in that region and elsewhere.

Prior to our Mekong experience we were involved in aspects of water resources management in Australia. Although infrastructure development commenced more recently in Australia than in Europe or North America, water resources management has been a major Australian concern, and was one of the drivers of the formation of the Australian Federation from six independent British colonies, and the establishment of the River Murray Commission in 1917 (Pigram, 2007; Royal Commission, 1902; Chapters 1 and 13).

The transfer and sharing of knowledge and expertise is a central pillar of science, but natural resource management is complex in that it operates at the interface between biophysical sciences, such as ecology, hydrology, and geology, and human societies. Water resources have been referred to as social-ecological systems (Ostrom, 2009). So, whereas sciences such as physics, chemistry, and mathematics can easily be transferred across geographic and cultural boundaries, that is not necessarily the case for natural resource management.

The transfer of ecological knowledge between biogeographical regions has been a substantial topic of discussion among ecologists over the years, with publications with titles such as "Are Australian ecosystems different?" (Dodson & Westoby, 1985), and "Are New Zealand stream ecosystems really different?" (Winterbourn et al., 1981) contributing to the discussion. For natural resource management, there is not only the complexity of transferring knowledge to an ecologically different location, but also the challenges of working in a politically and culturally different environment with the challenges and opportunities that presents. Indeed, we consider that the influence of politics and culture is a greater determinant of the success or otherwise of transfer of knowledge on natural resource management than are differences in biogeographical environments.

There are two common modes of knowledge transfer. The first is the transfer of specific skills, and the second is the transfer of more general experience. Specific skills transfer is the more straightforward, and usually arises as a request from the recipient organization or a third party such as an organization funding a development or aid project. Examples include training exercises in methods for ecological risk assessment, environmental flows techniques or fishery assessment methods. It is advantageous if the training exercises can incorporate local examples, and those running the training should be aware that the techniques may not be applicable in the circumstances of the recipient regions, even where they have been specifically requested.

The transfer of more general experience is more nuanced. It is generally an explanation of a series of activities that had been carried out in the developed country, an evaluation of the success, an identification of the difficulties, and an explanation of why particular approaches employed were selected. These types of transfers are generally better received when they take the form of discussing "what we did and why" rather than "what you should do." Rarely do those talking of their own experience have sufficient understanding and appreciation of the host situation to be prescriptive about lessons the host should draw and act on.

There are a number of well-recognized fields of potential conflict in natural resource management. One is between the goals of individuals or individual corporations and the broader community or the nation, and another is between short-term benefit and long-term sustainability of natural resources. In many relatively undeveloped countries, the political, social, and cultural imperatives may clash with long-term sustainable management of natural resources. Governments have poverty alleviation as their primary goal as a key social necessity as well as a necessity for their political survival. That may mean that wise environmental management is considered a lower priority, with polluting industries or excessive short-term exploitation of natural resources being accepted as part of the cost of improving the economic and social condition of the population.

In the remainder of this chapter, we discuss six key challenges to the transfer of water resources management knowledge between developed and less-developed regions, based mainly on our experience in the Mekong River Basin and elsewhere.

DIFFERENCES IN BIOPHYSICAL ENVIRONMENT

This is the most obvious difference, and the one that is most likely to be taken into account by those taking any sort of natural resource management between different global regions. Australian water resources differ from those in most other places because of the dry climate and the great variability in annual discharge displayed by most Australian rivers (e.g., McMahon et al., 1992). The entire Australian continent is highly prone to drought. Hydrologically, Australia is somewhat similar to southern Africa and California, but quite different from most other regions. This has important consequences for design and operation of water storages, which must be considerably larger in Australia to give the same security of supply as storages in regions with less variable rainfall and runoff.

Biologically, Australia differs from all other regions. The Australian freshwater fish fauna has few species (256—Unmack, 2013) compared with North America (about 1200—Page & Burr, 1991) or Asia (4411—Lêvêque et al., 2008, using data from Fishbase 2005). Australian freshwater fish species are more likely to be generalists, and perhaps less-sensitive ecological indicators than, for example, those of the Mekong Basin where there are approximately a thousand species in the single river basin (Valbo-Jorgensen et al., 2009). Furthermore, the Australian freshwater fish fauna is taxonomically distinctive (Humphries & Walker, 2013).

The Australian freshwater invertebrate fauna is also distinctive as is the vegetation, including the riparian vegetation. Williams and Campbell (1987) noted four broad components in the freshwater invertebrate fauna, a Gondwanan component, taxa with their closest relatives in South America, taxa of tropical-Asian origin, cosmopolitan taxa and a unique Australian group. Australian freshwater invertebrate assemblages are distinctive from those on other continents at the family level. The Australian vegetation, generally considered as having ancient origins, is about 30% endemic at the level of genus with a further 13% restricted to Australia and adjacent islands or Australia and South America (Specht, 1981). The riparian and floodplain vegetation is no less distinctive than that not dependent on fresh water (e.g., see Capon et al., 2016).

DIFFERENCES IN KNOWLEDGE STATUS

In developed countries, the documented scientific knowledge base is usually more substantial and more accessible than in developing regions. This knowledge base generally includes traditional, observational knowledge from previous generations, which has been amalgamated and rationalized with quantified, tested hypotheses explaining ecological functions, at least to the extent possible in inherently variable ecological systems. Conversely, the status of ecological knowledge in developing countries is still largely based on traditional knowledge of those who exploit the resources for their daily living. Often that knowledge has not yet been compiled and made widely available within the country, and incorporating that knowledge base with scientific observation and recording it is only just starting in many developing countries.

In addition, the knowledge base is more readily extended in developed economies, primarily because of the familiarity with English as the lingua franca. In the lower Mekong Basin, for example, there are four different countries, each with their own language, and each language with its own alphabet. Unlike Europe, where there are also many countries and languages, but a common alphabet, there are relatively few citizens who speak the language of a neighboring country, although there are rather more who speak their native language and English. The lack of common language skills is a substantial barrier to communication between developing countries, and the lack of strong English language skills is a barrier to accessing the international technical literature by natural resource managers in developing countries.

It is often difficult to discover what data and information is held within a country. A multitude of organizations, such as government agencies or research centers, nongovernment agencies or universities, may hold relevant data sets and documents. Frequently such documents or data sets are not indexed in accessible databases, so even discovering their existence can be problematic. When their existence is uncovered, the documents still must be located. If located they will often be in the local language and it may be slow, difficult and expensive to have them translated. Commercial translation services often struggle with technical documents.

Document and record storage has often been a relatively low priority in developing countries. In countries with high levels of poverty, governments have higher spending priorities than retention of records, particularly when the cost of storing paper documents in warm humid environments is high. Additionally, social and political disruption has often meant that storages were not maintained or were deliberately destroyed.

Often even the local "experts" are not aware of information that may be available concerning the natural resources of their own country. Libraries may have limited holdings and may be poorly cataloged. There may not be a system to collect the material published within the country. For example, many Thai universities publish their own science journals, which quite often contain material relevant to natural resource management. However, these journals are rarely indexed online, and no Thai university library has a complete collection of the journals published by the other universities in the country. In Cambodia, the national library was largely destroyed during the Khmer Rouge era, with the permanent loss of much irreplaceable material.

Universities in the lower Mekong region have very limited budgets for accessing technical journals either in print or online. Furthermore, the technical literature is largely written in English, which is difficult for most people in the lower Mekong to read. There is a weak culture in the region of reading and following technical and scientific developments, with most university students having large numbers of contact hours, mainly lectures, and very limited time devoted to project work or study modes that would require them to read technical material either in hard copy or online. Their primary information source is the notes provided by the teacher.

It is also not uncommon for the published or available knowledge base to be unreliable. This can occur for a number of reasons. First, it may simply be out of date. Developing countries are changing rapidly. There is often rapid population growth. Rapid changes in technology occur, such as the increased availability of cheap monofilament nylon fishing nets, which facilitate many more people fishing, and fishing more effectively. Data collected 5 years previously may be already outdated. Second, data collection methods may be inherently unreliable. Where formal or informal taxes are collected, those liable for the taxes may deliberately understate the amount of fish or forest products harvested. If there is a hierarchical informal taxation system, this understatement may occur repeatedly up the hierarchy, so that the final government statistic bears little relation to the true harvest. There may also be cases where officials use "creative" data to demonstrate compliance with national goals or 5-year plans and the like which may not have been achieved, or even have been achievable.

Underestimates of resource usage may be problematical where they lead governments to downgrade the significance of a resource to the community. The importance of the lower Mekong fishery was not appreciated in the 1990s. Early estimates of yield varied widely, but were generally around 500,000 tons/year, and it was not until the mid-1990s that it was suggested the yield could be in excess of 1 million tons/year (for a review of estimates, see Hortle, 2007).

Objective studies of fish consumption (Hortle, 2007) and yield per unit of wetlands (Hortle & Bamrungrach, 2015) have indicated the annual fishery yield in the lower Mekong is in excess of 2 million tons. There are a number of reasons why the early estimates diverged so widely from reality. Mainly it was the inherent difficulty in assessing yields from a geographically diffuse fishery employing numerous different fishing gears. But it was also a consequence of a lack of funding and effort to assess the fishery, brought about in part because those funding and driving development agendas were primarily hydropower engineers from developed countries where there were no substantial subsistence riverine fisheries. A riverine fishery such as that on the Mekong was simply too far outside their experience to even be considered. In contrast, other uses such as irrigation, hydropower production, and navigation were familiar and could be comparatively easily quantified in economic terms.

Accurate data are essential in allowing governments and communities to make the informed trade-offs necessary in natural resource management. However, accurate data may also prove "inconvenient" to those promoting particular agendas. For example, potential loss of a high-value river fishery may make the economic argument for a hydropower dam weak or unsustainable. In this regard, through the 2000s the Lao government used an estimate of 14,000 tons/year for inland capture fisheries (authors' observations), even though analysis of a nationwide expenditure and consumption survey undertaken by the National Statistics Centre (an agency of the Lao government) indicated a yield of 168,000 tons in the year 2000 (Hortle, 2007). More recently, the Lao government has increased the yield estimate for capture fisheries to 40,000 tons/year (SEAFDEC, 2015), although estimates based on wetland area assessments indicate that the capture fishery yield may be as high as 246,000 tons/year (Hortle & Bamrungrach, 2015).

In addition, in many developing countries it may be difficult to extend the knowledge base since there are often a limited number of local organizations or individuals interested in or capable of undertaking the research needed, even if there is sufficient funding to pay for the research.

DIFFERENCES IN CAPACITY

In many developing regions, capacity is severely limited. Access to primary and secondary schools has been limited in the recent past, and in places may still be so. Many universities are academically weak and poorly resourced. Often they have very poor access to scientific literature because of limited funding for subscriptions (either hard copy or online), and most of the literature is written in English, which is not the first language of either staff or students.

Education systems in southeast Asia tend to emphasize content at the expense of skills development, and there is little in the way of a research culture, particularly in biophysical sciences and natural resources management. In such environments, highly competent people may be swamped by demand. In addition, people who receive good training, either through education in a foreign country, or in their own country, are often quickly promoted to the positions where they are unable to utilize the skills developed. For example, academics sent to another country to complete a Ph.D. degree, which trains them to do research, may be promoted to administrative positions, such as dean or deputy dean positions, as soon as they return, and thus have little opportunity to conduct research. The problem is that, unless there is an upgrading of the education system, the younger staff coming through will have the same lack of capacity that their promoted predecessors had.

Capacity building has become a high-profile component of most aid projects over the past few decades. However, building capacity is a long-term process, requiring decades to achieve, while at the same time aid programs in many countries have become more and more short term. Projects are often required to demonstrate "measurable outcomes" within 3 years or thereabouts to satisfy donor country political cycles. It is important to be able to demonstrate results from aid projects, but, with the exception of disaster assistance, meaningful outcomes are rarely achieved within a 3-, 4- or even 5-year project.

Many international aid projects have recognized that within a limited, relatively short-term project they will have little impact on the capacity of the country receiving assistance. A solution to this shortcoming, which is widely applied within the aid community, is the "training of trainers" approach. The intention here is to obtain a multiplier effect by training trainers, who will then train others within their own country. Unfortunately, training of trainers is rarely successful beyond the duration of the project cycle.

There are a number of factors that may lead to the failure of training of trainers projects. First, those being trained are often not experienced teachers. They are often nominated staff from government agencies with no prior experience working as trainers. In our experience, they are frequently not staff normally mandated within their organization to work as trainers when they return from their own training. Second, even where those being trained were academics and presumably had some training expertise, and were in a position to train others, our experience has been that success was elusive or short-lived at best. The primary reason for not meeting expectations is that institutions or donor agencies fail to provide the necessary follow-up support, with funding for salaries, travel, and materials, which might allow the trained trainers to carry out their role effectively.

DIFFERENCES IN SOCIOECONOMIC CONDITIONS

There are usually huge differences in socioeconomic conditions between the populations of developed countries and those of less-developed countries. In less-developed countries there are often many people who are subsistence users of wild natural resources. They may be fishers or they may forage in wetlands or forests. In Cambodia, for example, 80% of the animal protein consumed by the human population consists of fish and other aquatic animals such as shrimp, crabs, mussels and snails (Hortle, 2007). In many cases, people have a small-scale farm growing rice and vegetables for their own consumption, which they supplement with wild-caught fish.

Thus, the loss of a fishery, or a forest, can mean that many people are displaced from the traditional lands and consequently lose their entire livelihood. Many of those people have very few other options available to them. They earn their livelihoods as their parents and grandparents did, often having little education and no other skills. If they lose the fishery or the forest, they struggle to earn a living as dryland farmers even if they are provided with land. In addition, their community is often dispersed as well, so their local community support structures can also be lost.

In developed countries landowners are generally more highly skilled, literate, and living in a society where there is more likely to be legal recourse, and better compensation, should they lose their property or livelihood. So, the social consequences of changes in access to land are usually far more severe for at least some of the population in developing than in developed countries.

Another stark difference between developed and developing countries is the relative salary rates. In general, western cultures do not openly discuss levels of remuneration, but it is a common topic and indeed interest in some southeast Asian countries. *Ipso facto*, an average salary paid to a scientist or technical specialist from a developed economy may seem exorbitant from the perspective of colleagues in developing economies, which can lead to jealousies within organizations and criticism about the employment of "highly paid foreign experts." In our experience, such internal issues over salary differences can usually be negated by mutual empathy, good will, and understanding. Many local staff in resource management organizations in developing countries are as well or better remunerated as the foreign specialists. Attracting well-qualified midcareer foreign specialists to short-term contract positions in developing countries is difficult, and often exacerbated by salaries perceived by them to be inadequate considering the inherent disruption to career paths.

DIFFERENCES IN CULTURE

Differences in culture are the most complex set of challenges to meet in any project aimed at transferring knowledge. For any given project, they will be unique, depending both on the cultures of the transferee and the recipient countries as well as the particular characteristics of the individuals involved. It is not possible to be comprehensive in the context of a single book chapter. However, there have been many books written on local cultural mores for foreigners moving to particular countries (e.g., Culotta, 2013; Mikes, 1974; Segaller, 2006). Following, we discuss some of the cultural differences, which we encountered sufficiently often for them to be significant in our daily work.

We believe there are two overriding factors, which are essential prerequisites for the success of any knowledge-transfer project. The first is selection of staff with empathy and respect for cultural differences. The second is to facilitate long-term partnerships, which allow participants to build trust and develop personal relationships. Both factors are inherent in the following discussion.

IMPORTANCE OF AGE/SENIORITY

In southeast Asia age is synonymous with wisdom. Older or more senior people must be respected and not challenged, even in private. Younger people must defer to them. It follows that a person is not considered an "expert" with valuable opinions unless they are of sufficient age, usually at least 50. We have seen highly qualified and experienced job applicants rejected for positions purely on the basis that they were in their late 30s, and thus could not be employed in positions where advice to senior, older managers was required.

This cultural feature also makes it difficult to review or criticize any document written or published by a senior person. Any criticism, no matter how logical, would be viewed as unacceptably rude.

IMPORTANCE OF AGREEMENT

In many regional fora, agreement by all the parties is considered essential for decision making, even for contentious issues. This could bring great pressure on any party with a differing point of view, and lead to decisions being deferred, sometimes indefinitely, even if almost all the parties agreed. It is always important to be sure that the parties are all in genuine agreement and comfortable with carrying out the agreed actions. Feigned agreement is not uncommon, where parties publicly agree but then ignore any follow-up requirements for implementation.

IMPORTANCE OF SOCIAL NETWORKS

Social networks may have important positive and negative roles. In a positive sense, development of a strong personal relationship between the technology donor and the recipient is often crucial. In the absence of a strong personal relationship and trust between the two, it is very difficult to get timely and open feedback. Recipients may agree to undertake activities that they know in advance they will be unable to undertake, for example because of lack of personnel or other demands on their time and resources, but to admit these issues up front may cause loss of face. Where a sufficiently strong relationship has been established, both participants will be aware of the limitations and constraints, and consequently be able to constructively discuss at the outset of the program what activities are realistically possible.

The negative aspect of social networks is nepotism, a long-standing historical practice, which can work well when those being appointed have the right skill sets. It was, for example, one reason for the success of the British Navy in the 18th century, where the vast majority of the ships' officers and the commanders in every successful battle had gained their appointments through the "interest" of influential relatives (Rodger, 1986). However, in that situation it was rather important that those appointed had the necessary skills and ability, and those who were thought to be lacking were presumably not recommended for appointment. Social networks are important to the process of locating potential staff and consultants everywhere, but in some places they are the only technique. That can have serious repercussions if the staff members or consultants appointed are not suitable. It leads to poor performance of the necessary tasks, as well as an inability to adjust the work program or remove the consultant because of their connections. The project fee must be paid in full regardless of the quality of the work and/or report, because the consultant is the friend of somebody senior. In this context, the use of independent external referees is particularly useful, as it removes the evaluation of the work from the purview of the project managers. Implementation of a system of external independent peer review of consultant reports is, in general, a relatively cheap, and cost-effective quality assurance tool, which is used far too infrequently where consultants are employed.

REJECTION OR SUPPRESSION OF REPORTS THAT PRODUCE THE "WRONG" ANSWER

This is an issue throughout the world, particularly if a piece of work suggests that a current government policy may not be having the desired results. This is possibly more an issue where those transferring technology or expertise are academics dealing with government officers for the first time, as is often the case in technology-transfer projects. Instances of the "unwanted answer" abound in natural resource management. For example, the inability of fish passage structures on hydropower dams in the Mekong to maintain fish migrations and hence fisheries resources is widely accepted by fisheries scientists who have worked in the region (Halls & Kshatriya, 2009), but the structures are still universally proposed as a "solution" to allow fish migration by agencies proposing to build dams on the mainstream and major tributaries. Within Australia, research and investigation reports that either established that environmental impacts would be substantially greater, or that identified potential economic benefits would be far smaller, or nonexistent, than suggested by proponents of dams were ignored in the cases of the Lake Pedder dam (Lake Pedder Committee of Enquiry, 1974) and the Ord River scheme (Economists at Large, 2013).

CORRUPTION

Corruption is a problem everywhere, but is often more pervasive in developing countries. It may directly or indirectly impact projects. It can range from project delivery companies diverting project funds for other purposes, to project staff being forced to pay kickbacks to obtain or maintain their jobs, to nondelivery of project components that have been paid for, to training activities needing to be conducted in locations that will allow participants to obtain travel allowances. It is often a vexing problem for aid programs, as practices viewed as corruption by donor agencies might be viewed as normal operating practices in recipient countries. Projects must, at the outset, endeavor to establish structures and procedures that preclude corruption, which requires input from project managers with previous experience of working within the same cultural and political environment.

LOSS OF FACE

Loss of face (equivalent to "saving face" in western societies) is an omnipresent aspect of working relationships in Asia. It is fundamental to all relationships and everyday dialog. Colleagues must not be disrespectful during discussions where disagreements or even contrary opinions are expressed. In part, it is avoidance of loss of face that ensures that there must always be agreement after discussions about controversial issues. If any party disagreed with the final outcome, then they would be losers who also lose face.

DIFFERENCES IN POLITICAL CIRCUMSTANCES

In large international river basins, there are multiple levels of government: national, provincial, and local. Specialists from developed regions such as North America and Australia are rarely experienced in dealing with issues requiring international negotiation. In Europe there is far more experience, but even there the presence of strong international frameworks such as the EU results in many Europeans considering both a Eurocentric and a nationalist point of view for any issue.

The situation is different in many developing regions, such as much of Africa and southeast Asia. Neither of these regions has a regional organization with the same level of power and support as the EU, and in both regions there has been open warfare between countries or ethnic groups in the relatively recent past. Thus, there is often substantial mistrust and misunderstanding between nations, and political imperatives for leaders not to be seen as being "weak" when dealing with countries that may have been viewed as traditional rivals.

There is also often a desire by leaders of developing countries for their country to be seen as "modern," or making progress towards modernity, which can lead to pressure to identify and build "modern" large infrastructure projects such as international airports, large buildings, and dams.

Relationships between the power nations—such as China and the United States—and developing countries can also be influential. Depending on their location, most countries would prefer to maintain cordial relationships with both of these powers. In the Mekong, China is a very strong influence, controlling the headwaters of the river and being a major aid donor to several of the lower Mekong countries. But history can make for tense relationships. For example, Vietnam fought a border war with China in 1979, and China and Thailand were on opposite sides during the regional wars in Laos, Cambodia and Vietnam.

CONCLUSIONS

Transferring expertise and knowledge between countries and regions and developing capacity in places where existing capacity is low are challenging. Providing training in specific techniques is the most straightforward aspect of technology transfer, but even so there may be obstacles. Techniques may not be applicable, or able to be implemented by the recipients. Broader approaches to water resource management are often difficult to transfer because of the different historical, social, political, economic and biophysical circumstances faced by the recipients. People everywhere tend to be resistant to being told what they should do by outsiders.

Knowledge transfer can only be successful if the knowledge is wanted and needed by the recipients. Nevertheless, while being wanted or needed is necessary, it is not sufficient in itself for transfers to be successful. Potential recipients may enthusiastically request knowledge transfer activities even if they do not believe them to be potentially useful, possibly because the activities come with financial support that can be at least partially directed to broader institutional imperatives.

Knowledge transfer processes are often quite slow. With new approaches and techniques, those wishing to learn and adopt them will often take time to develop the confidence to implement them without external support. In other cases, especially when junior staff learn the new techniques, implementation may not be supported by senior staff, and consequently may be delayed until senior staff retire or move on.

In transferring knowledge or techniques, it is essential to be sensitive to the biophysical, and especially cultural and political, differences between regions. In discussion of management approaches that have been successful in Australia (or elsewhere), it is important to place them in context by explaining the biophysical, historical, and political environment in which they were implemented, as well as their strengths and weaknesses. Good managers are happy to "borrow" ideas from elsewhere and wise enough to adapt them to their local circumstances. Those offering insights into their own successful management strategies need to accept that their audiences will appropriate and adapt the components that appear to them to be most useful, given their own circumstances.

REFERENCES

Capon, S., James, C., & Reid, M. (2016). *Vegetation of Australian riverine landscapes*. Collingwood, Australia: CSIRO Publishers.

Culotta, N. (2013). *They're a Weird mob*. Australia: Text Classics.

Dodson, J. R., & Westoby, M. (1985). Are Australian ecosystems different? *Proceedings of the ecological society of Australia: Vol. 14*. Carlton, Australia: Ecological Society of Australia and Blackwell Scientific Book Distributors.

Economists at Large. (2013). *Rivers, rivers, everywhere. The Ord River Irrigation Area and the economics of developing riparian water resources*. Online at: http://www.ecolarge.com/wp-content/uploads/2014/10/Rivers-Rivers-Everywhere-Developing-Northern-Australia-The-Ord-River-Irrigation-Area-Ecolarge-FINAL.pdf (Viewed 5 July 2016).

Halls, A., & Kshatriya, M. (2009). *Modelling the cumulative barrier and passage effects of mainstream hydropower dams on migratory fish populations in the lower Mekong Basin: MRC technical paper no. 25*. Vientiane: Mekong River Commission [104 pp.].

Hortle, K. (2007). *Consumption and the yield of fish and other aquatic animals from the lower Mekong Basin: MRC technical paper no. 16*. Vientiane: Mekong River Commission [87 pp.].

Hortle, K. G., & Bamrungrach, P. (2015). *Fisheries habitat and yield in the lower Mekong Basin: MRC technical paper no. 47*. Phnom Penh: Mekong River Commission.

Humphries, P., & Walker, K. F. (Eds.), (2013). *Ecology of Australian freshwater fishes*. Collingwood, Australia: CSIRO Publishing.

Lake Pedder Committee of Enquiry. (1974). *The flooding of Lake Pedder: Final report of Lake Pedder Committee of Enquiry*. Canberra: Australian Government Publishing Service.

Lêvêque, C., Oberdorff, T., Paugy, D., Stiassny, M. L. J., & Tedesco, P. A. (2008). Global diversity of fish (Pisces) in freshwater. *Hydrobiologia, 595*, 545–567.

McMahon, T. A., Finlayson, B. L., Haines, A. T., & Srikanthan, R. (1992). *Global runoff—Continental comparisons of annual flows and peak discharges*. Cremlingen, Germany: Catena Press.

Mikes, G. (1974). *How to be inimitable*. London: Coronet.

Ostrom, E. (2009). A general framework for analysing the sustainability of social-ecological systems. *Science, 323*, 419–422.

Page, L. M., & Burr, B. M. (1991). *A field guide to freshwater fishes of North America North of Mexico*. Boston: Houghton and Mifflin Co.

Pigram, J. (2007). *Australia's water resources: From use to management*. Collingwood, Victoria: CSIRO Publishing.

Rodger, N. A. M. (1986). *The Wooden World. An anatomy of the Georgian navy*. London: Collins.

Royal Commission. (1902). *Report of the commissioners*. Melbourne, Australia: Interstate Royal Commission on the River Murray. Sands and MacDougall.

SEAFDEC. (2015). *Fishery statistical bulletin on Southeast Asia 2013*. Bangkok: Southeast Asian Fisheries Development Center [142 pp.].

Segaller, D. (2006). *Thai ways*. Bangkok: Silkworm Books.

Specht, R. L. (1981). Major vegetation formations in Australia. In A. Keast (Ed.), *Ecological biogeography of Australia*. Vol. 1. The Hague: Dr. W. Junk.

Unmack, P. J. (2013). Biogeography. In P. Humphries & K. F. Walker (Eds.), *Ecology of Australian freshwater fishes*. Collingwood, Australia: CSIRO Publishing.

Valbo-Jorgensen, J., Coates, D., & Hortle, K. (2009). Fish diversity in the Mekong Basin. In I. C. Campbell (Ed.), *The Mekong. Biophysical environment of an international river basin* (pp. 161–196). Amsterdam: Academic Press.

Williams, W. D., & Campbell, I. C. (1987). Major components and distribution of the freshwater fauna. In G. R. Dyne (Ed.), *Fauna of Australia. Vol. 1A general articles*. Canberra: Australian Government Publishing Service.

Winterbourn, M. J., Rounick, J. R., & Cowie, B. (1981). Are New Zealand stream ecosystems really different? *New Zealand Journal of Marine and Freshwater Research, 15*, 321–328.

BUILDING EXPERTISE IN RIVER BASIN MODELING—TRANSFER OF KNOWLEDGE FROM AUSTRALIA TO INDIA

17

C.A. Pollino*, P.J. Wallbrink*, A.K. Parashar*, G.M. Podger*, R. McLoughlin[†]

CSIRO Land and Water, Canberra, ACT, Australia* Commonwealth Department of Agriculture and Water Resources, Canberra, ACT, Australia[†]

CHAPTER OUTLINE

Decision Making in Water Resources Policy and Management. http://dx.doi.org/10.1016/B978-0-12-810523-8.00018-5

INTRODUCTION

Commonly accepted Integrated Water Resource Management (IWRM) practices promote taking a basin-scale approach for water planning, considering the basin as a system, with multiple drivers and pressures, and with diverse stakeholders (Global Water Partnership, 2009). Such an approach needs to focus on the physical, social, and political dynamics of water, from understanding the physical properties of a basin through to the relevant governance and institutional frameworks that support and enable resource use within them.

In recent years, India has been on a water reform journey, including revisions of its policy, legal, governance, and institutional arrangements. Like India, Australia also has a federated model of government; however, its reform has been ongoing for a significantly longer time. It has had ongoing negotiations with its states and territories over relevant water planning institutions and instruments over a period of a century. Starting in the 1990s, specific reforms responded to water climate extremes and recognized the widespread overallocation of water resources, particularly in the Murray-Darling Basin (Chapter 2). This culminated in a series of actions including: (a) a comprehensive policy blueprint to water reform known as the National Water Initiative (2004), (b) legislation to support this policy through the Australian governments' *Water Act (2007)* (Aust Govt, 2014), and (c) the tabling of the Murray-Darling Basin Plan (Aust Govt, 2012). In combination, these transformed the governance and institutional water management arrangements in the Murray-Darling Basin (Hart, 2016) and sought to chart the basin on a course of sustainable water management over the coming decades.

The challenges in managing water in multijurisdictional basins are well recognized and, where negotiated, solutions take substantial time, money, and political will to resolve (see Chapter 3). Australia's experiences in the Murray-Darling Basin can provide some useful insights on how to navigate water reform across jurisdictions and the associated risks and responses to this (see Chapter 13).

Recently, Australia has been working with central and state water agencies within the Government of India to build their capacity in managing water at basin scales. This relationship was undertaken under the auspices of a Memorandum of Understanding (MoU) on Water Resource Management signed between the Government of Australia and the Government of India in 2009 and subsequently renewed for 5 years in 2014. The intention of the MoU is to enhance cooperation on water resources development and management through policy and technical experiences, considering development and management of surface and groundwater resources, river basin management and impact of climate variability and change. This chapter describes a project that was established to meet the objectives of the MoU.

WATER IN INDIA

Throughout the South Asia Region, water security concerns are a major issue, with India facing a daunting water security challenge (Asian Development Bank, 2013; Briscoe & Malik, 2006; Vorosmarty et al., 2010). Sufficient and clean water resources are vital for supporting people and their livelihoods, economies with some water dependence (e.g., agriculture, industry, and energy) and the health of river systems. Changing demographics, rapid urbanization, agricultural development and climate change have the potential to impact India's ability to achieve future water security requirements (Riviere, 2015), as does a lack of institutional capacity to manage increasing competition for water within and between state borders, and deliver government priorities for Basin-level water management outcomes.

While India's average rainfall (1200 mm/year) is higher than most other countries of similar size (Rodell et al., 2009), it is unevenly distributed over the country both spatially and temporally. To counter this variability, India has sought to increase the reliability of water supply through large investments in water infrastructure (Briscoe & Malik, 2006). Such investments have had economic benefits nationally and regionally, by growing the national economy and reducing the incidence of poverty (Briscoe & Malik, 2006). However, despite this investment, the average surface water storage per capita is vastly lower in India (~0.2 ML per person) as compared to Australia (~5 ML per person).

While there is a desire to increase surface-water storage, various factors such as physical constraints, the capital investment required, social and environmental impacts, and marginal economic returns suggest that building large storages in India is not a sufficient option alone (Ashley & Cashman, 2006; Briscoe & Malik, 2006). Furthermore, estimates of current spending on maintenance of existing water infrastructure is lower than that required, with criticisms being that the allocated resources are either misdirected, not available or are unspent (Ashley & Cashman, 2006; Briscoe & Malik, 2006). This criticism is supported by evidence that many aging irrigation systems are rapidly deteriorating (Poddar et al., 2014). The financial liabilities of deferred maintenance of irrigation infrastructure subsequently impacts on the quantum of public funds available to finance new infrastructure (Briscoe & Malik, 2006), although there has been a recent focus on growing investment in completing "ongoing" irrigation projects that are in various stages of completion (Times of India, 2016).

In India, groundwater is regarded as a more reliable water resource than surface water, contributing to India's economic development (Ashley & Cashman, 2006). The limitation of surface-water delivery systems within the basins has further created incentives to tap into groundwater as the primary source of irrigation water (Shah, 2010). Consequently, India has developed a high dependency on groundwater, with up to 60% of the population in some regions being dependent on groundwater alone (Ashley & Cashman, 2006). Recent evidence at a national scale has demonstrated the consumption of groundwater by irrigation and other uses is unsustainable in some parts of the country, particularly in the northeast (Rodell et al., 2009). This is further evidenced by loss of surface water connectivity and soil salinization in parts of the country (Ashley & Cashman, 2006). The problem has been compounded by subsidized power for pumping, contributing to greater use of groundwater (Poddar et al., 2014), and a lack of investment or incentivizing of water-use efficiency (Briscoe & Malik, 2006).

WATER FOR AGRICULTURE

One of the highest users of water in India is agriculture. It has been estimated that 70% of the country's grain comes from irrigated agriculture (Gandhi & Namboodir, 2009). Through investments in water infrastructure, agriculture is now a significant contributor to the nation's economy, with the sector now contributing close to 16% of the nation's gross domestic product. As well as having social benefits, water infrastructure is seen as a way of lifting communities out of poverty (some 60% of the population are dependent on this sector). By contrast, in dryland areas, water resources are often degraded by pollution, and there are high levels of poverty and poor food security (Reddy et al., 2002). Approximately 30% of the population in India live in such areas (Bouma et al., 2007).

In irrigation areas, the costs of access to surface and groundwater are largely negligible. This is accompanied by inefficient irrigation systems and low agricultural yields (Briscoe & Malik, 2006). It has been estimated that the national average surface-water use irrigation efficiency is about 38%, while groundwater irrigation efficiency is estimated to be 70%–80% (Poddar et al., 2014). This compares with around 70%–90% efficiencies for a typical surface- and groundwater irrigation delivery system in Australia's Murray-Darling Basin.

CURRENT AND LOOMING CHALLENGES

India's water resources are under growing pressure, which manifests in symptoms such as water scarcity and decline in river and aquifer health. Recently, India has experienced two successive years (2015–16) of low monsoonal rains, accompanied with high dry-season temperatures. Exacerbated by overextraction of ground and surface waters, this has resulted in one of the most severe droughts on record, leading to impacts on approximately 330 million people. The drought has caused displacement of people, huge costs for water transportation to meet basic water needs (The Guardian, 2016) and conflicts between states where there are shared water resources.

As early as 1990, a national report estimated that by 2050 national water demands will exceed all available sources of water supply (NCIWRD, 1999). To manage the supply aspects of the water resource, there has been growing momentum in India for the interlinking of rivers to divert rivers to drought-prone areas. Unfortunately, there is little momentum in managing the demand-side aspects, such as increasing water efficiencies and/or strategic allocations of water to high-value uses. Part of the reason for this is that the agricultural sector absorbs around 60% of the population, hence reforming it comes with the challenge of shifting people out of the agricultural sector to other productive sectors. This is in stark contrast to Australia's agriculture and related sectors, which directly employs 2.6% of the employed population[1] and produces 2.2% of GDP (Commonwealth of Australia, 2015).

The deepening water crisis in India is manifesting itself in calls for increases in water demand through "supply side" technical solutions rather than accompanying this with demand management (Bharucha et al., 2014). This ignores the downstream impacts, with water demands not being met through the system (Briscoe & Malik, 2006).

Another complicating factor is the growing reliance on water for energy production in India. India as a signatory of the Paris Climate Agreement, and as a nation with the fourth largest carbon emissions globally, has committed to reduced carbon emissions. Coal-fired power stations are both key emitters and energy sources within India, supporting a growing population and industries. To meet their obligations in reducing carbon emissions, India will need to seek alternatives for energy production. While early agreements have focused on solar, another potential energy source that may compensate for reduced coal use is hydropower. However, hydropower dams, while nonconsumers of water, can have substantive impacts on river systems, by acting as longitudinal barriers to river flows and in changing flow regimes. Barriers have both spiritual and ecological impacts, with "uninterrupted" flows being a stated objective for many river systems.

INDIA WATER POLICY AND RIVER BASIN PLANNING

Under the Indian Constitution, water is managed at the state level within a federated system, similar to the Australian water policy environment. Although the constitution does allow the central government to regulate and develop interstate rivers and river valleys when declared by parliament as a matter of

---

[1]http://www.aph.gov.au/About_Parliament/Parliamentary_Departments/Parliamentary_Library/pubs/rp/rp1516/Quick_Guides/EmployIndustry—Accessed 20 October 2016.

public interest, this power is rarely exercised. To establish a national approach to water governance, the National Water Policy was first formulated in 1987, and reviewed and updated in 2002 and 2012. The National Water Policy of 2012 outlines a framework for the creation of a system of laws and institutions, and a plan of action for a unified national perspective considering surface and groundwater (Government of India, 2012). The framework advocates an integrated approach to management of water resources, superseding the 2002 Policy, which established a priority approach for sectorial water rights and entitlements.

The 2012 Policy recognizes the need for comprehensive legislation for interstate rivers and river valleys to facilitate interstate coordination of the planning of land and water resources at a basin or subbasin unit. The Policy includes aspects of water pricing, water-use efficiency, water sanitation, databases, monitoring and evaluation, conservation, climate change, and institutional arrangements.

The Policy also advocates an IWRM approach to planning, development and management, advocating a multidisciplinary approach. There are aspirations to legislate this framework as law in India. Under current arrangements, interstate tribunals are established to negotiate outcomes for interstate conflicts. However, these can take years to complete and there are limited enforcement mechanisms for tribunal findings (Briscoe & Malik, 2006; Saleth, 2016).

RIVER BASIN PLANNING: A WAY FORWARD?

While not a panacea, basin-planning processes can promote a more holistic approach to planning of water resources, exploring opportunities for sustainable water resource development as well as restoring the balance in overdeveloped basins. This is particularly the case where cross-border water management arrangements need to be implemented. Embedded above a basin-planning approach is the perspective of IWRM. Indeed, the Indian National Water Policy views basin planning as a key part of a broader IWRM approach.

As such, basin planning takes a broad view of water as an integrated resource, considering its use by a range of water sectors and how they interact, and is underpinned by a scientific evidence base. The outcomes seek to (Global Water Partnership, 2009):

- Be participatory, engaging stakeholders in defining objectives and in exchanging knowledge,
- Set out a clear and transparent process for decision making,
- Streamline and standardize planning processes,
- Promote intergovernmental cooperation,
- Balance outcomes for industry, people, and the environment,
- Be refined as new knowledge is gained.

In India, basin planning has the potential to promote a strategic approach to water resource planning, prioritizing investments in existing and new infrastructure, exploring demand and supply, and managing the system as a hydrological management unit, rather than by jurisdictional boundaries. To be successful, a basin approach needs the establishment of institutions, such as a river basin organization that can manage the water resource at the scale of the basin, and a matching set of incentives to facilitate adoption. Currently, however, there are few examples of institutional arrangements in India that support a basin-planning environment.

The current governance structures for water in India promote a tradition of command-and-control, demand-and-supply water management approaches, focused on state-level water management and allocation problems. This challenges the adoption of a basin-planning approach to water resource management, which places an emphasis on broader integrated outcomes, including to the environment and people's livelihoods. In water-constrained basins, local and state-centric approaches to water planning and management can lead to problems, including interstate water-sharing issues, water-quality problems, and a decline in the overall health of the water resource.

Although outside of the scope of this chapter, a significant and ongoing risk to be managed in advancing India's water management planning is the need to manage perceptions of who pays the costs of water reform and who ultimately benefits. There are strong socioeconomic arguments involved in debates on this subject, which center on concepts of fairness and equity in the supply of affordable water to communities. While this challenge is also experienced in Australia, the context is quite different. India has significant water, energy and food security issues and widespread poverty in rural areas.

INDIA-AUSTRALIA COOPERATION

As overviewed previously, the intention of the 2009 MoU was to enhance cooperation in water resources development and management of surface- and groundwater resources achieved by the sharing of policy and technical experiences as well as direct engagement and field visits.

The MoU is overseen by a Joint Working Group, which directs and monitors joint activities. The MoU aims to build on the experience of Australian organizations in the planning and management of basin water resources, particularly in the Murray-Darling Basin.

The nature of the relations established under the MoU includes the supporting of water resource planning through capacity building. This has been achieved through the use of water-planning tools, such as Australia's hydrological modeling platform eWater "Source,"[2] as well as flood and rainfall forecasting.

In the next section, we review one aspect of the program, specifically the building of capacity in basin planning with the focus being on a demonstration basin case study.

CAPACITY BUILDING: BASIN DEMONSTRATION

One of the capacity constraints in India is the technical skill required to evaluate water resources at basin scales, consider the requirements of an integrated basin-planning process, and consider the requirements to meet policy requirements.

To meet the capacity-building requirements at a technical level, a collaborative approach was taken to building expertise within India and to improving Australia's skill set to meet India's needs. To do this, it was agreed early on that the Australia-India project teams would take a partnering approach, and that there would be shared obligations and deliverables as part of a "project." The project commenced in mid-2013 and concluded in mid-2016.

Project efforts were focused on a case study basin, the Brahmani-Baitarni Basin, which has two subbasins that join at the river delta. The focus of the technical work of the Australian-led project team was on the Brahmani subbasin while the focus of the Indian-led teamwork was the Baitarni subbasin.

---

[2]http://ewater.org.au/products/ewater-source/—Accessed 27 July 2016.

Both subbasins were treated as hydrologically disconnected in modeling, as the models did not extend to consider tidal influences at the delta; rather, the focus was on water sourced from runoff within the catchment.

The basin is located in the eastern part of India (Fig. 1) and crosses three Indian states, these being Chhattisgarh, Jharkhand, and Odisha. The basin was selected for use in this case study as there were no interstate water-sharing issues identified between states. Using a digital elevation model,[3] the rivers of the Brahmani subbasin have a total length of 799 km and those of the Baitarni subbasin are 365 km long.

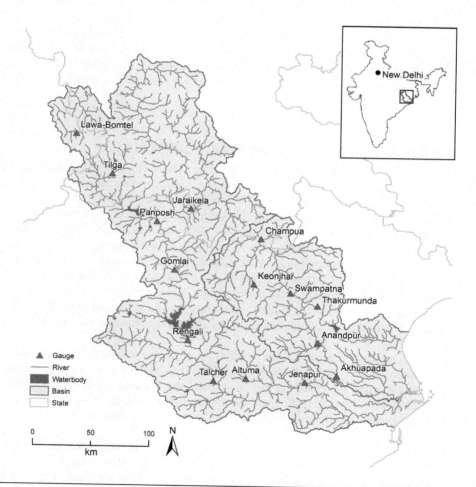

FIG. 1

Brahmani-Baitarni Basin, showing subbasin boundaries and river gauge stations. The inset shows the location of the basin on the Indian subcontinent.

---

[3]hydrosheds.cr.usgs.gov/—Accessed 10 September 2016.

The basin is bounded by the Chhotanagpur Plateau on the north, by the ridge separating it from Mahanadi Basin on the west and the south, and by the Bay of Bengal on the east. The physiography of the basin is defined by four regions: the northern plateau, the eastern ghats, the coastal plains, and the central tableland. The first two regions are hilly and forested, and the coastal plains consist of a fertile delta area. The main soil types found in the basin are red and yellow soils, red sandy and loamy soils, mixed red and black soils, and coastal alluvium.

The Brahmani River is formed by the confluence of the South Koel and Sankh Rivers, which largely flow through the State of Jharkhand. The Sankh River headwaters originate in Chhattisgarh, but this is only a small proportion of area of the basin. The confluence of the rivers is in the State of Odisha, where the Brahmani is formed and the water availability is higher. The Baitarni River is contained in the State of Odisha, with only a small proportion of area in Jharkhand. The land use of the basin is a mix of forest and agriculture, with agricultural land covering approximately 50% of the basin area; the bulk of this land use is in the State of Odisha where the coastal plains dominate. The forested land area is facing rapid degradation, particularly in the State of Jharkhand, which still has extensive areas of uncleared forest.

CONTRASTING INDIAN AND AUSTRALIAN POLICY CONTEXTS

The Brahmani-Baitarni Basin has a total area of 52,000 km^2. The average annual rainfall is 1488 mm and it has a population of 8.94 million. By way of contrast, the Australian Murray-Darling Basin is 1 million km^2 in area, with an average annual rainfall of 468 mm and a population of 2.1 million; it is an overall much drier system with an average end-of-system flow of 5100 GL (Chapter 13).

The Brahmani-Baitarni Basin has a tropical monsoonal climate, where rainfall is dominated by the southwest monsoon, between Jun. and Oct., with 80% of annual precipitation occurring during these months. The average maximum rainfall is approximately 3000 mm. The temperature ranges from 4°C to 47°C. The average end-of-system flow is 28,500 GL. The rural population makes up approximately 80% of the total population, with large tribal communities in the basin.

By way of contrast, flows towards the end of the Murray River and the end of the Brahmani River are shown in Fig. 2. The Brahmani River is dominated by an annual monsoonal pattern, which typically floods the delta region.

The Murray-Darling Basin presented Australia with the challenge of how to restore flows of water to meet environmental needs, within a socioeconomic setting and a highly variable climate over decadal scales. Within India, the challenge is how to meet the drinking water and sanitation needs of a growing population, and provide sufficient water to secure food supplies within an annual cycle of variability. There is also the opportunity to enhance development to increase food production and thus potentially rural livelihoods.

Australia has embarked on establishing limits on water resource use and a water market to encourage efficiencies of water resource use and water shifting to higher-value uses. India has sought to meet supply by securing additional water resources to meet the growing needs for water. Irrigated agricultural opportunities remain theoretical at this stage.

In this study, the major challenge in transfer of knowledge was how to apply the lessons learned from modeling water resources in the Murray-Darling Basin to the Brahmani-Baitarni Basin, where the physical and social contexts and the knowledge base are vastly different.

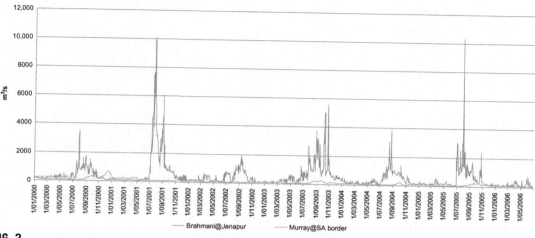

FIG. 2

Comparison of modeled daily flows from the Brahmani River at Jenapur and the Murray River at the South Australian border.

TRANSFER OF KNOWLEDGE

The knowledge transfer activities within the Brahmani-Baitarni project had a technical focus. Knowledge development was aimed at: constructing a river system model; using global, national and state datasets as inputs to the model; building a baseline to represent current water use and availability; and developing and evaluating scenarios exploring change in water resources as a consequence of water management.

This technical focus acknowledges the roles of river system models as fundamental analytical tools used to support basin planning, water management and water policy development. The river system model has two key components: a geographic component to represent the rainfall-runoff process, and a node-link river system model to show water distribution and use within the basin.

River system models have the potential to improve the understanding of water resources and their potential change with climate and land use, and enable assessments of change as a consequence of water management and use, including understanding how water resource developments in upstream environments can impact downstream. The form and focus of the model depends on the objectives of basin stakeholders as well as the available information that can be used to support its development.

Through the development of scenarios, river system models can also be used to explore the interactions between surface and groundwater, environmental flows, and human water use and infrastructure, under current and possible future scenario states. These scenarios are designed to enable water planners to gain a better understanding of the impacts of changes, such as those related to management, investments, and climate, as well as to inform decisions on infrastructure development and management, water allocations, water sharing, and sustainable limits of water use. In this project, the scenarios focused on development of new, and management of existing, infrastructure for increased irrigated food production, energy production, and water use across water sectors.

CAPACITY-BUILDING PROGRAM

A range of capacity-building activities supported the transfer of technical knowledge. Trainees included employees of the central and state governments, as well as members of nongovernment organizations who contribute to water planning. Training activities were aimed at three participant types:

- *Trainers*—those who would gain enough skills and knowledge of manipulating and building models to become a trainer in river modeling (so-called "train the trainer" approach)
- *Users*—those who would gain skills to be able to use data and river system models, but with limited skills to build models independently
- *Influencers*—those who would gain skills to understand the use of models and their outputs, but have limited direct interactions with models.

Over the life of the project, training activities undertaken included:

- Joint workshops and meetings, to develop plans and focus project objectives (all groups);
- Five basin trips, to better understand the basin physical and policy environment as well as water issues, focus the problem description through engagement with basin stakeholders, understand existing water, infrastructure, and irrigation management and operations (all groups);
- Three formal hands-on training workshops, with a mix of theory and practice, to undertake directed exercises in river modeling (trainers and users);
- Parallel modeling teams in India and Australia, with joint meetings face-to-face and remotely (trainers).

The joint workshops and basin trips were valuable in establishing the relationships and communicating the progress of project activities. They also enabled the project outcomes to be grounded in meeting local needs, reflecting local practices, and using the best available local knowledge. The basin engagement activity was also targeted at a broad base of people, including policy makers and state-based engineers, as well as engaging with farming communities where possible.

The initial formal training event was framed around the Murray-Darling Basin, and gave some level of familiarity of what was possible in river system modeling. The subsequent training workshops were focused on the development of the Brahmani and Baitarni subbasin models. Using a case study approach on a local basin was essential to focus the use of the model on the Indian contexts, using Indian data. Consequently, the training outcomes were targeted towards meeting Indian training needs, with a special focus on irrigation modeling.

Over the life of the training program, we observed that there was an attrition of skills between training events without dedicated practice on the skills obtained in the intervening periods. To counter this, we established parallel modeling teams within Australia and India.

The parallel teams each focused on a subbasin. The scope of the task, which was to produce tangible scenarios to explore improving livelihoods in the basin, was set by a senior water manager in the central Indian government. Working in parallel, model-building activities included: data collation, synthesis and evaluation; defining model objectives and developing conceptual models; developing model structure and parameters; model evaluation and scenario testing; and assessment of outcomes to inform basin planning and reporting.

PROJECT OUTCOMES

The outcomes from this project include building technical expertise in India in the conceptualization, data requirements, evaluation, and set-up of models that can be used in a basin-planning context. Further outcomes were through demonstration of model applications themselves, with river models in the subbasins of the Brahmani and the Baitarni Rivers being used to support subsequent scenario analysis.

From the project inception, we sought a strong focus on collaboration and joint activities rather than the traditional approach of delivering training events. An outcome of this has been in establishing a stronger relationship between organizations, and strengthening the ties established under the MoU. This relationship allows for an enhanced understanding of the governance processes for water in India, enabled the modeling team to tailor their approach to be fit for purpose for the Indian context and in defining the role of the central government in supporting basin-planning processes and the states in planning for water resources.

The models for the Brahmani and Baitarni subbasins represent current water resource use (the baseline), the level of change in water resources considering no water resource development, and scenarios that explore development of new infrastructure and agriculture projects, and the impacts of this across different water sectors. These models represent a synthesis of the best data and knowledge collected at reach scale or by individual states to represent the basin's water resources. In India, there have been few instances of basin-scale analysis.

In terms of delivering expertise, the participants in the parallel India team have since held independent training events within India, and the project outcomes are being used to frame future investments in basin planning. The lasting capacity-building outcomes are weaker in the states than in the central government. This is explored further in the challenges outlined in the following section.

Despite this demonstration of new capacity, it is worth noting that building capacity often requires a concerted effort over longer timeframes for there to be an enduring legacy. Capacity building requires fostering these skills and applying these in real-world contexts and sharing of knowledge and experiences at the policy level to enhance governance processes and institutions to support the development and implementation of a basin plan, moving beyond a technical exercise, as undertaken here.

PROJECT CHALLENGES

A number of challenges were experienced over the life of the project. These are synthesized here.

KNOWLEDGE, DATA, AND INFRASTRUCTURE

- At a basin level, knowledge of the systems was limited, with most of the required information located within different agencies and in various levels of quality. Such data included: the location and size of storages, land-use information, and location of cultural and environmental assets. Generally, the data sourced from the central government was of higher quality than other datasets provided. Central government data provenance, including collection methods, was more easily obtained. Some data-quality assurance had also been undertaken.
- Access to current practice for water management remains a challenge, with some information on water use and operations at the basin and the farm scale being unavailable. In the study

basin, relevant collation methods were still evolving and in many cases the underlying data required had not been digitized.

- There was limited cross-sectoral knowledge of water use and management. This was particularly so in understanding water use by agriculture (i.e., crop types, crop areas, groundwater use), the environmental and cultural values of the basin, as well as the design of current and planned basin infrastructure and the operational decisions in water management. To address this gap, assumptions were made in the modeling. These would need to be revisited if the models were used in a decision-making context.

- A greater emphasis is needed on groundwater to address the whole water cycle, in particular the usefulness of existing available data in exploring ground- and surface-water connectivity and conjunctive use. Within this project, we concluded that the available data was not fit for purpose for conceptualizing or quantifying surface and groundwater interactions. If there is a need to understand the groundwater as a resource, this requires investment in aquifer characterization and an understanding of groundwater recharge and use, and a fit-for-purpose modeling solution for groundwater.

- Information sharing was critical in the project. Basin states have their own operational and planning processes and interests in water development. Focusing the problem and analysis at the basin scale required openness and transparency in data and information sharing. This challenge was largely overcome through the project process.

- Building from an existing approach for river system modeling and making this fit for purpose for India's basin-planning context was important given the limited project timeframe. We were able to demonstrate the model within the climate and surface-water resource challenges in the basin, but we were unable to extend it to groundwater and water-quality issues. Both are pressing issues in India and represent a substantial challenge in providing water to meet human and environmental needs. Further efforts are needed to build a modeling framework to extend to these issues.

- For technical trainees, who undertook hands-on learning, analytical, and computing skills were limited. We observed this to be more so at the state level. As with the Brahmani Basin, simple models are needed that establish a baseline of water distribution and use, and that enable simple scenarios of water management. This suggests there is a need for a less technically complex modeling approach to basin planning in India.

- Basic computing infrastructure was only available to some participants taking part in the training events. This limited their ability to use the technologies away from the training venue. The fitness for purpose of the approach needs to be further demonstrated through application, which includes use of local computing data and infrastructure for use of software tools.

GOVERNANCE

- Technical capacity in river system modeling needs to be developed with the oversight of water managers and policy makers to ensure outputs are focused and fit for purpose.
- Once the project was advocated strongly at the policy level, the project activities and project buy-in from the center and the states was enhanced, and the project momentum increased substantially.
- Focusing the project scenarios on investment types and their potential benefits (and limitations) resonated with a policy audience.

- While the focus was modeling at a basin scale, representing the benefits across jurisdictions resonated with the basin state policy audiences.
- Future efforts in the basin require a focus on the establishment of interbasin institutions and governance to avoid and resolve potential conflicts in managing the shared water resource.

LESSONS LEARNED

Following is a synthesis of the lessons learned over the life of the project, which will hopefully assist in future projects with similar objectives.

The policy oversight from within the Government of India Water Ministry was critical for maintaining project momentum and focusing activities to be of direct relevance to the policy environment in India.

The interaction with the states, through field visits and joint workshops, was critical in defining a clear pathway for the technical components of the project. This was in terms of both defining and implementing models. An observation we made in the initial part of the project was the need for a tighter governance model. The benefits of this would include a greater clarity on water planning and policy objectives to align the project objectives and deliverables.

The project would have also been well served by the establishment of a technical oversight group, to provide legitimacy, review the proposed methods and deliverables for consistency, and to ensure these were fit for purpose.

We suggest that future basin-planning efforts should include other ministries as key stakeholders. These include agriculture, environment, forests, climate change, energy, finance, and possibly transport. We also suggest that basin-planning processes should have representation from nongovernment stakeholders, particularly if broader outcomes are needed to shape water policy.

A critical aspect is the ongoing maintenance of the skills and knowledge gained in the project. This needs to be fostered through ongoing practical application, with the focus being on real and tangible basin outcomes rather than theoretical ones. As project training events lacked a real-world purpose in water management, they were sometimes regarded as an "academic" exercise. Empowering trainees through allocation of time and grounding these as necessary core skills is also needed to maintain genuine engagement between government agencies. The challenge for India is ensuring skills are maintained in the short term, and that a larger cohort of people is created in the medium and longer term to create an ongoing critical mass. It is also important for this cohort to see that their endeavors are being utilized in genuine policy, planning, and operational decision-making processes.

THE LEGACY

India is increasingly experiencing issues of water scarcity, as evidenced by the recent 2016 drought. This drought was exacerbated by overpumping of groundwater. To meet India's water-planning challenges, water management requires an integrated approach, considering surface and groundwaters, improving efficiencies in water use, more effectively managing demands through formal water allocations and water markets in contested basins, and effectively managing and maintaining current infrastructure, and strategically developing new infrastructure to harness further water resources. Much of this can be achieved within an analytical basin-planning software framework. To embed this within planning and operational practice will require both developing and strengthening the skills for analysis and the institutional means of governing, and guiding water resource development.

Through this project, a series of strong working relationships have been built both with, and between, the three basin states and central government project officers at the technical level and the policy level. Within Australia and India, building these connections across government hierarchies is a challenge and this close cooperation was enabled under the MoU. The creation of trust, and consequent improved relationships, has also enabled a higher level of information sharing by the states with each other. This has included site and field trips by the basin states and central government officers. The project has also provided a forum for basin discussions, where the focus is on the basin, and at the basin scale, rather than on specific state issues.

This project sought to achieve a transfer of technical skills from Australia to India and with Australian partners getting a deeper understanding of the drivers of the Indian water sector. The Australian water reform was largely a reform of the surface water; India, however, has the challenge of reforming surface and groundwater, which until recently have had different management responses.

One clear message is the integrated nature of capacity building that India requires, not just one of surface- and groundwater management, but also coupling that with water-quality management to ensure a holistic response by policy makers. With crafting of these skills to meet India's water challenges, this objective was met with the legacy of this project in technology transfer being reflected in the in-country "train the trainers." There is potential for the skill base of these trainees to be further advanced through ongoing activities and further experience through application linking to investments and planning.

REFERENCES

Ashley, R., & Cashman, A. (2006). The impacts of change on the long-term future demand for water sector infrastructure. *Infrastructure to 2030: Telecom, land transport, water and electricity* (pp. 241–349). Paris: OECD.

Asian Development Bank. (2013). *Asian water development outlook 2013: Measuring water security in Asia and the Pacific.* Mandaluyong, Philippines: Asian Development Bank.

Aust Govt. (2012). *Water Act 2007—Basin Plan 2012, extract for the Federal Register of Legislative Instruments.* Canberra: Australian Government, Office of Parliamentary Counsel [245 pp.].

Aust Govt. (2014). *Water Act 2007 (with amendments).* Canberra: Australian Government, Office of Parliamentary Counsel [571 pp.].

Bharucha, Z. P., Smith, D., & Pretty, J. (2014). All paths lead to rain: Explaining why watershed development in India does not alleviate the experience of water scarcity. *The Journal of Development Studies, 50,* 1209–1225.

Bouma, J., Van Soest, D., & Bulte, E. (2007). How sustainable is participatory watershed development in India? *Agricultural Economics, 36,* 13–22.

Briscoe, J., & Malik, R. P. S. (2006). *India's water economy.* Washington, DC: The World Bank.

Commonwealth of Australia. (2015). *Australian Industry Report.* Canberra, Australia: Department of Industry, Innovation and Science.

Gandhi, V. P., & Namboodir, N. V. (2009). *Groundwater irrigation in India: Gains, costs and risks, No. 2009-03-08.* Ahmedabad: Indian Institute of Management.

Global Water Partnership. (2009). *A handbook for integrated water resources management in basins*. Sweden: Global Water Partnership and International Network of Basin Organizations.

Government of India. (2012). *National water policy*. India: Ministry of Water Resources.

Hart, B. T. (2016). The Australian Murray-Darling Basin Plan: Factors leading to its successful development. *Ecohydrology & Hydrobiology*, *16*, 229–241. http://dx.doi.org/10.1016/j.ecohyd.2016.09.002.

NCIWRD (1999). *Integrated water resource development: A plan for action* (Report of the National Commission for Integrated Water Resource Development (NCIWRD)). Ministry of Water Resources, Government of India.

Poddar, R., Qureshi, M. E., & Shi, T. (2014). A comparison of water policies for sustainable irrigation management: The case of India and Australia. *Water Resources Management*, *28*, 1079–1094.

Reddy, V. R., Reddy, Y. M., & Soussan, J. (2002). *Water and poverty: A case of watershed development in Andhra Pradesh, India*. Manila: Asian Development Bank Water and Poverty Initiative.

Riviere, D. (2015). *The thirsty elephant—India's water security challenges: A test for regional relations over the next decade*. Canberra: Commonwealth of Australia.

Rodell, M., Velicogna, I., & Famiglietti, J. S. (2009). Satellite-based estimates of groundwater depletion in India. *Nature*, *460*, 999–1002.

Saleth, R. M. (2016). Water rights and entitlements in India. In *Indian water policy at the crossroads: Resources, technology and reforms* (pp. 179–207). Switzerland: Springer.

Shah, T. (2010). *Taming the anarchy: Groundwater governance in South Asia*. Washington, DC: Routledge.

The Guardian. (2016). Indian drought 'affecting 330 million people' after two weak monsoons. *The Guardian*.

Times of India. (2016). Budget 2016: Irrigation focus waters parched hopes at farms. *Times of India*.

Vorosmarty, C. J., McIntyre, P. B., Gessner, M. O., Dudgeon, D., Prusevich, A., Green, P., et al. (2010). Global threats to human water security and river biodiversity. *Nature*, *467*, 555–561.

DESIGN OF A NATIONAL RIVER HEALTH ASSESSMENT PROGRAM FOR CHINA

C. Gippel*, Y. Zhang[†], X.D. Qu[‡], N. Bond[§], W.J. Kong[†], C. Leigh*, J. Catford[¶], R. Speed**, W. Meng[†]

Griffith University, Nathan, QLD, Australia* Chinese Research Academy of Environmental Sciences, Beijing, People's Republic of China[†] China Institute of Water Resources and Hydropower Research, Beijing, People's Republic of China[‡] Latrobe University, Wodonga, VIC, Australia[§] University of Southampton, Southampton, United Kingdom[¶] Badu Advisory Pty Ltd, Brisbane, QLD, Australia**

CHAPTER OUTLINE

INTRODUCTION

Over the last two decades, reform of water laws has been initiated in a number of countries in response to societal concerns about the ecological status of rivers, security of water supplies, inefficient use of water and inequitable distribution of water. For example, in Australia, reforms that followed a set of ecosystem principles set out by ARMCANZ & ANZECC (1996) culminated in changes to legislation that provided water specifically for protection of the ecological values of aquatic ecosystems (Godden, 2005; Tan, 2002). To meet the biological integrity objective of the US Clean Water Act, first specified in the 1972 Federal Water Pollution Control Act Amendments, the importance of bioassessment is now made explicit in US EPA (Environmental Protection Agency) policy and guidelines (Bauer & Ralph, 1999). In the European Union, following the Amsterdam Treaty in 1997, a new Water Framework Directive (WFD) was introduced that required monitoring of river health and achievement of standards by given deadlines (Barth & Fawell, 2001; Chave, 2001). Recently, comparable developments in water law, policy, and community attitudes in China have created the need for consistent nationwide assessment of river health.

Design of a national river health assessment program for China can draw on considerable local as well as global experience. In this respect, being a close neighbor of China and a leader in development and application of river health assessment technologies, Australia is well positioned to make a significant contribution. As with all technology transfer endeavors, the challenge is to adapt familiar methods to be compatible with China's social, administrative, and biophysical situation (see Chapter 16). There are many aspects to a river health assessment program.

This chapter discusses the rationale behind the framework for a national program to assess the health of China's rivers, developed collaboratively over the period 2010–12 by Australian and Chinese scientists, and supported by the Australian Agency for International Development (AusAID) as part of the Australia China Environment Development Partnership (ACEDP) (Speed et al., 2012).

BACKGROUND
RIVER HEALTH ASSESSMENT KNOWLEDGE AND PRACTICE IN AUSTRALIA

Catchment-scale bioassessment of rivers began in the United States in the 1980s, with Australia soon showing interest in this approach. Subsequent development of river health assessment policies and technologies in Australia borrowed from, or influenced, similar work happening in other countries. Early US studies focused initially on benthic macroinvertebrates (Karr, 1981; Karr et al., 1986), and these remain the most commonly used indicator group to this day (Rosenberg & Resh, 1993; Verdonschot et al., 2000). A debate ensued about the relative merits of multimetric (Karr & Chu, 1997) versus multivariate (Norris & Hawkins, 2000) approaches to bioassessment, but both remain in widespread use.

In the early days of bioassessment, concerns were raised about use of the term "river health" to describe the ecological status of rivers (e.g., Calow, 1992; Kapustka & Landis, 1998; Suter, 1993; Wicklum & Davies, 1995), but it has since gained widespread traction as a powerful metaphor for communicating complex scientific information about ecosystem structure and function to a broad audience (e.g., Boulton, 1999; Davies et al., 2010; Fairweather, 1999; Karr, 1999; Karr & Chu, 1997; Meyer, 1997; Norris & Hawkins, 2000; Norris & Thoms, 1999; Ryder, 1990).

The traditional view is that river health is mainly about the biological integrity of stream channels, indicated by the diversity and relative abundance of aquatic biota (usually fish and benthic macroinvertebrates) measured in the field at times of baseflow (Karr, 1999; Novotny et al., 2005). However, the contemporary understanding of river health has broadened to include physical and chemical aspects, as well as a wide range of biological aspects, and often now encompasses environments other than shallow perennial channels, such as wetlands, riparian zones, large rivers, temporary streams, lakes, and estuaries (e.g., Albert & Minc, 2010; Boulton, 1999; Cui, 2002; Davies et al., 2010; Flotemersch et al., 2006; Gamito et al., 2012; Ladson et al., 1999; Norris & Thoms, 1999; Novotny et al., 2005; Peng & Chen, 2009; Steward et al., 2012). Some authors have also argued for inclusion of the social and economic dimensions of rivers within the scope of river health assessment (e.g., Feng et al., 2012; Meyer, 1997; Rogers & Biggs, 1999), but little attention has so far been paid to development of the theoretical and practical basis of river-dependent social indicators.

River health assessment programs provide information to help guide policies and on-ground actions within an adaptive management framework, provide evidence for the success (or otherwise) of management decisions, and promote participation of the wider public in policy development and implementation. The programs are designed to match the scale of the resource management jurisdiction that they service. Thus, river health assessment programs exist at the scales of catchment/region, state/province, major river basin (often trans-boundary), and nation/confederation.

Australian examples of this jurisdictional scale of programs are the Ecosystem Health Monitoring Programme (EHMP), which covers the South East Queensland region (Bunn et al., 2010); the Index of Stream Condition (ISC), which covers the state of Victoria (Ladson et al., 1999); the Sustainable Rivers Audit (SRA), now discontinued, which covered the Murray-Darling Basin (Davies et al., 2010); and the National Framework for the Assessment of River and Wetland Health (FARWH), which was proposed to cover the entire nation (Norris et al., 2007).

A comprehensive national-scale program could potentially meet the needs of all jurisdictions through the advantage of consistent indicators and metrics that would be easily compared across jurisdictions without requiring intercalibration. However, there are challenges in establishing a universal method that would work satisfactorily in all physiographic, ecological, and sociopolitical settings. Another potential problem is that the indicators, site density, site locations, and sampling frequency appropriate to efficiently meet the objectives of a nationwide program could be inadequate to meet the objectives of a regional-scale program. For example, in developing FARWH, Norris et al. (2007) recognized that development of policy to address problems of national interest "…generally does not require site-specific information but does need assessment at regional, state or national levels." In comparison, the regional-level EHMP is primarily a tool for a partnership of government and nongovernment resource managers, local community groups and parochial stakeholders to not only assess current river health status and trends, but also to guide management actions (Bunn et al., 2010).

A national assessment program could also be scale-independent, such that the data it produces could satisfy objectives relevant to the smallest scale of interest (the catchment or region) up to the largest scale of interest (the nation or confederation of member states). It is doubtful that any existing national-scale program uses consistent methods and informs all scales of interest. For example, the WFD requires member states to follow a consistent framework to accommodate broad-scale reporting, but freedom in choice of methods means that most results require intercalibration, and some aspects of programs are not readily compared across the EU (Heiskanen et al., 2004; Kelly et al., 2008;

Schmutz et al., 2007). The EMAP-West and WSA programs both used consistent methods across the vast areas that they covered, and provided information useful for national policy formulation (Hughes & Peck, 2008), but the sites were too widely spaced to have relevance to local river management activities.

BACKGROUND TO DEVELOPMENT OF A NATIONAL-SCALE RIVER HEALTH ASSESSMENT PROGRAM IN CHINA

River health assessment is undertaken in a number of Asian countries, but Morse et al. (2007) identified some obstacles to progress: lack of knowledge about fauna and their tolerance values, scarcity of research programs and training opportunities, shortage of equipment, and limited institutional interest and support. In China, these obstacles are being overcome, and currently there is a high level of enthusiasm for development and implementation of a national river health assessment program.

River health assessments have been undertaken in various catchments throughout China, but mainly as a research activity. China currently has no ongoing river health assessment programs linked to river management policy operating at any jurisdictional level. This presents an opportunity to develop a program that uses nationally consistent methods and is scale-independent, such that it would meet the needs of local authorities responsible for implementing appropriate river restoration actions, as well as supplying the type of information required to assist national policy development.

Large, ambitious programs like the WFD, WSA, SRA, and the South African River Health Programme (RHP) (Hohls, 1996; Roux, 2001) were able to draw on many years of experience in biological research and assessment. As well as having trained field and laboratory staff available to carry out the work, these programs had access to considerable knowledge of the local fauna and biogeography to aid program design. In China, pilot studies could provide useful information and build local capacity, but definition of appropriate ecoregions (the sampling and reporting units) and refinement of the methods and sampling strategy requires knowledge of the variability in physical, biological, social, and institutional conditions across the country.

The preliminary framework for a national river health assessment program in China was one outcome of a collaborative effort between Australian and Chinese scientists, students, river managers, and policy makers. The work first identified how such a program might be constrained or shaped by factors of management objectives, timelines, funding, and institutional capabilities (cf. Hughes & Peck, 2008). Within these constraints and opportunities, and in consideration of the biophysical history and setting of China's rivers, methods were preferentially chosen from existing published methods that were familiar in China, known to work in China, or had a high prospect of transferability to China. The Australian contributors resisted the temptation to expediently transfer technologies they were familiar with and knew worked well in Australia. Rather, the knowledge transfer took the form of using Australian and Chinese experience to collaboratively explore ways of tailoring a solution specific to China.

The program framework outlined in the following applies to freshwater rivers. It is also important to monitor the health of permanent lakes and reservoirs, nonpermanent wetlands, and estuaries, but they are sufficiently physically, chemically, and biologically different from freshwater rivers that separate frameworks will be required.

A NATIONAL-SCALE RIVER HEALTH ASSESSMENT PROGRAM IN CHINA—CONSTRAINTS AND OPPORTUNITIES

ADMINISTRATIVE STRUCTURE AND POLICIES

Water resources are managed in China at the scale of major basins (Fig. 1) and smaller regional boundaries. The Water Law gives the Ministry of Water Resources (MWR), under the State Council, primary responsibility for the unified administration and supervision of water resources (Shen, 2009, 2012). According to the 2008 Water Pollution Prevention and Control Law, water pollution control plans are also based around the river basin and/or regional unit (Shen, 2009, 2012). Water quality is mainly the responsibility of the Ministry of Environmental Protection (MEP) with Environmental Protection Bureaus (EPBs) under the MEP at each administrative level. EPB responsibilities include monitoring water quality. Under the current arrangements, there is a degree of overlap in responsibility for monitoring and managing rivers between the MWR and MEP (Qiu & Li, 2009; Shen, 2009, 2012; Yan et al., 2006), so both ministries have an interest in river health assessment.

The 2002 amendment to the 1988 Water Law of the People's Republic of China (National People's Congress, 2002) specifically requires that the ecological environment be protected and improved when developing and using water resources. In 2008, the State Council approved a National Key Scientific Programme—Water Pollution Control and Treatment Programme (Zhou et al., 2011). In 2011, the annual agricultural policy priority document of the Central Committee and the State Council (the "Number 1 document"), for the first time focused on water conservation, citing alleviating drought, mitigating floods, improving water-use efficiency, and protecting and improving ecosystem health. The MEP National Biodiversity Conservation Strategy and Action Plan (2011–30) includes freshwater ecosystems as one of the priority areas (Li et al., 2011; Ministry of Environmental Protection, 2011).

In Australia, the complex political and administrative water management landscape presents a barrier to agreement on a nationally consistent river health assessment program. In contrast, the centralized and hierarchical administrative structure in China presents an opportunity to implement a standard river health assessment method across the nation. Recent water law reform, restructuring of the environmental ministry, and water-environment related policy announcements, will create urgent demand for river health information.

KEY MANAGEMENT ISSUES

Gleick (2008) referred to China's water resources as overallocated, inefficiently used, and grossly polluted by human and industrial wastes, largely as a result of more than 20 years of rapid economic growth. China faces serious constraints on water availability, due to high demand and uneven distribution of water resources across the country (Gleick, 2008; Shen, 2009). Recent institutional reforms and policy announcements recognize that declining environmental health threatens China's economic growth, sustainable development, public health and rural stability (Qiu & Li, 2009).

The main stressors limiting good river health in China are hydrological alteration and impaired water quality, with nutrient contamination from fertilizer application a major problem. These issues are manageable through implementing environmental flows and emphasizing contaminant source control rather than dilution (ironically, one of the main functions of environmental flows in China).

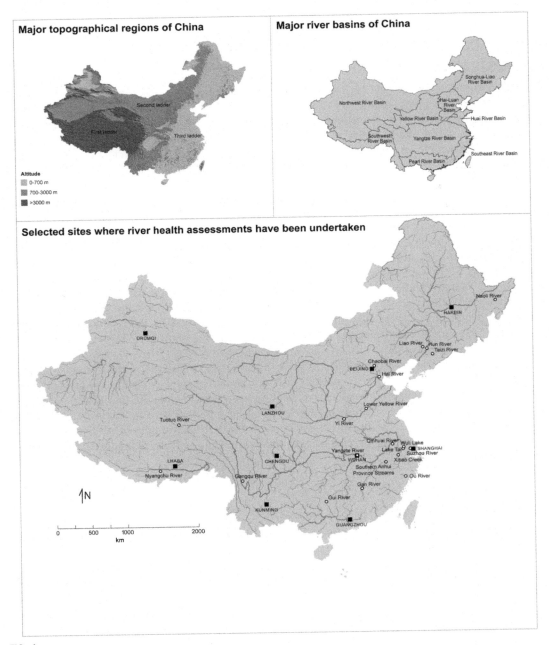

FIG. 1

Topography of China, major river basins, and the distribution of sites where a literature review found that river health assessments have been undertaken.

Rivers in China are subjected to other stressors that will also require amelioration if significant gains are to be made in river health. Many rivers have been channelized (straightened and hard-lined), and while dikes lessen the risk of flooding for cities and rural populations, they alienate river channels from their floodplains. As well as being regulated by large reservoirs, the longitudinal continuity of many rivers is interrupted by numerous weirs and hydropower dams. Widespread extraction of sand and gravel from active riverbeds to supply the construction industry directly disturbs the channel form and causes elevated turbidity. Riparian vegetation has long been cleared from most midland and lowland river reaches in favor of farmland, and this has reduced habitat availability and shading of channels. These issues can be addressed by physical restoration of channels and wetlands, installation of fishways (Chang, 2008), controls on sand and gravel mining, and planting of riparian revegetation. Thus, a river health assessment program that identifies the most appropriate management actions should include indicators that characterize the magnitudes and distributions of these stressors.

TECHNICAL CAPACITY AND EXISTING APPROACHES

The history of bioassessment of rivers in China extends back to the late 1950s, when an ecological survey was undertaken of major rivers and lakes. From the late 1970s to the 1980s, quantitative methods and various biotic indices were used to monitor water quality of a number of major rivers (Wang, 2002, cited by Morse et al., 2007). Chinese scientists began to take a strong interest in river health assessment from the early 2000s. This coincided with research funding opportunities associated with increasing emphasis on environmental protection in central policy, and a rapidly growing body of global literature on the topic.

The published Chinese river health assessment literature can be divided into three main categories: (a) review of concepts and methods, and recommendations for program design, (b) case study application of river health assessment based on the multimetric Index of Biotic Integrity (IBI) approach (cf. Hering et al., 2006; Karr, 1981, 1999; Karr & Chu, 1997), and (c) case study application of river health assessment using the holistic multiindicator approach that incorporates multimetric indexes representing a number of environmental components, such as fish, macroinvertebrates, algae, vegetation, hydrology, water quality, and physical form (cf. Chave, 2001; Davies et al., 2010; Ladson et al., 1999).

As part of the ACEDP, a collaborative review of published river health studies from China (in English and Chinese languages) was undertaken by Chinese and Australian scientists to establish the current level of technical knowledge and experience available. Six key observations emerged from the review of existing studies, all of which might be regarded as unofficial pilot studies for a national program:

Capacity building is required within local jurisdictions

The level of understanding of the concept of river health is advanced in China, and many sophisticated field assessments have been undertaken. There is clearly a high capacity for undertaking technical work, but the expertise for biological assessment largely resides within universities and government research institutes. Implementation of comprehensive assessment of river health will require capacity building within local EPBs.

The IBI method is not suitable for initial inclusion in a national program

Although the IBI has proved popular in catchment-scale assessments in China, it could be problematic to apply this method in a consistent way at the national scale. A range of subjective approaches has been applied to identification of reference sites in applications of the IBI method, and the reality in China is that most streams are impaired to some degree. However, a bigger problem is that the reference sites were invariably located in the uplands, and their characteristics were then used to judge lowland sites, ignoring natural environmental gradients. For the IBI method to have a role in a national program, it would have to be applied in a consistent way and at the spatial unit of ecoregions. Initially at least, there will be insufficient technical capacity and biogeographical knowledge available in China to include the IBI in a national river health assessment program.

The concept of holistic river health assessment is widely supported

There appears to be widespread support among academics in China for a holistic multiindicator approach that links the assessment process with the process of setting priorities for, and implementing, management action. Various indicators and metrics have been tried, but the case studies lack comparison and critical evaluation of methods, so there is no consensus on which are the most appropriate indicators and metrics for China.

Probabilistic sampling needs to be considered

In the Chinese studies reviewed, a nonprobabilistic approach was used to select the site locations and number of sites within each study area, which would not suit the main objectives of a national program. Probabilistic sampling designs locate sites randomly within the study area (which might be stratified into smaller areas such as ecoregions), and site locations are rerandomized for each survey (Stein & Bernstein, 2008; Theobald et al., 2007). This approach provides reliable, unbiased estimates of aggregate regional ecological conditions and is appropriate when the objective is a broad-scale audit of river health, as for a national program.

In nonprobabilistic sampling, sites are selected using nonrandom criteria, most often for convenience (ease of access for the field team, which reduces costs) or to target specific management activities or types of disturbance. The main limitation of nonprobabilistic sampling is that the data cannot be used to make inferences about regional patterns of river health (Keizer-Vlek, 2012; Stoddard et al., 1998; Theobald et al., 2007). It is not surprising that Chinese assessments have not dwelt on the issue of sample site location and density, because they were mostly one-off surveys of fairly small areas and simply reported individual site data. Stein and Bernstein (2008) suggested that a hybrid sampling design incorporating a mix of fixed target sites and probabilistically selected sites would satisfy programs having multiple objectives. This approach could suit China, where probabilistically sampled sites could be used to report at the national scale, and targeted sites could report on issues of local interest, as required.

The concept of reference condition requires review

The prevailing understanding of reference river condition in China is one of ecosystem integrity, idealized as being unimpaired by human disturbance, which is consistent with most programs around the world. Defining this state often relies on finding unmodified reference sites, which are rare in China. This problem can potentially be overcome by developing environmental filter models (Chessman, 2006; Chessman & Royal, 2004; Chessman et al., 2006; Stranko et al., 2005). In China, development of filter models would first require an ecological survey of the nation's rivers, also measuring relevant environmental variables.

In some situations, perhaps including China, reference states other than unimpaired may be more relevant (Stoddard et al., 2006). In heavily utilized rivers, the community may be willing to trade river health for social and economic benefits, such that the agreed (or implicit) vision for river health could be quite distant from an unimpaired river (e.g., Liu & Liu, 2009). If the local river health-management target was distant from unimpaired condition, then river health scores measured relative to an unimpaired state could remain discouragingly low, despite river managers achieving local targets through investment in appropriate actions. One solution would be to use dual reference scales, one based on an unimpaired condition that would allow a consistent comparison of rivers nationwide, and the other relative to management targets, which would have more local relevance.

Whatever approach is taken to defining reference conditions in China, it is apparent that it will be a technically challenging task that is best addressed by research institutes with access to national data sets. In the meantime, a national river health survey can be undertaken, with local jurisdictions reporting ecological health relative to interim criteria that are based on expert judgement and published standards.

The concept of social indicators of river health has strong intuitive appeal

It is not uncommon for Chinese researchers to make the case for inclusion of social indicators in a river health assessment program. The high level of competition for scarce water resources in most Chinese rivers means that managers must routinely balance the likely social and economic benefits against potential ecological benefits when making water allocation decisions. Availability of data on indicators of flood risk, drought risk, navigability, food and fiber production from irrigation, fisheries production, electricity generation, tourism, angling, and water quality for human uses (recreation, industry, agriculture, drinking water supply), together with ecosystem health data, would inform the trade-off between ecological and social benefits. The concept of social indicators of river health has strong intuitive appeal in China, but at this stage the concept is too immature to be considered for inclusion in a national river health assessment program.

EXISTING DATA

Although the quality of China's rivers is principally the responsibility of the MEP, water quality is also routinely monitored by the MWR. Together, they administer thousands of monitoring sites across China, including a network of automatic water quality monitoring stations in each of the major river basins. The MWR monitors river flows using a network of thousands of hydrological stations, and many of these double as water quality monitoring stations.

It would seem reasonable that a national assessment of river health could be made on the basis of data from the existing monitoring networks, given the recognition that the main human disturbances to rivers in China impact both hydrology and water quality. Also, in many places it would be possible to establish reference conditions, associate trends in water quality and hydrology with human disturbances, and assess the effectiveness of rehabilitation efforts implemented to date.

Data from existing hydrological and water quality monitoring networks is used in a number of countries as an input to annual State of Environment (SoE) reports (including Australia and China), but it is usually ignored in river health monitoring programs. The reasons could be the low density of sites, nonprobabilistically selected and fixed site locations, difficulty in accessing data, or lack of interest in hydrology and water quality as indicators of river health.

Most river health monitoring programs were set up principally as bioassessment exercises aimed at collecting data at times of baseflow from sites spread evenly over catchments. In contrast, hydrological stations tend to be located on the main stems of rivers and major tributaries, or near dams. In Australia, the ISC (Ladson et al., 1999) and the SRA (Davies et al., 2010) were able to incorporate hydrology as an indicator by using a distributed network of sites having modeled hydrology data.

River flow and water quality are so variable in time and space that infrequent snapshot field measurements made in association with biological sampling are not normally very meaningful. The main reason for measuring water quality or river flow parameters in association with biological sampling would be to provide the input data to an environmental filter or other multivariate predictive biological model. Thus, incorporating hydrology and water quality into a multiindicator river health assessment program will require use of data from existing monitoring networks that measure at frequent intervals. In China, difficulties in accessing raw data mean that the method should expect monthly, rather than mean daily, hydrology data, and water quality data converted to grades, rather than parameter concentration data.

BIOPHYSICAL CHARACTERISTICS OF CHINA'S RIVERS

China has a highly diverse geography, with streams present over a wide range of climate, altitude, and geological zones. A nationally consistent river health assessment method would need to have a high degree of flexibility. There may be a need to tailor some aspects of the methodology to suit ecoregions. Li et al. (2012) found that macroinvertebrate indicators defined ecoregion boundaries that corresponded with the three main topographic zones (called geographic ladders) of China (Fig. 1). These ladders correspond with the main geographic regions of China: the Tibet-Qinghai plateau (first ladder), the Xinjiang-Mongolia central mountains and plains (second ladder), and the eastern coastal plains or monsoon region (third ladder). The eastern region has a strong north-south increasing rainfall gradient. Most Chinese river health assessments reported in the literature have been undertaken in the low elevation, monsoon region, with a few in the Tibet-Qinghai plateau, and none in the vast second ladder region (Fig. 1). Although high mountain areas may be difficult to access, routine river health assessment methodologies can be applied in such environments (e.g., Hartmann et al., 2010; Shah & Shah, 2012).

It is not clear whether the existing mapped aquatic ecoregions of China (Liang et al., 2010; Yin et al., 2005; Zhu et al., 2011) are suitable for setting reference standards for, and reporting, river health at the national scale. It might be necessary to classify rivers (rather than land areas) as highland, midland, and lowland types, with each having a distinctive character defined by geomorphological, hydrological, and ecological variables. Further separation of these broad river types by climate and altitude could also be required. For example, forested riparian zones would not be expected in alpine or arid areas, so different riparian vegetation metric standards would apply there compared with humid and temperate regions.

Duan et al. (2008) demonstrated that benthic macroinvertebrates were very sensitive to bed material particle size and interstitial space, with taxa richness and density much higher in cobble and gravel compared to coarse and fine sand beds. This suggests that bed material size might be an important variable in classifying ecoregions. As bed material size data are generally unavailable, it could be included as an adjunct variable in a national river health survey. Also, it might be possible to model expected bed material size, in which case it would become a candidate indicator of physical form.

Globally, most methodologies for river health assessment apply to small, wadeable streams rather than larger rivers. This is in part due to the sampling difficulties and other complexities posed by larger river systems, but also in recognition of the many thousands of kilometers of tributary streams that are directly influenced by land use, which in turn influence downstream water quality. China has tens of thousands of reservoirs that offer aquatic habitat, many lowland rivers have associated large lowland lakes, and, compared with Australia, China's lowland rivers are numerous and large (Leigh & Gippel, 2012). Large rivers are the focus of human settlement, are critical for supporting the livelihoods of millions of people, act as important transport routes, and have traditional cultural significance. Not surprisingly, large rivers are the focus of river health concerns in China. Field sampling of large water bodies, compared with smaller streams, often involves considerably more effort per site. Remote sensing is often more efficient than field sampling, but some indicators cannot be measured by remote sensing. Thus, design of a river health assessment program that includes China's large water bodies must trade off the expediency of remote sensing against the detail and additional variables that are possible with field-based sampling (Leigh & Gippel, 2012).

Most river health assessment programs exclude temporary (intermittent and ephemeral) streams. Temporary rivers are globally common (Steward et al., 2012), and China is no exception. Steward et al. (2012) argued that dry riverbeds should be included in river health assessments because they are valuable habitats in their own right, with peculiar physical and biological characteristics (Steward et al., 2011; Williams, 2006; Wishart, 2000). Sampling drought refuges is one possible solution (Robson et al., 2011). However, given the traditional cultural and ongoing economic significance of large perennial rivers in China, temporary streams are likely to remain low priority for inclusion in a national river health assessment program for some time.

COLLABORATIVE AUSTRALIA-CHINA RIVER HEALTH ASSESSMENTS

Crucial to the success of the Australia-China collaboration (Speed et al., 2012) was undertaking pilot river health assessments across a range of jurisdictions and river types, as a test of potential methodologies and as a practical way of transferring, and adapting as necessary, existing technologies. Pilot studies were undertaken in three catchments: the Gui River, a subcatchment of the Pearl River (Bond et al., 2011); the lower Yellow River (Gippel et al., 2012; Liu et al., 2016); and the Liao River (Leigh et al., 2011; Qu et al., 2016; Zhang et al., 2013) (Fig. 1).

These assessments all followed a holistic methodology, although there were differences in the suite of variables measured and the way the reference condition was defined. Indicator groups included fish, benthic macroinvertebrates, riparian vegetation, water quality, hydrology, and physical form. Each assessment followed a similar framework (Gippel & Speed, 2010), whereby a team of scientists, representing a wide range of expertise: (a) selected appropriate variables based on the physical and biological characteristics of the river and the program objectives, (b) selected sampling sites, (c) conducted field sampling, laboratory analysis and office-based measurements, (d) developed indicators and a way of calculating a score relative to reference, and then (e) published the results in the form of a technical report and a report card (Fig. 2).

The Gui River is considered one of the least impaired in China. A reasonably high density of sites was sampled (752 km^2 per site) (Fig. 2), but it was found that this small catchment, with high rainfall and a relatively low gradient of disturbance, might require an even a higher sampling density in order to effectively inform management at this scale (Bond et al., 2011). In common with the other pilot study areas, the Gui River team prepared a report card to communicate the results of the assessment. The style and format of the report card was based on the team's experience with creating effective report cards in Australia, adjusted for local preferences. The Gui River report card included separate graphics to represent the distributions of catchment disturbance, riparian vegetation, physical form (barriers to movement), water quality, algae, benthic macroinvertebrates, fish, hydrological alteration, and overall river health (Fig. 2).

The lower Yellow River study (Gippel et al., 2012; Liu et al., 2016) highlighted the need to develop unique local indicators to assist with management of locally special issues (in this case, severe bed

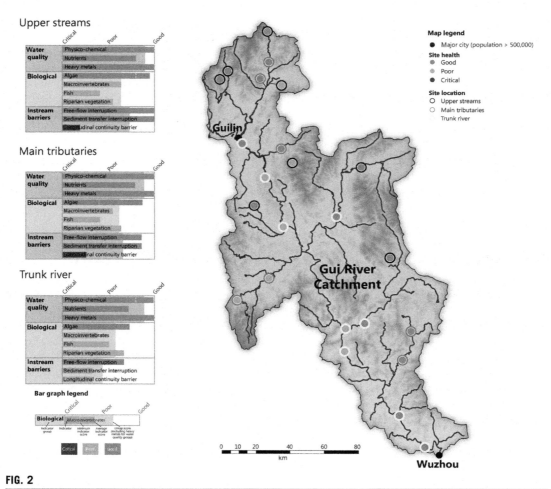

FIG. 2

River health scores for sites sampled on the Gui River, China, in Mar.–Apr., 2010.

sedimentation and the associated flood risk) and highly valued ecological assets (in this case, the Yellow River delta). Data from historical records were used for reference. The scoring system used weighting to adjust for differences in the importance of the chosen indicators and the relative efficiency of the sampling effort.

This study remains one of the few that has attempted to incorporate social indicators of river health. Social river health means the condition of a river with respect to the cultural, social, and economic values that are derived directly from the river. The social targets for the lower Yellow River center on flood protection, no interruption to water supply, and meeting water quality standards for human needs.

In China, policies for managing river-derived social benefits are often integrated with those for achieving ecological health. This is partly because rivers with a high level of competition for scarce water resources require managers to routinely balance the likely social benefits against potential ecological benefits when making water-allocation decisions, so they need to have an awareness of the trade-off involved. The other reason for the close association between social and environmental river values in the lower Yellow River is the very long history of human dependency on the river's resources for protein (fishery, now virtually nonexistent), domestic water supply, agriculture and transport (and more recently for industry). The capacity of the river to provide these resources was partly dependent on its ecological health. The risk of flooding has always been an overriding concern of people living near the lower Yellow River. Using readily available data, Gippel et al. (2012) developed a suite of social indicators comprising: flood risk, due to both storm event and ice jam; drought risk; water consumption (as a secondary indicator of the economic value of production); hydropower production; and navigability.

In the Liao River, Leigh et al. (2011) selected metrics based on their sensitivity to a human disturbance gradient. They first sampled 70 sites in the Taizi River subcatchment for water quality, benthic macroinvertebrates, fish, algae, and aquatic and riparian vegetation, and also analyzed long-term hydrology data. An associated outcome of this study was development of an approach to ecoregional classification, which was applied to the Taizi River catchment (Kong et al., 2013). Later, Leigh et al. (2011) sampled a further 176 sites spread across the Liao River basin and assessed river health using the same indicators. A high site density of 198 and 293 km^2 per site was achieved for the Taizi River subcatchment and Laio River basin, respectively. These sampling densities are an order of magnitude greater than those typically used in national-scale programs.

KEY CHARACTERISTICS OF A RIVER HEALTH ASSESSMENT PROGRAM FOR CHINA
OBJECTIVES

The Australia-China collaborative research culminated in the development of a framework for a national river health assessment program tailored to suit China. The agreed overarching objective of the program was to provide reliable information on the health status of rivers to guide rational river management policy at all jurisdictional levels. When implemented, the assessment should provide a periodic audit of the nation's river assets using consistent data. These data should assist development of investment policy for river management by identifying priority rivers, priority issues, and priority management actions. Data collected through the program would also be used to evaluate the effectiveness of management actions in maintaining or improving river health.

SAMPLING

Both probabilistic and fixed sampling sites are required to meet program objectives. Probabilistically selected sites should be spatially balanced and stratified within predefined ecoregions. The number of sites needs to be sufficient to characterize the aggregate health of streams in each ecoregion and to detect change from one sampling period to the next. The mean catchment area represented per site would ideally be in the range 1000–2000 km^2, which would amount to 5000–10,000 sites nationally. Fixed sites would be located at established hydrological and water quality monitoring stations, and at other locations deemed important by local river management jurisdictions. For local jurisdictions interested in the detailed spatial distribution of river health, a site density one order of magnitude higher than the national-scale sites would probably be required.

In China, the standard reporting year begins with the onset of winter (usually Dec.) and ends at the close of autumn. Field sampling should be undertaken at least once in the reporting year in spring (generally Mar. to May) during a period of relatively constant baseflow. Sampling a second time, in autumn (generally Sep. to Nov.), would be optional.

REFERENCE STANDARDS

To enable implementation of a river health assessment framework across China, a set of interim standards, with the highest score reflecting unimpaired condition, was established using information from the literature and expert opinion. Establishing reference standards for metrics is a research task best undertaken centrally in a consistent way at a large spatial scale, using data collected in the first national survey.

INDICATORS AND METRICS

Indicators are readily measured attributes of river systems that reflect human disturbances and which have meaning to scientists, policy makers, river managers, and the wider community. River health indicators fall within the three main categories of physical, water quality, and biological. The recommended indicator groups were: hydrology, physical form, water quality, fish, benthic macroinvertebrates, algae, and riparian vegetation. Indicators can be quantified using a number of different river health metrics, and these will likely vary between rivers and ecoregions. The research found difficulty in devising universal and objective indicators and metrics of physical form. At this stage, physical form variables might make a useful contribution to characterization of the catchment disturbance gradient, rather than being indicators of river health at the point scale. This might also apply to water quality, hydrology, and vegetation indicator groups.

SCORING AND REPORTING

In the pilot studies, metrics were calculated by comparing the raw value of a measured variable against the expected value, according to the selected reference standard. The reference was interpreted in a number of ways, depending on the situation. A national program will also need flexibility in setting reference standards, but initially fixed interim standards should be used. To be consistent with most programs around the world, the metric scores should be standardized over a scale between 0 and 1, divided into five equal classes to facilitate presentation of results. In the pilot studies, indicator scores were a composite of metric scores, and indicator group scores were a composite of the scores for the indicators belonging to the group. A combined overall river health index score was derived by combining the indicator group scores. The overall index would be mainly of value at the national scale. In the first implementation of the river health assessment framework across China, scores should be combined using a simple unweighted averaging procedure. This method is open to review, with other options being expert rules (Davies et al., 2010), or basing the overall river health score on biological indicators, and using the stressor indicators to guide management priorities. For reporting at the basin and national scales, the site results should be aggregated by ecoregions.

CONCLUSION

China's commitment to rapidly improve river health will demand objective data to inform policy, and local EPBs will require information additional to water quality data to help them undertake appropriate river rehabilitation works. In China, the current absence of provincial river health assessment programs, and the centralized administrative structure, is conducive to the establishment of a nationally consistent river health assessment program. Many catchment-scale river health assessments have been undertaken in China as research projects, and there is little point in commissioning more unless they contribute directly to the implementation of a coordinated national program.

The ACEDP China-Australia collaborative research project facilitated the rapid and effective design of a framework for national river health assessment that would meet the needs of all jurisdictions and was sensitive to the particular constraints and opportunities present in China.

The most appropriate approach in China would be to collect biological data on fish, algae, and benthic macroinvertebrates, supported by data that relate directly to the manageable aspects of the environment that affect biological health, and which are commonly impaired by human activities, namely hydrology, water quality, physical form, and riparian vegetation. This approach will assist in deciding the high-priority rivers for protection or rehabilitation, and which disturbances should be addressed. The program would also serve as a way of monitoring the performance of the agencies responsible for improving river health.

ACKNOWLEDGMENTS

The contributions of Christopher Gippel and other ACEDP team members were funded by the Australian Agency for International Development (AusAID) as part of the Australia China Environment Development Partnership (ACEDP).

REFERENCES

Albert, D. A., & Minc, L. D. (2010). Plants as regional indicators of Great Lakes coastal wetland health. *Aquatic Ecosystem Health & Management, 7,* 233–247.

ARMCANZ., & ANZECC. (1996). *National principles for the provision of water for ecosystems. Sustainable Land and Water Resources Management Committee, Subcommittee on Water Resources: Occasional paper SWR no. 3.* Canberra: Agriculture and Resource Management Council of Australia and New Zealand, Australian and New Zealand Environment and Conservation Council, Commonwealth of Australia.

Barth, F., & Fawell, J. (2001). The Water Framework Directive and European water policy. *Ecotoxicology and Environmental Safety, 50*, 103–110.

Bauer, S. B., & Ralph, S. C. (1999). *Aquatic habitat indicators and their application to water quality objectives within the Clean Water Act: EPA-910-R-99-014.* Seattle, WA: US Environmental Protection Agency, Region 10.

Bond, N. R., Liu, W., Weng, S. C., Speed, R., Gippel, C. J., Catford, J. A., et al. (2011). *Assessment of river health in the Pearl River Basin (Gui sub-catchment).* Brisbane: River Health and Environmental Flow in China Project, Pearl River Water Resources Commission and International Water Centre.

Boulton, A. J. (1999). An overview of river health assessments: Philosophies, practice, problems and prognosis. *Freshwater Biology, 41*, 469–479.

Bunn, S. E., Abal, E. G., Smith, M. J., Choy, S. C., Fellows, C. S., Harch, B. D., et al. (2010). Integration of science and monitoring of river ecosystem health to guide investments in catchment protection and rehabilitation. *Freshwater Biology, 55*(Suppl. 1), 223–240.

Calow, P. (1992). Can ecosystems be healthy? Critical consideration of concepts. *Journal of Aquatic Ecosystem Health, 1*, 1–5.

Chang, J. B. (2008). Construction of fish passages in China: An overview. H. Rosenthal, P. Bronzi, M. Spezia, & C. Poggioli (Eds.), *Passages for fish: Overcoming barriers for large migratory fish* (pp. 22–29). *World sturgeon conservation society, special publication no 2* Norderstedt: Books on Demand.

Chave, P. (2001). *The EU Water Framework Directive—An introduction.* Colchester, Essex: IWA Publishing.

Chessman, B. C. (2006). Prediction of riverine fish assemblages through the concept of environmental filters. *Marine and Freshwater Research, 57*, 601–609.

Chessman, B. C., & Royal, M. J. (2004). Bioassessment without reference sites: Use of environmental filters to predict natural assemblages of river macroinvertebrates. *Journal of the North American Benthological Society, 23*, 599–615.

Chessman, B. C., Thurtell, L. A., & Royal, M. J. (2006). Bioassessment in a harsh environment: A comparison of macroinvertebrate assemblages at reference and assessment sites in an Australian inland river system. *Environmental Monitoring and Assessment, 119*, 303–330.

Cui, B. S. (2002). Establishing an indicator system for ecosystem health evaluation on wetlands I. A theoretical framework. *Acta Ecologica Sinica, 22*, 1005–1011 [in Chinese with English abstract].

Davies, P. E., Harris, A. J. H., Hillman, T. J., & Walker, K. F. (2010). The Sustainable Rivers Audit: Assessing river health in the Murray-Darling Basin, Australia. *Marine and Freshwater Research, 61*, 764–777.

Duan, X. H., Wang, Z. Y., & Tian, S. M. (2008). Effect of streambed substrate on macroinvertebrate biodiversity. *Frontiers of Environmental Science & Engineering in China, 2*, 122–128.

Fairweather, P. G. (1999). State of environment indicators of 'river health': Exploring the metaphor. *Freshwater Biology, 41*, 211–220.

Feng, Y., Kang, B., & Yang, L. P. (2012). Feasibility analysis of widely accepted indicators as key ones in river health assessment. *Journal of Geographical Sciences, 22*, 46–56.

Flotemersch, J. E., Stribling, J. B., & Paul, M. J. (2006). *Concepts and approaches for the bioassessment of non-wadeable streams and rivers.* Washington, DC: U.S. Environmental Protection Agency.

Gamito, R., Pasquaud, S., Courrat, A., Drouineau, H., Fonseca, V. F., Gonçalves, C. I., et al. (2012). Influence of sampling effort on metrics of fish-based indices for the assessment of estuarine ecological quality. *Ecological Indicators, 23*, 9–18.

Gippel, C. J., Jiang, X. H., Fu, X. F., Liu, X. Q., Chen, L., Wang, H. Z., et al. (2012). *Assessment of river health in the lower Yellow River. Australia-China Environment Development Partnership, River Health and Environmental Flow in China.* Brisbane: Yellow River Conservancy Commission and International Water Centre.

Gippel, C. J., & Speed, R. (2010). *River health assessment framework: Including monitoring, assessment and applications: Australia-China Environment Development Partnership, River Health and Environmental Flow in China.* Brisbane: International Water Centre.

Gleick, P. H. (2008). China and water. Chapter 5 In P. H. Gleick, H. Cooley, M. Cohen, M. Morikawa, J. Morrison, & M. Palaniappan (Eds.), *The world's water 2008–2009: The biennial report on freshwater resources: Pacific Institute for Studies in Development, Environment and Security* (pp. 79–100). Washington, DC: Island Press.

Godden, L. (2005). Water law reform in Australia and South Africa: Sustainability, efficiency and social justice. *Journal of Environmental Law, 17*, 181–205.

Hartmann, A., Moog, O., & Stubauer, I. (2010). "HKH screening": A field bio-assessment to evaluate the ecological status of streams in the Hindu Kush-Himalayan region. *Hydrobiologia, 651*, 25–37.

Heiskanen, A. -S., van de Bund, W., Cardoso, A. C., & Noges, P. (2004). Toward good ecological quality of surface waters in Europe—Interpretation and harmonisation of the concept. *Water Science and Technology, 49*, 169–177.

Hering, D., Feld, C. K., Moog, O., & Ofenböck, T. (2006). Cook book for the development of a multimetric index for biological condition of aquatic ecosystems: Experiences from the European AQEM and STAR projects and related initiatives. *Hydrobiologia, 566*, 311–324.

Hohls, D. R. (1996). *National biomonitoring programme for riverine ecosystems: Framework document for the programme: NBP report series no. 1.* Pretoria: Institute for Water Quality Studies, Department of Water Affairs and Forestry.

Hughes, R. M., & Peck, D. V. (2008). Acquiring data for large aquatic resource surveys: The art of compromise among science, logistics, and reality. *Journal of the North American Benthological Society, 27*, 837–859.

Kapustka, L. A., & Landis, W. G. (1998). Ecology: The science versus the myth. *Human and Ecological Risk Assessment, 4*, 829–838.

Karr, J. R. (1981). Assessment of biotic integrity using fish communities. *Fisheries, 6*, 21–27.

Karr, J. R. (1999). Defining and measuring river health. *Freshwater Biology, 41*, 221–234.

Karr, J. R., & Chu, E. W. (1997). *Biological monitoring and assessment: Using multimetric indexes effectively: EPA 235-R97-001.* Seattle, WA: University of Washington.

Karr, J. R., Fausch, K. D., Angermeier, P. L., Yant, P. R., & Schlosser, I. J. (1986). *Assessing biological integrity in running waters: A method and its rationale: Special publication 5.* Champaigne, IL: Illinois Natural History Survey.

Keizer-Vlek, H. E., Verdonschot, P. F. M., Verdonschot, R. C. M., & Goedhart, P. W. (2012). Quantifying spatial and temporal variability of macroinvertebrate metrics. *Ecological Indicators, 23*, 384–393.

Kelly, M., Bennett, C., Coste, M., Delgado, C., Delmas, F., Denys, L., et al. (2008). A comparison of national approaches to setting ecological status boundaries in phytobenthos assessment for the European Water Framework Directive: Results of an intercalibration exercise. *Hydrobiologia, 621*, 169–182.

Kong, W., Zhang, Y., Meng, W., Gippel, C., & Xu, X. (2013). A freshwater ecoregion classification approach in Taizi River Basin, northeastern China. *Ecological Research, 28*, 581–592.

Ladson, A. R., White, L. J., Doolan, J. A., Finlayson, B. L., Hart, B. T., Lake, P. S., et al. (1999). Development and testing of an index of stream condition for waterway management in Australia. *Freshwater Biology, 41*, 453–468.

Leigh, C., & Gippel, C. J. (2012). *Ecological health assessment of large freshwater rivers, lakes and reservoirs: Study design options, sampling methods and key issues: Australia-China Environment Development Partnership, River Health and Environmental Flow in China.* Brisbane: International Water Centre.

Leigh, C., Qu, X. D., Zhang, Y., Kong, W. J., Meng, W., Hanington, P. J., et al. (2011). *Assessment of river health in the Liao River Basin (Taizi sub-catchment). Australia-China Environment Development Partnership, River Health and Environmental Flow in China.* Brisbane: Chinese Research Academy of Environmental Science and International Water Centre.

Li, F. Q., Cai, Q. H., Qu, X. D., Tang, T., Wu, N. C., Fu, X. C., et al. (2012). Characterizing macroinvertebrate communities across China: Large-scale implementation of a self-organizing map. *Ecological Indicators, 23*, 394–401.

Li, G., Wu, X. P., Luo, Z. L., & Li, J. S. (2011). Establishing an indicator system for biodiversity assessment in China. *Biodiversity Science, 19*, 497–504 [in Chinese with English abstract].

Liang, J. J., Zuo, Q. T., Xia, J., Shi, X. X., & Dou, M. (2010). Theoretical method and scheme of aquatic ecological regionalization in China. *Journal of Irrigation and Drainage, 29*, 47–51 [in Chinese with English abstract].

Liu, X. Q., Gippel, C. J., Wang, H. Z., Leigh, C., & Jiang, X. H. (2016). Assessment of the ecological health of heavily utilized, large lowland rivers: Example of the lower Yellow River, China. *Limnology:18*, (pp. 17–29). http://dx.doi.org/10.1007/s10201-016-0484-9.

Liu, C. M., & Liu, X. Y. (2009). Healthy river and its indication, criteria and standards. *Journal of Geographical Sciences, 19*, 3–11.

Meyer, J. L. (1997). Stream health: Incorporating the human dimension to advance stream ecology. *Journal of the North American Benthological Society, 16*, 439–447.

Ministry of Environmental Protection. (2011). *China national biodiversity conservation, group strategy and action plan*. Beijing: China Environmental Science Press [in Chinese].

Morse, J. C., Bae, Y. J., Munkhjargal, G., Sangpradub, N., Tanida, K., Vshivkova, T. S., et al. (2007). Freshwater biomonitoring with macroinvertebrates in East Asia. *Frontiers in Ecology and the Environment, 5*, 33–42.

National People's Congress. (2002). *Water Law of the People's Republic of China (Order of the President no.74). In: Adopted at the 24th Meeting of the Standing Committee of the Sixth National People's Congress on January 21, 1988, revised at the 29th Meeting of the Standing Committee of the Ninth National People's Congress on August 29*. www.gov.cn.

Norris, R. H., Dyer, F., Hairsine, P., Kennard, M., Linke, S., Merrin, L., et al. (2007). *Australian Water Resources 2005. A baseline assessment of water resources for the National Water Initiative. Level 2 assessment. River and wetland health theme. Assessment of river and wetland health: A framework for comparative assessment of the ecological condition of Australian rivers and wetlands*. Canberra: National Water Commission [Australian Capital Territory].

Norris, R. H., & Hawkins, C. P. (2000). Monitoring river health. *Hydrobiologia, 435*, 5–17.

Norris, R. H., & Thoms, M. C. (1999). What is river health? *Freshwater Biology, 41*, 197–211.

Novotny, V., Bartošová, A., O'Reilly, N., & Ehlinger, T. (2005). Unlocking the relationship of biotic integrity of impaired waters to anthropogenic stresses. *Water Research, 34*, 189–198.

Peng, T., & Chen, X. H. (2009). Assessment of ecosystem health for typical estuary in Haihe River Basin. *Engineering Journal of Wuhan University, 42*, 631–634 [in Chinese with English abstract].

Qiu, X., & Li, H. L. (2009). China's environmental super ministry reform: Background, challenges, and the future. *Environmental Law Reporter, 39*, 10152–10163.

Qu, X. D., Zhang, H. P., Zhang, M., Liu, M., Yu, Y., Xie, Y., et al. (2016). Application of multiple biological indices for river health assessment in northeastern China. *International Journal of Limnology, 52*, 75–89.

Robson, B. J., Chester, E. T., & Austin, C. M. (2011). Why life history information matters: Drought refuges and macroinvertebrate persistence in non-perennial streams subject to a drier climate. *Marine and Freshwater Research, 62*, 801–810.

Rogers, K., & Biggs, H. (1999). Integrating indicators, endpoints and value systems in strategic management of the rivers of Kruger National Park. *Freshwater Biology, 41*, 439–451.

Rosenberg, D. M., & Resh, V. (Eds.), (1993). *Freshwater biomonitoring and benthic macroinvertebrates*. London: Chapman Hall.

Roux, D. J. (2001). Strategies used to guide the design and implementation of a national river monitoring programme in South Africa. *Environmental Monitoring and Assessment, 69*, 131–158.

Ryder, R. A. (1990). Ecosystem health, a human perception: Definition, detection, and the dichotomous key. *Journal of Great Lakes Research, 16*, 619–624.

Schmutz, S., Cowx, I. G., Haidvogl, G., & Pont, D. (2007). Fish-based methods for assessing European running waters: A synthesis. *Fisheries Management and Ecology, 14*, 369–380.

Shah, R. D. T., & Shah, D. N. (2012). Performance of different biotic indices assessing the ecological status of rivers in the Central Himalaya. *Ecological Indicators, 23*, 447–452.

Shen, D. (2009). River basin water resources management in China: A legal and institutional assessment. *Water International, 34,* 484–496.

Shen, D. (2012). Water quality management in China. *Water Resources Development, 28,* 281–297.

Speed, R., Gippel, C., Bond, N., Bunn, S., Qu, X., Zhang, Y., et al. (2012). *River Health and Environmental Flow in China Project: Final report and summary of findings.* Brisbane: International Water Centre.

Stein, E. D., & Bernstein, B. (2008). Integrating probabilistic and targeted compliance monitoring for comprehensive watershed assessment. *Environmental Monitoring and Assessment, 144,* 117–129.

Steward, A. L., Marshall, J. C., Sheldon, F., Harch, B., Choy, S., & Bunn, S. E. (2011). Terrestrial invertebrates of dry riverbeds are not simply subsets of riparian assemblages. *Aquatic Sciences, 73,* 551–566.

Steward, A. L., von Schiller, D., Tockner, K., Marshall, J. C., & Bunn, S. E. (2012). When the river runs dry: Human and ecological values of dry riverbeds. *Frontiers in Ecology and the Environment, 10,* 202–209.

Stoddard, J. L., Driscoll, C. T., Kahl, J. S., & Kellog, J. P. (1998). Can site-specific trends be extrapolated to a region? An acidification example for the northeast. *Ecological Applications, 2,* 288–299.

Stoddard, J. L., Larsen, D. P., Hawkins, C. P., Johnson, R. K., & Norris, R. H. (2006). Setting expectations for the ecological condition of streams: The concept of reference condition. *Ecological Applications, 16,* 1267–1276.

Stranko, S. A., Hurd, M. K., & Klauda, R. J. (2005). Applying a large, statewide database to the assessment, stressor diagnosis, and restoration of stream fish communities. *Environmental Monitoring and Assessment, 108,* 99–121.

Suter, G. W. (1993). A critique of ecosystem health concepts and indexes. *Environmental Toxicology and Chemistry, 12,* 1533–1539.

Tan, P. L. (2002). *Legal issues relating to water use: Issues paper no. 1.* Canberra: Institute for Rural Futures, University of New England, Armidale, Murray-Darling Basin Commission, Australian Capital Territory.

Theobald, D. M., Stevens, D. L., Jr., White, D., Urquhart, N. S., Olsen, A. R., & Norman, J. B. (2007). Using GIS to generate spatially balanced random survey designs for natural resource applications. *Environmental Management, 40,* 134–146.

Verdonschot, P. F. M., Nijboer, R. C., Wright, J. F., Sutcliffe, D. W., & Furse, M. T. (2000). Typology of macrofaunal assemblages applied to water and nature management: A Dutch approach. In J. F. Wright, D. W. Sutcliffe, & M. T. Furse (Eds.), *An international conference on assessing the biological quality of fresh waters: RIVPACS and other techniques* (pp. 241–262). Oxford: Freshwater Biological Association.

Wang, B. X. (2002). *Water quality biomonitoring using benthic macroinvertebrates in China* (Ph.D. thesis). Nanjing: Nanjing Agricultural University (in Chinese).

Wicklum, D., & Davies, R. W. (1995). Ecosystem health and integrity? *Canadian Journal of Botany, 73,* 997–1000.

Williams, D. D. (2006). *The biology of temporary waters.* New York, NY: Oxford University Press.

Wishart, M. J. (2000). The terrestrial invertebrate fauna of a temporary stream in Southern Africa. *African Zoology, 35,* 193–200.

Yan, F., He, D. M., & Kinne, B. (2006). Water resources administration institution in China. *Water Policy, 8,* 291–301.

Yin, M., Yang, Z. F., & Cui, B. S. (2005). Eco-hydrological regionalization of river system in China. *Acta Scientiae Circumstantiae, 25,* 423–428 [in Chinese with English abstract].

Zhang, Y., Zhao, R., Qu, X. D., & Meng, W. (2013). The researches of integrated river health assessment of Liaohe River basin. *Engineering Science, 15,* 11–18 [in Chinese with English abstract].

Zhou, Y., Xi, B. D., & Su, J. (2011). *Water pollution control and treatment in China: Geophysical Research Abstracts 13, EGU2011-4407.* EGU General Assembly 2011.

Zhu, D. S., Zhang, J. Y., Li, Y., & Shi, X. X. (2011). Key technologies for aquatic ecosystem protection and rehabilitation planning. *Water Resources Protection, 27,* 59–64 [in Chinese with English abstract].

CHALLENGES FOR THE FUTURE

FUTURE CHALLENGES

19

B.T. Hart*,†, J. Doolan‡, S.E. Bunn§, A. Horne¶, C.A. Pollino**, R. Rendell††, A. Webb¶

Water Science Pty Ltd, Echuca, VIC, Australia* Monash University, Melbourne, VIC, Australia† University of Canberra, Canberra, ACT, Australia‡ Griffith University, Brisbane, QLD, Australia§ University of Melbourne, Melbourne, VIC, Australia¶ CSIRO Land and Water, Canberra, ACT, Australia** RM Consulting Group, Bendigo, VIC, Australia††

CHAPTER OUTLINE

INTRODUCTION

Australia's approach to water resource policy, planning and management has evolved considerably over the last 100 or so years. Throughout this evolution, a dominating theme has been the need to secure water supplies and buffer the effects of drought, which is a constant feature of our highly variable climate.

The Australian water resource reform journey commenced in the late 1800s under Alfred Deakin. In establishing a federated Constitution, Deakin ensured that ownership of water was vested with the Crown and that the management of water resources was the responsibility of state governments. Following this, state laws were enacted governing basic water allocation and water delivery.

Following this establishment phase, the key focus was on building water infrastructure to provide cities and towns with reliable water supplies, and developing irrigation networks to grow food and promote regional development. This focus on development—the "build and supply" phase—lasted until the late 1970s and it was during this period that most of the infrastructure that currently services Australia's urban and rural communities was constructed. Even in this early phase where the focus was on infrastructure development, water resource managers of the time showed considerable initiative and vision in the legislative frameworks and management systems that they established to allocate and administer the resources they then controlled. Water systems were constructed using the best contemporary understanding of the hydrology of the water systems involved, water allocation systems were put in place that reflected the variability of water availability, and basic arrangements for water measurement were put in place (e.g., using the Dethridge wheel).

This phase of Australia's water history was governed by a different set of social and political imperatives from those that exist today, and it reflected the philosophy and practices of an immature water management system. And while this early phase provided a firm foundation for contemporary water resource management, it also left a number of significant legacy issues, including large government debt, poor pricing policies, service delivery challenges and unaccounted externalities, represented as widespread environmental damage. This triggered governments, in the 1980s and 1990s, to dramatically transform the way water is managed in Australia, shifting the focus from development to pursuing productivity and efficiency within capped water allocations. This saw the emergence of a new paradigm of economically and environmentally sustainable water resource management that sought to provide a secure basis for future investment, and to yield high returns to the community.

Doolan (2016a) summarized the aims of the recent Australian water reform agenda as being "to increase the productivity and efficiency of Australia's water use and ensure the health of river and groundwater systems, whilst servicing rural and urban communities." Furthermore, she identified the following as the key areas being pursued as part of this reform agenda:

- Transforming water allocation and establishing water markets,
- Improving environmental management,
- Reforming pricing of water services,
- Modernizing institutional arrangements,
- Ensuring community and stakeholder engagement in all reform processes,
- Improving water information and water knowledge.

As documented in this book, there has been substantial, wide-ranging reform over the past three decades in Australia in all of these key areas, with clear and demonstrable outcomes.

However, despite the major gains that have been made over the last 30 years, it is clear that there is still a need to maintain the focus, commitment and pace of water reform if we are to manage water to support a healthy environment, a prosperous economy and thriving communities into the future (ATSE, 2015).

Australia is facing a number of challenges that will continue to have major implications for water resource management. In our view, the five biggest challenges to water management over the next three decades to 2050 will be: *climate change, population growth, water-energy interactions, increasing community expectations and demands,* and *maintaining affordability.* These are not the only challenges. Nevertheless, it is inarguable that these five areas will present significant challenges over the coming decades.

These challenges will demand new trade-offs and create new policy dilemmas that will require the same attention and effort in decision making that has prevailed for the past 30 plus years. And they will create a sixth challenge—that of *maintaining the impetus for and commitment to water reform in the future.*

In the sections following, these challenges are briefly described to provide an indication of the scale and seriousness of the issues faced. We then discuss how the challenges will influence future decision making in rural, urban and environmental water management, as managers and researchers search for solutions to these pressing issues. In our view, the responses to these challenges will form the basis for the next wave of water reform in Australia.

KEY CHALLENGES FOR AUSTRALIA
CLIMATE CHANGE

Australia, in common with much of the rest of the world, is starting to feel the effects of a changing climate. Over the next 30 years to 2050, we face the prospect of increased variability as *climate change* starts to really bite. The most recent climate change projections predict that Australian average temperatures will continue to increase in all seasons, with more hot days and warm spells (CSIRO & BOM, 2015, 2017). This will be accompanied by a continuation in the current trend of decreasing winter rainfall. In southern and eastern Australia, there will also be changes in the seasonality of rainfall with predicted decreases in spring rainfall and possible changes in summer and autumn. In these areas of the country, there will be a consequential significant decrease in stream flows and increase in the frequency and severity of droughts, accompanied by an increase in the intensity of storms (BOM, 2016; CSIRO & BOM, 2015, 2017; Leblanc et al., 2012). It should also be noted that recent research has shown that postdrought the volume of runoff generated from a rainfall event does not necessarily return to the predrought relationship (Saft et al., 2015, 2016). Also of relevance for coastal cities and towns will be the continued rise in sea levels and the increase in the height of extreme sea-level events (Milly et al., 2008).

The specter of less runoff, warmer temperatures, more frequent, more intense and longer droughts, more intense storms and potentially less frequent but bigger floods will raise significant challenges for the environment, agriculture (particularly irrigation) and urban centers in Australia. This will be accompanied by a predicted increase in wildfires and changes in vegetation distribution as the climate changes, which will also modify runoff and stream flows.

The Millennium drought (1997–2009), which was followed by some of the biggest floods on record for much of the country, provided Australia with "real life" experience of the challenges that climate change will create for water management in each of these sectors (Kiem, 2013; van Dijk et al., 2013). Some have argued that the reason irrigated agriculture in southeastern Australia handled this drought relatively well was a direct result of the reforms that occurred prior to and during the drought (Doolan, 2016b, 2016c).

For urban water management, the Millennium drought made clear the increased uncertainty regarding the reliability of future water sources and the need to develop new and innovative ways to ensure water security (Head, 2014). Also it made clear the problems with using conventional predictive models based on historical rainfall/runoff data that perform poorly in capturing extremes (see Chapter 5).

POPULATION GROWTH

Australia's population is projected to increase to between 36.8 million and 48.3 million by 2061, and to reach between 42.4 million and 70.1 million in 2101, representing an increase of 1.4% over this period (ABS, 2016).

These increases in population will be mainly felt in the capital cities and regional centers. For example, the population of greater Melbourne is predicted to increase from 4.5 million in 2015 to somewhere between 7.6 million and 9.8 million by 2061 (ABS, 2016). The Victorian government, in their recently released Victorian Water Plan, is planning for a state population of 10.1 million by 2050 (DELWP, 2016a, 2016b).

The expected population increases over the next 30 years, coupled with climate change, will challenge water managers in several ways. Urban water managers will be required to at least maintain the level and standard of water supply, wastewater management, flood protection and environmental outcomes that they currently provide to cities and towns that will be growing rapidly. However, as these urban areas grow, communities are also increasingly appreciating and valuing "liveability." This has real implications for urban water management because water plays a critical role by providing for green spaces, recreational areas, wetlands and streams, and urban cooling—all of which have been shown to be important in maintaining mental and physical health and well-being (DELWP, 2016b).

Population growth will also have significant implications for rural water management, as there will be an inevitable increased demand for more food to meet the needs of the larger domestic market. This is expected to drive increased agricultural activity in the Murray-Darling Basin and the growth of other key irrigation areas across the country.

WATER-ENERGY INTERACTIONS

In Australia, the water sector uses about 45 petajoules of energy for transporting water, representing about 1.4% of all Australia's energy use (Apostolidis, 2012). This use of energy results in the emission of greenhouse gases through two pathways:

- *Indirect emissions* through the use of energy to pump water around, to treat water and wastewater to safe standards, and to heat water in houses;
- *Direct emissions*, particularly from wastewater treatment plants.

In Victoria, the water sector emits almost 1 million tonnes of carbon dioxide equivalents each year and is responsible for almost 25% of the emissions from all government activities in that state (DELWP, 2016b).

As governments seek to meet international obligations and reduce carbon emissions, there will be requirements for the water sector to do its part. As part of this, there are now major moves to reduce both energy use and costs in the water sector (WSAA, 2012). For example, the Victorian government has committed to a long-term target of net-zero greenhouse emissions by 2050. In line with this, the recent *Water for Victoria* plan requires the four metropolitan water authorities to examine a pathway to achieving net-zero emissions by 2030, with all other water authorities plotting pathways to meet the 2050 target (DELWP, 2016b).

The urban water sector has already reduced emissions by developing more energy-efficient equipment, capturing biogas for energy generation, and investing in renewable energy generation, particularly where these have been cost-effective business decisions. However, as they seek in the future to augment water supplies with alternative, non–climate dependent water sources (e.g., desalination plants), they will be faced with a dilemma in that the energy use associated with these sources is often greater and more costly than that for more traditional water sources.

In irrigated agriculture, as irrigators seek to improve their water efficiency by using technology, such as pumping and systems like drip irrigation, they will find that these use more energy. In this area of water-energy interactions, there is often a trade-off between improving water efficiency and increasing carbon emissions, and a further trade-off between reducing carbon emissions and the cost of service provision.

INCREASING COMMUNITY EXPECTATIONS AND DEMANDS

Water managers in the future will have to deal with an increase in the expectations and demands of the community. The Millennium drought in Australia showed very clearly the social dependence on and importance of water in both urban and rural environments. During the drought, local lakes and streams dried up in rural areas (often for the first time) and urban communities were on stringent water restrictions for extended periods (sometimes years) with very limited water available for parks and private and public gardens (Doolan, 2016c; Lindsay et al., 2016). The communities in all major capital cities responded very positively to the need for water conservation and changed their expectations around what was an appropriate use of water (Lindsay et al., 2016). Also during this period, rural communities became increasingly aware of how important those environments were for amenity, regional tourism and recreation, and how much influence they had on local economies, liveability and social well-being.

It is inevitable that communities will expect more from their water services as they look towards a future with less water. In urban areas, this will include the capacity to maintain green spaces, recreation areas and water environments, even during periods of extended drought. Thus, as our cities become larger and more highly urbanized with population growth, the effective use of water will become even more important for maintaining liveability and social well-being.

In rural areas, there is often pressure to use environmental water to maintain environments of local economic or social importance. In many cases, this can be compatible with good environmental outcomes, but this is not always the case. On occasions, managers will be required to work through some difficult trade-offs.

These increasing expectations and demands will require innovation from all water managers—urban, rural, and environmental—to get the utmost benefit from an increasingly scarce and valuable resource. Solutions will need to meet their primary requirements, but also maximize secondary environmental, social or economic benefits for their communities. As we move into the Information Age, this will be achieved through improved access to data, both at the site and national scales, which can lead to real-time decision making and greater transparency of the risks and benefits of water to both managed and unmanageable parts of the system. The challenge for water authorities will be to provide sufficient amenity value from reduced amounts of water, and to do it in a way that does not increase the cost to do this.

A final area where there is a significant increase in community expectations and requirements is from our indigenous communities. To date, Traditional Owners have had very little involvement in decision making on water management despite having a significant dependence on and great knowledge about water in Australia. In recent years, there has been a major drive from these communities to take a greater role in water management, seeking access to both the consumptive and nonconsumptive pools to improve cultural, social and economic outcomes for indigenous communities (MLDRIN, 2009; FVTOC, 2014; NNTC, 2014). While work in this area has commenced with recent Victorian government commitments (DELWP, 2016b), the granting of cultural access licenses in NSW water sharing plans,[1] and initiatives such as the National Cultural Flows Research Project[2] and the Living Murray Indigenous Partnerships Program,[3] progress has been slow and it remains an ongoing issue for the future.

Thus, a further challenge will be how to deliver indigenous access to water for cultural reasons and economic development, and to increase the participation of Traditional Owners in water planning and management over the coming decades.

AFFORDABILITY

A key objective of water reform in Australia has been to ensure that the water industry is financially sustainable, achieved through the introduction of cost-reflective water pricing (see Chapter 1). Thirty plus years of water reform have seen this objective largely achieved in metropolitan areas, mostly achieved in other urban areas and significant progress made in irrigation in relation to the costs of service provision. However, there is still some debate about what this actually means, e.g., the current policy settings do not include full accounting of environmental externalities or any form of resource rent and there are still differences about how assets should be valued. How this is done is still a challenge, with techniques such as nonmarket benefit valuation and ecosystem services showing some promise.

Regardless of any future policy directions, current policy settings have seen water prices increase. In particular, prices in urban areas have significantly increased over a short period of time to pay for large-scale augmentations (e.g., desalination plants) undertaken in response to drought and population growth. This has put significant pressure on households that are also dealing with cost-of-living increases associated with rising cost of electricity and gas, and other cost increases.

Prices for irrigation services have also increased, although not to the same extent as in urban areas. Moreover, as new water-efficient technologies for farms and irrigation systems are introduced, water savings occur but recurrent costs increase. For new irrigation enterprises, the initial purchase of their water entitlement is also a consideration, as water entitlements must now be bought on the water market. Water entitlements are now a key capital cost for irrigated agricultural businesses, with the value of water being around 50% of a rice property, 40% of a dairy farm and 30%–35% of a horticulture farm (Rob Rendell, RMCG, Pers. Comm. Jan. 2017).

Given the preceding discussion, affordability has become a real focus for governments and, as a result, continues to be a significant challenge for the water industry as it looks to deal with the challenges of climate change, population growth and increasing community expectations.

FUTURE DECISION MAKING TO ADDRESS THESE CHALLENGES

Water resource managers are adaptively dealing with the challenges of implementing the existing national water reform agenda, responding to a variable climate, changing community expectations and political environments. The challenges described in this chapter add new and emerging issues and new dimensions to water resource decision making. In the sections following, we describe the key issues and decision-making challenges that will be posed for each of the urban, rural and environmental water management sectors.

---

[1] www.water.nsw.gov.au/watersharing plans.
[2] culturalflows.com.au/.
[3] www.mdba.gov.au/aboriginalpartnershipprograms.

URBAN WATER MANAGEMENT

Over the next 30 years, urban water managers will need to provide water supply, sewerage and liveability services for rapidly growing cities and towns in a much drier climate, while cutting carbon emissions and with their water prices under strict scrutiny.

This raises several key issues for urban water policy and management, including:

- *The need to develop a diverse range of water sources while making the most from existing water supplies,* e.g., using variably treated water to suit alternative uses. Current policy settings and community attitudes do not always support this approach, with community debates about what should be allowable uses for stormwater and recycled water. As housing density increases in cities, the opportunities for cost-effective, fit-for-purpose use decrease (with smaller gardens or no gardens at all). In large coastal cities, desalination is one climate-independent option for augmentation of urban water supplies, but the excessive energy requirements pose a related challenge. Managed aquifer recharge of stormwater is also being explored at scale, but pretreatment costs can be high. Increasingly, cities and towns are looking to groundwater or surface water trading for their next augmentations, particularly where available catchments and aquifers are at their sustainable diversion limits.

- *The role of the water industry in providing water to enhance liveability*:
  - **(a)** *Water availability*—There is real potential to manage stormwater to more effectively provide green spaces and improved water environments. However, this is frequently a costly option and it is difficult to retrofit older urban areas. Stormwater management is also complicated by boundary issues between planning departments, local governments and water managers. Rather than broad-scale approaches, local recycled water systems can augment water supplies. As these schemes are developed, it will require governments to define roles and responsibilities in the new area of provision of "liveability" services, determining cost-effective and efficient mechanisms of augmenting water supplies, including being prepared to allow higher prices or higher rates to pay for them
  - **(b)** *Integrated planning*—In most urban centers, land use planning and urban water planning are not (or poorly) integrated, with the result that opportunities for innovative, resilient, and cost-effective water system development in the (extensive) new growth areas are lost. Significant institutional and regulatory reform is needed to address existing barriers to integrated planning, taking into account objectives across multiple domains such as water security and housing affordability.

- *Setting standards for services* particularly during drought, including understanding the role of water conservation and restriction policies in the provision of water supply. Once again, this will result in a trade-off for communities between water security and water price.

- *Moving to carbon neutrality* will involve initially making sensible business decisions to reduce electricity consumption or to generate their own electricity. Many of these "low-hanging fruit" decisions have already been taken. In future years, low carbon operations will require innovation and will probably cost more to run, again reflecting the need for careful consideration of implementation strategies to mitigate impacts on water prices.

Given all of this, there will be an ongoing imperative for urban water managers to improve their operational and economic efficiency. This includes examining the issue of the role of the private sector in providing urban water services. Currently, urban water services in Australian urban centers are largely provided by government entities, although they may contract out significant components.

While much of this future will be highly challenging, it will also create an environment for real innovation, which will see the emergence of new technologies and improved local integrated solutions to water and liveability issues.

RURAL (IRRIGATION) WATER MANAGEMENT

Over the next 30 years, irrigators will need to increase productivity to provide for a growing population, but in a drier climate and within established limits on the amount of available water (sustainable use limits). The imperative will be on the irrigation industry to continue to become more efficient, carrying on the trend of "producing more with less water." This is entangled with correctly valuing agricultural products, ensuring primary industries, such as the dairy industry, can remain profitable and continue to invest in best practice of water use and treatment.

In the Murray-Darling Basin, it seems likely that successful irrigation industries will need to optimize the mix of enterprise types with an eye to maximizing production in a highly variable and drying climate, and to provide sufficient flexibility to operate within wetter and drier climatic periods. For example, in drier years, the market will drive the available water to enable the higher-value enterprises (e.g., horticulture) to be sustained, but at the temporary expense of lower-value crops. This will result in irrigation enterprises needing to make smart business decisions that trade off between crop type, crop yield and water reliability. Technology, including access to real-time sensor network data and reliable forecasting, has a great potential to help rural industries make better decisions.

In the Murray-Darling Basin, it is highly likely that the value of water will rise in line with the value of production. This will create a greater driver for efficiency in all aspects of water management, including river operations and environmental water management. It is inevitable that the system will be managed adaptively, and knowledge gaps will be addressed where possible. As well, multiple benefits of water use and management will be sought and communities will become empowered as part of decision processes.

One likely future is that irrigation authorities in the Murray-Darling Basin will have to cope with the challenge of delivering less water with existing infrastructure in an increasingly variable climate. This will put increased pressure on the price for irrigation services, the customer's capacity to pay, and the financial viability of irrigation enterprise. This is a trade-off that has been a constant in Australian water management for the past 100 years, but is likely to become more acute.

Increasingly, there will be pressure to better manage catchment water sources (both surface and groundwater), storages and delivery pathways for consumptive, environmental and cultural (indigenous) uses. This will inevitably see the development of new methods for balancing the needs and desires of irrigators, local communities, Indigenous groups and the environment. It will require better understanding of community values and perspectives, and allowing them to participate in decision making.

Outside the Murray-Darling Basin, there will be pressure to grow production from other major irrigation schemes, which remain unprofitable at this stage. In addition, there will be a push to achieve higher environmental standards and meet targets for diffuse pollution in critical catchments, such as those that discharge to the Great Barrier Reef.

Once again, while much of this is challenging, it is an area where innovation and improvements in forecasting, smarter river operations, farm and system delivery technology and crop types and yields, will make a substantial difference.

ENVIRONMENTAL WATER MANAGEMENT

The next 30 years will pose significant challenges for water, people and the environment. Climate change will mean that there is less water within the landscape overall. Unregulated rivers and lake systems will dry more frequently. Hart et al. (2017a) have shown that, given the current policy settings (at least in the Murray-Darling Basin Plan), the environment will bear the brunt of a drier climate, receiving disproportionately less water. Additionally, the sequencing of wet and dry spells, both preceding and within a drought, can significantly influence ecological outcomes during the drought period, but results in only minor changes to consumptive allocations (Wang et al., 2017). In regulated systems, where environmental water managers hold water entitlements, there will be more opportunity to mitigate the impacts of climate change. Groundwater dependent ecosystems will also be subject to change with reduced recharge of aquifers. Moreover, there will be significant and increasing pressure to get as many social and other community benefits as possible from the use of this water. This could include water for recreational and amenity purposes.

In urban systems, as cities and towns grow, the challenge will be to maintain or improve stream and wetland health. This will be particularly challenging in existing urban areas, but offers many opportunities in new "greenfield" growth areas. In the latter, new ways to capture, treat and utilize stormwater and domestic wastewater for environmental benefit are already occurring, and are likely to expand (City West Water, 2016).

Climate change will bring several difficult issues for environmental water policy and management including:

- *Determining environmental objectives for managing rivers, streams and wetlands under climate change*. With the pressures placed on many of our river and wetland systems due to climate change, it will be necessary to become far more sophisticated in setting management objectives to meet future community needs. It will mean managing systems to ensure they have the greatest chance of adapting to a drier climate rather than looking backwards at what they were (e.g., pre-European state). This will be challenging in resolving the balance between environmental values and the need to provide for the community's social, cultural and economic requirements. This is an anthropocentric view, but it does represent the reality of contemporary environmental management.
- *Improving the management of river systems for environmental outcomes*. Despite some improvements, the highly regulated sections of the Murray-Darling Basin are still managed largely to optimize consumptive water outcomes. The future challenge will be to operate these river systems with a view to optimizing the outcomes for both irrigators and the environment. For environmental water, this will require the development of new modeling tools that provide support for active management decisions and allow representation of how management decisions might change under a drying climate (Horne et al., 2017). Current models focus on long-term planning and flow targets, and not on supporting active management. Most do not perform well in estimating changes to low flows, which often have high environmental significance (see Chapter 6).

- *Demonstrating to communities that the environmental water entitlements are being managed to maximum effect.* The Australian government has made a significant investment in providing water for the environment (see Chapter 13), but not without considerable community concern about the impact this has had on irrigation-dependent communities. Thus, there is an imperative for environmental water managers, and particularly the Commonwealth Environmental Water Office and the Murray-Darling Basin Authority, to demonstrate that the investment in environmental water has benefited the environment. This will be a significant challenge given that environmental recovery can take a long time to become apparent.
- *Improving the methods for measuring the social, human-health and indigenous benefits of environmental flows.* There is some research around the benefits of healthy environments for mental and physical health of resident populations. But currently, these values are not seriously considered in the assessment of benefits of environmental watering. The challenge will be to develop improved methods for assessing the broader societal benefits of environmental water, and with this to improve the case for retaining and, where necessary and agreed, increasing the volumes of environmental water.
- *Ensuring that associated complementary tasks of habitat restoration, water quality management and catchment management are maintained to maximize the value gained from environmental water.* There is potential for these essential activities to be downplayed compared with the focus on the provision of water for the environment, and particularly as the amount of that environmental water is reduced due to climate change. It will be a challenge to maintain these complementary management activities into the future to ensure the desired environmental outcomes are achieved.
- *Developing an understanding within the community and governments that these environmental assets underpin the economy and well-being of Australian communities.* If this can be achieved, the case for long-term management and investment regimes, similar to those existing for other water infrastructure assets (e.g., reservoirs), will be easier to achieve. In Victoria, an environmental contribution is legally required from water authorities to assist in the long-term funding of waterways management and monitoring.[4]

Again, while these are challenging issues, there are also opportunities. There are real opportunities for river managers and other water users to work together to optimize water delivery and get the best environmental outcomes and efficient delivery of irrigation water. Already investment in environmental "works and measures" has resulted in enhanced environmental outcomes with less environmental water[5] (DPIE, 2013; NSW-OW, 2013), although these investments are not without some criticism (Pittock & Finlayson, 2013). Additionally, the Australian and state governments have developed a "constraints management strategy" aimed at reducing some of the major constraints to more efficient environmental watering in the Murray-Darling Basin (MDBA, 2013). A key plank in this strategy is the minimization of constraints (e.g., levees, flooding of minor bridges, flooding of agricultural land) that are preventing more water getting onto floodplains. Again, this strategy has run into considerable local opposition in some regions.

---

[4]www.depi.vic.gov.au/water/rivers-estuaries-and...implementation-and-monitoring.
[5]www.agriculture.gov.au/water/mdb/programmes/.../environmental-works-measures.

A FINAL CHALLENGE—MAINTAINING THE MOMENTUM AND COMMITMENT TO WATER REFORM

The preceding sections show clearly that the need for ongoing effort in water reform in Australia is still required, even after three decades of concerted effort. Indeed, there is unlikely to ever be an "end" to water reform, with a constantly evolving human and natural environment requiring an ongoing effort to continuously reform and improve water management. This, in itself, will be a challenge. As Australian water managers have found, water reform is complex, hard, politically challenging and resource intensive and can only proceed at a pace that the community can adapt to (Chapter 1; Doolan, 2016a).

In some areas of Australia, the combination of water reform (including the Murray-Darling Basin Plan), the Millennium drought, increasing fuel and energy costs, and fluctuating commodity prices have created conditions where there is considerable community hardship that is being attributed to the Basin Plan specifically and water reform more generally.

Conversely, however, the Millennium drought created the conditions where communities and politicians were more prepared to make the effort to make major advances in water policy and management. These greatly increased the efficiency of water use. However, since the breaking of the drought in 2010, there has been some increase in per capita water use in urban environments (although nowhere near predrought levels), and politicians arguing that environmental allocations need review. It is clear that the drought provided a major point of focus (a crisis) that, with the groundwork laid over the previous two decades, allowed further progress in water reform to be made.

The challenge therefore is to maintain momentum for water reform, particularly during wetter periods when the need for reform is not so obvious, and when the increased costs that might come with a different approach to water management might not appear justified to those focused on the short-term availability of water.

Additionally, community controversy and backlash can affect the political will for further reform, and because of this water reform can easily be moved to "the back burner" for governments. All parties can be (either separately or mutually) fatigued by the ongoing conflict, resulting in a decreased appetite for any further reform.These issues highlight the importance of ongoing community and stakeholder engagement, obviously in the development of policies, but also in their implementation where community attitudes can change depending on their circumstances. This requires a capacity to adapt to and manage such changes.

Further, governments have spent significant amounts of money on water initiatives. For example, the Australian government is spending around $13 billion on the development and implementation of the Basin Plan alone as well as other water-related initiatives. State governments also invested heavily during the Millennium drought on water efficiencies, water for the environment, new infrastructure and drought relief. In these times of economic uncertainty and budget austerity, and without the immediate imperative of drought (a crisis), there is always a risk that governments will view water reform as "having had its turn."

Another potential issue is that the current generation of water managers and key water stakeholders who were responsible for developing and implementing the reform agenda over the last 30 years are retiring and moving on. The key agencies are also being "hollowed out" due to restructuring and downsizing. It is imperative that this huge body of institutional knowledge, as well as "the fire in the belly" for reform, is passed on to the next generation of water managers to ensure that the economic, social and environmental gains from past reforms are not lost, but are built on for greater future benefits. There are some excellent signs that this is occurring, with the emergence of a younger generation of water planners and managers.

Despite the difficulties, governments are recognising the need for continued reform. The Victorian Government recently released its *Water for Victoria—Strategy* (DELWP, 2016b), which outlines water reform over the next 10 years in that state. It is to be hoped that 13 years after the National Water Initiative (NWI), the other state and Australian governments will not only maintain their commitment to the existing national water reform framework but will also recognize the need to review, renew and extend their own water reform programs and the collective national agenda.

A key test of this for the Australian government and the northern states of Queensland, Western Australia and Northern Territory will be to ensure that the new developments planned for northern Australia (Aust Govt, 2015) are undertaken within the policy framework of the National Water Initiative (NWI). Hart et al. (2017b) have stressed the crucial importance of water in underpinning any sustainable irrigated agricultural activities in the north, and have discussed the key aspects that will need to be addressed in developing a robust, transparent and coordinated water resource planning process to support sustainable irrigated agriculture in northern Australia, based on NWI principles. Others have also commented on the need for strong governance regarding these developments (Humphries et al., 2017; Stephens et al., 2015).

CONCLUSIONS

This book has sought to summarize the key aspects of the Australian water reforms over the past 30 years, and to capture the learnings from them, particularly those elements that have contributed to improved decision making. This collective review and analysis of the Australian experience should assist in advancing the decision-making process in water resources policy, planning and management, both in Australia and overseas.

As a developed country, with demonstrated experience in decision making in water resources management, Australia has some capacity to assist other countries, particularly developing countries, in their water resources management journeys. This may become more useful, as our region becomes drier and hotter, with less water, and therefore greater water scarcity.

Australia also has an opportunity to contribute to water reform on the world stage. In a recent (December 2016) and important move, the United Nations declared the *2018–2028 Water for Sustainable Development* decade to raise awareness of the critical state of water resources around the world and to inspire more action.[6] This represents a significant opportunity for Australia to play a role in helping other countries achieve their sustainable development targets for water.

The preceding sections in this chapter have identified the key issues that in our view will need to be addressed in the next wave of water reform in Australia. Working through these issues and trade-offs will require the same attention and effort in decision making that has prevailed for the past 30 years. It will require the same or greater level of commitment to community and stakeholder engagement.

---

[6]http://www.tajikistan-un.org/2-uncategorised/104-adoption-of-a-resolution-entitled-international-decade-for-action-water-for-sustainable-development-2018-2028.

It will require a much greater level of partnership with Traditional Owners than we have had previously and it will require continued investment in knowledge and information to ensure that future decision-making processes are based on the best evidence available.

Australia has had considerable success in the past with water reforms. What is required in the future is the acknowledgement from governments and communities of the benefits realized, and the social, political and legislative license to press on with the difficult and ongoing task of further water reform.

ACKNOWLEDGMENTS

We are most grateful for the comments received from Chris Barlow, Jon Brodie, Joanne Chong, Amber Clarke, Campbell Fitzpatrick, Chris Gippel, Tony Jakeman, Sarina Loo, Poh-Ling Tan and Peter Wallbrink.

REFERENCES

ABS. (2016). *Population projections, Australia, 2012 to 2101*. Canberra: Australian Bureau of Statistics. www.abs.gov.au/ausstats/abs@.nsf/Lookup/3222.0main+features52012%20%28base%29%20to%202101.

Apostolidis, N. (2012). *Australian experience in water and energy footprints*. Rockingham, WA: National Centre of Excellence in Desalination, Murdoch University. http://desalination.edu.au/2012/12/australian-experience-in-water-and-energy-footprints/#.WHb7lGY1pLU.

ATSE. (2015). *Australia's liquid assets: Meeting our water reform challenges*. Melbourne: Australian Academy of Technological Sciences and Engineering. 60pp.

Aust Govt. (2015). *Our north, our future: White paper on developing northern Australia*. Canberra: Australian Government. 201pp.

BOM. (2016). *Recent rainfall, drought and southern Australia's long-term rainfall decline*. Bureau of Meteorology. www.bom.gov.au/climate/updates/articles/a010-southern-rainfall-decline.shtml Accessed 01.12.16.

City West Water. (2016). *Integrated water cycle management strategy—Summary document*. Melbourne: City West Water. 22pp.

CSIRO and BOM. (2015). *Climate change in Australia: Information for Australia's natural resource management regions*. Technical report Melbourne: CSIRO and Bureau of Meteorology. 212pp.

CSIRO and BOM. (2017). *Climate change in Australia*. Melbourne: CSIRO and Bureau of Meteorology. www.climatechangeinaustralia.gov.au/ Accessed 11.01.17.

DELWP. (2016a). *Victoria in future 2016: Population and household projections to 2051*. Melbourne: Department of Environment, Land, Water and Planning. 20pp.

DELWP. (2016b). *Water for Victoria—Strategy*. Melbourne: Department of Environment, Land, Water and Planning. 192pp.

Doolan, J. (2016a). The Australian water reform journey: An overview of three decades of policy, management and institutional transformation. In *Paper prepared for the Australian Water Partnership, Canberra, August 2016*. 32pp.

Doolan, J. (2016b). Building resilient economies and communities through the effective management of water scarcity and drought: A framing paper for the high-level panel on water. In *Paper prepared for the Australian Water Partnership, Canberra, June 2016*. 11pp.

Doolan, J. (2016c). Building resilience to drought: The Millennium drought and water reform in Australia. In: *Paper prepared for the Australian Water Partnership, Canberra, August 2016*. 31pp.

DPIE. (2013). *Victorian Murray Darling Basin environmental works and measures feasibility program—Final report*. Melbourne: Department of Primary Industries and Environment. 21pp.

FVTOC. (2014). *Victorian traditional owners' water policy framework*. Melbourne: Federation of Victorian Traditional Owner Corporations. http://fvtoc.com.au/wpcontent/uploads/2013/10/Water-Policy-Framework.pdf.

Hart, B. T., McLeod, A., & Neave, I. (2017a). The Murray-Darling Basin plan and climate change. *Australasian Journal of Water Resources*, [submitted].

Hart, B. T., O'Donnell, E., & Horne, A. (2017b). Water resources planning for Northern Australia. *International Journal of Water Resources Development*, [submitted].

Head, B. W. (2014). Managing urban water crises: Adaptive policy responses to drought and flood in Southeast Queensland, Australia. *Ecology and Society*, *19*(2), 33.

Horne, A., Kaur, S., Szemis, J., Costa, A., Webb, J. A., Nathan, R., et al. (2017). Using optimization to develop a "designer" environmental flow regime. *Environmental Modelling & Software*, *88*, 188–199.

Humphries, F., Anton, D. K., Tan, P. -L., Akhtarkhavari, A., Butler, C., & England, P. (2017). Ecological governance and the development plan for Northern Australia. *Australian Environmental Review*, [submitted].

Kiem, A. S. (2013). Drought and water policy in Australia: Challenges for the future illustrated by the issues associated with water trading and climate change adaptation in the Murray–Darling Basin. *Global Environmental Change*, *23*, 1615–1626.

Leblanc, M. J., Tweed, S. O., Van Dijk, A. I. J. M., & Timbal, B. (2012). A review of historic and future hydrological changes in the Murray-Darling Basin. *Global and Planetary Change*, *80–81*, 226–246.

Lindsay, J., Dean, A. J., & Supski, S. (2016). Responding to the Millennium drought: Comparing domestic water cultures in three Australian cities. *Regional Environmental Change*. http://dx.doi.org/10.1007/s10113-016-1048-6.

MDBA. (2013). *Constraints management strategy 2013–2024, Publication No. 28/13*. Canberra: Murray-Darling Basin Authority.

MLDRIN. (2009). *Echuca declaration*. Murray and Lower Darling Rivers Indigenous Nations. www.savanna.org.au/nailsma/publications/downloads/MLDRIN-NBAN-ECHUCA-DECLARATION-2009.pdf.

Milly, P., Betancourt, J., Falkenmark, M., Hirsch, R., Kundzewicz, Z., Lettenmaier, D., et al. (2008). Stationarity is dead: Whither water management? *Science*, *319*, 573–574.

NNTC. (2014). *Recognising indigenous water interests in water law—A submission by the National Native Title Council to the 2014 review of the Water Act 2007*. National Native Title Council. www.agriculture.gov.au/SiteCollectionDocuments/water/63-national-native-title-council.pdf.

NSW-OW. (2013). *NSW environmental works and measures feasibility program—Evaluation report*. Sydney: NSW Office of Water. 40pp.

Pittock, J., & Finlayson, C. M. (2013). Climate change adaptation in the Murray-Darling Basin: Reducing resilience of wetlands with engineering. *Australian Journal of Water Resources*, *17*, 161–169.

Saft, M., Peel, M. C., Western, A. W., Perraud, J. -M., & Zhang, L. (2016). Bias in streamflow projections due to climate-induced shifts in catchment response. *Geophysical Research Letters*, *43*, 1574–1581.

Saft, M., Western, A. W., Zhang, L., Peel, M. C., & Potter, N. J. (2015). The influence of multiyear drought on the annual rainfall-runoff relationship: An Australian perspective. *Water Resources Research*, *51*, 2444–2463.

Stephens, A., Oppermann, E., Turnour, J., Brewer, T., & O'Brien, C. (2015). Identifying tensions in the development of Northern Australia: Implications for governance. *Journal of Economic and Social Policy*, *17*(1), 23.

van Dijk, A. I. J. M., Beck, H. E., Crosbie, R. S., de Jeu, R. A. M., Liu, Y. Y., Podger, G. M., et al. (2013). The Millennium drought in southeast Australia (2001–2009): Natural and human causes and implications for water resources, ecosystems, economy, and society. *Water Resources Research*, *49*, 1040–1057.

Wang, J., Horne, A., Nathan, R., Peel, M., & Neave, I. (2017). Vulnerability of water management objectives to the sequencing of wet and dry spells prior to and during drought conditions, in preparation.

WSAA. (2012). *Cost of carbon abatement in the Australian water industry*. Occasional Paper 28Melbourne: Water Services Association of Australia. 53pp.

Index

Note: Page numbers followed by *f* indicate figures, *t* indicate tables, *b* indicate boxes, and *np* indicate footnotes.